VIEW FROM THE TOP
∾
Forest Service Research

by

R. Keith Arnold

M. B. Dickerman

Robert E. Buckman

Edited by

Harold K. Steen

Forest History Society ∾ Durham ∾ North Carolina

The Forest History Society is a nonprofit, educational institution
dedicated to the advancement of historical understanding of human
interaction with the forest environment. It was
established in 1946. Interpretations and conclusions in FHS
publications are those of the authors; the institution takes
responsibility for the selection of topics, the competence of the
authors, and their freedom of inquiry.

Work on this book and its publication were supported by
grants to the Forest History Society.

Cover photo: Table Rock from Wiseman's View,
Nantahala National Forest, North Carolina.
Courtesy of the United States Forest Service.

Library of Congress Cataloging-in-Publication Data

Arnold, R. Keith (Richard Keith), 1913-
    View from the top: Forest Service research / by R. Keith Arnold, M. B. Dickerman,
Robert E. Buckman; edited by Harold K. Steen.
    p. cm.
    ISBN 0-89030-049-6
    1. Arnold, R. Keith (Richard Keith), 1913—Interviews.
2. Dickerman, M. B. (Murlyn Bennet), 1912—Interviews. 3. Buckman, Robert E. (Robert
Erwin), 1927—Interviews. 4. Foresters—United States—Interviews. 5. United States. Forest
Service—Officials and employees—Interviews. 6. Forests and forestry—United States—
History. I. Dickerman, M. B. (Murlyn Bennet), 1912- . II. Buckman, Robert E. (Robert Erwin),
1927- . III. Steen, Harold K. IV. Title.
SD127.A76 1994
634.9'092'2—dc20
[B]   93-46520
                                                                        CIP

CONTENTS

ᘯ

R. KEITH ARNOLD

⌘

M. B. DICKERMAN

ROBERT E. BUCKMAN

✑

Introduction

Having its own research arm makes the Forest Service an atypical federal agency. Having a research arm independent of the agency's management responsibilities makes it even more atypical. But what follows is not an account of differences; instead, three former deputy chiefs for research narrate their careers and the politics and practice of science. The agency's research findings are widely available throughout the worldwide forestry community. Therefore, this is much more than one examination of Forest Service research, it is a study of forest science at home and abroad.

The three interviews appear in the same sequence as R. Keith Arnold, M. B. Dickerman, and Robert E. Buckman served as deputy chief for research. Previous interviews with Clarence L. Forsling, E. I. Kotok, Verne L. Harper, and George M. Jemison, plus this latest three, add up to a half century (1937-86) of Forest Service research leadership on tape.

This book was jointly sponsored by the Forest Service and the Forest History Society. Special thanks go to our three narrators, who gave freely of their time and hospitality, and to Jerry Sesco, current deputy chief for research, who thought the whole thing was a good idea.

Richard Keith Arnold was born on November 17, 1913, in Long Beach, California. He earned a bachelor of science degree in forestry from the University of California in 1937 and a masters degree in forestry from Yale University in 1938; in 1950 he would receive a Ph.D. from the University of Michigan. A four-year stint in the Navy during World War II, plus graduate school, plus teaching at the Berkeley forestry school, fairly well rounded out the 1940s.

In 1951 Keith began working in fire research at the Forest Service experiment station, also in Berkeley, and in the forestry school building. During the interview I commented to Keith that I had difficulty sorting out when he worked for the Forest Service and when he taught at the forestry school—he was constantly on loan from one to the other. He confided that it was confusing to him, too, but nonetheless it worked well because the assignments were very compatible. The reader is thus cautioned.

One of his most challenging assignments was as manager of a research team for the Armed Forces Special Weapons Project, a multi-agency undertaking. Central to the effort was the use of nuclear devices and the development of instrumentation to measure ignition and combustion at various distances from ground zero. Of all the scientific groups involved, Keith's forestry team was the most efficient and effective. An impressed military asked if he was interested in taking over similar instrumentation needs for their weapons testing in the South Pacific. He laughed as he remembered how the officer's face fell when he learned that Keith's Ph.D. was in forestry, not nuclear physics. It was okay, though, Keith wasn't interested anyway. As noted above, his career has not been typical.

Keith was named chief of the Division of Fire Research in 1954, and in 1957 he moved up to director of the station. He moved again in 1963, this time across the nation to be director of forest protection research in the Washington Office. Now he was responsible for research on fire, insects, and disease—topics that were intellectually unrelated but made sense to the land manager. He believes that these three moves resulted from him being "in the right place at the right time," a situation that would occur more often in his life than in yours or mine.

His career path took an abrupt jog in 1966, when he accepted the deanship at the University of Michigan School of Natural Resources, where he had earned his Ph.D. a decade and a half earlier. One of his goals was to make the school more important to the university, a process that became rather controversial with alumni and others. Here, he relates publicly for the first time what happened and why.

The next "right time" was in 1969 when Forest Service chief Ed Cliff called, asking if he would be interested in being deputy chief for research. It seems there was an unexpected vacancy, and the chief was concerned that the secretary of agriculture would bring in an outsider. Apparently Keith was the only one with the necessary experience and credentials to quickly pass civil service muster for the position. He accepted the offer, and his career path jogged back to its Forest Service center.

Keith discusses and describes the budget process, testifying in Congress, relationships with the Agricultural Research Service, and minority hiring. President Nixon had announced that the 1970s was to be the environmental decade, and Forest Service research responded in several ways, including studying fire in wilderness areas and methods to manage natural areas. There were international issues to address and a need to achieve balanced programs. And there was more—much more—as he recounts what it was like to be deputy chief.

Another career jog occurred in 1973, when his good friend Steve Spurr (who was also president of the University of Texas) called. Would he like to be director of the Division of Natural Resources? As it turned out he would, and soon he was a professor in the LBJ School of Public Administration. The word intrigue takes on fuller meaning when following Keith's account of life in Austin, which he nonetheless thoroughly enjoyed.

It was at this same time that he became active in the Society of American Foresters at the national level. He was president when a major bequest made it possible for the SAF to own its headquarters, and spiral downward into an extraordinarily divisive—and expensive—conflict. There were also antitrust implications in the Code of Conduct, which apparently restricted competition. The code was appropriately revised.

Finally, 1980 was the time to retire. But there were enjoyable things yet to do, and for a few more years Keith was visiting professor at forestry schools in Colorado, Oregon, Maine, and Arkansas. By 1992 he seemed truly retired but is keeping busy with family, stepping off his front porch into his boat, and playing championship seniors tennis. He lives in Hot Springs Valley, he says, because it is half-way between Austin and Washington, and it was his favorite stop-over point for the many, many trips he made between the two towns.

Murlyn Bennet Dickerman was born on July 29, 1912, in Hamden, Connecticut. He earned a bachelor of science degree from the University of Connecticut in 1934 and a masters degree in forestry from the University of California in 1936. From 1937 to 1941, Dick was assigned to the Northeastern Forest Experiment Station and was involved with cooperatives for managing forests, producing and marketing timber.

He transferred to the Lake States Forest Experiment Station and continued marketing studies. Then World War II caused severe civilian budget cuts and reassignments to war-related activities. Dick found himself in Washington, D.C., on loan to the War Production Board and Office of Price Administration. During this period of price and wage control, Dick worked to see that the price of lumber was at a proper level. He was frustrated by the futility of the assignment, and when he had a chance to go as a civilian to liberated Italy to help reestablish a forestry program, he jumped at it. In 1945 and 1946 he was first with the Allied Control Commission in Rome and then with the United Nations Relief and Rehabilitation Administration in Athens.

At war's end, patriotism became less passionate for some, and the chief's office began asking Dick when he was going to report back for work. An especially terse inquiry from Assistant Chief Edward C. Crafts brought Dick home in 1947, but this time he was assigned to the Northern Rocky Mountain Forest and Range Experiment Station. He would stay in Missoula for four years, and it was then that he became an avowed westerner and collected the objects now lining his den.

Dick was in charge of forest inventory and economics programs. However, the director was frequently on other assignments, and Dick spent much of his time as acting director. He found that he liked research administration, and was pleased to move to St. Paul in 1951 to be the director of the Lake States Forest Experiment Station. He would remain there for fifteen years. As director he added programs in entomology, pathology, genetics, utilization, watershed, engineering, and recreation. Also, his station was the first to have its own headquarters building.

He liked work and life in St. Paul, but after a decade and a half the time seemed right for a change. Therefore, he was glad to move to Washington, D.C., in 1965 to be a staff member for the Research Program Development and Evaluation Group. The goal was to examine research efforts throughout the Department of Agriculture and develop efficiencies, standards, and coordination. The net effect was production of a long-range research program.

Dick's next assignment came in 1968 as associate deputy chief for research under George Jemison and Keith Arnold. Now he was responsible for a broad program in forest research, including planning, budgeting, and review and inspection. He gave much attention to various specific issues, such as minority hiring and the need for new facilities.

In 1973 Chief John McGuire asked Dick to be deputy chief, overseeing an eighty million dollar research budget with three thousand employees. As deputy, he would testify to Congress and perform other high-level tasks—not everything can be delegated to one's associate. Research on endangered species and selected species of insects that were far from being rare received emphasis, as did international forestry with the upgrading of the Puerto Rican experiment station into the Institute of Tropical Forestry.

It had been a long and productive career, and Dick was ready to retire in 1975. However, his proximity to Washington, D.C., has made it convenient for him to remain active in professional programs. From 1976 to 1978 he served as science advisor to the Society of American Foresters and also a brief stint as acting executive vice president. He played a key role in reinvigorating the International Society of Tropical Foresters, as director and as secretary from 1980 to 1983. At this writing, he still goes "into town;" he has yet to fully retire.

Robert E. Buckman was born on June 28, 1927, in Superior, Wisconsin. He earned a bachelor of science degree in 1950 and a master of forestry degree in 1953 from the University of Minnesota, and a Ph.D. in 1959 from the University of Michigan. Following military service in World War II and in Korea, and with his formal education well along, he began with the Lake States Forest Experiment Station in 1955, and he would remain there for a decade.

For Bob, the Lake States years were among his happiest, and he would have been content to stay. He spent much of his time studying red pine silviculture, along with prescribed burning and related topics. A plus was having M. B.

Dickerman as station director; ironically, it would turn out that Bob would be Dick's successor as deputy chief for research. But that would be much later.

In 1965 Bob was transferred to Washington, D.C., to work in timber management research. He was awarded a mid-career sabbatical to earn a masters of public administration at Harvard. He wrote a major research paper, "Evolution of Science Policies in the Forest Service," showing an interest in history that remains strong.

When Bob arrived in Washington, Les Harper was deputy chief for research, to be shortly succeeded by his associate deputy George Jemison. Keith Arnold would be deputy when Bob was reassigned in 1971. Thus he was in Washington during a time of transition; Harper's long and seminal tenure was followed by a series a relatively brief appointments. Too, the civil rights movement prompted a shift in personnel priorities, as did the environmental movement cause a reexamination of research projects. During this period, the experiment stations were reorganized so that fewer individuals reported to the director. Toward the end of this Washington assignment, Bob was responsible for overall research budget preparation and coordination.

In 1971 Bob was named director of the Pacific Northwest Forest and Range Experiment Station in Portland, Oregon. There he was responsible for Forest Service research in Washington, Oregon, and Alaska with nine laboratories hosting about one hundred scientists. Some of the studies included the tussock moth, spruce budworm, prescribed fire, forest ecology, and hydrology.

His final Forest Service career move came in 1975; Deputy Chief Dickerman asked him to return to Washington to be his associate deputy. The next year, Dickerman retired and Chief John R. McGuire named Bob as successor. As it would turn out, Bob was deputy chief for ten years, as long as the combined tenures of his three, immediate predecessors. Only Les Harper (1952-65) and Earle Clapp (1915-35) served longer in that capacity.

As deputy chief, Bob was responsible for eight hundred scientists at seventy-five laboratories organized into eight experiment stations and the Forest Products Laboratory. Planning, coordination, and execution of a broad array of research topics was now his domain. He was also in the top agency leadership, participating in discussions and decisions on virtually all major policies concerning programs, budgets, and personnel. He would meet with top-ranked members of the administration and testify before Congress. His watch included the transition from President Carter to President Reagan, with the attendant budget cuts of the latter administration.

New statutes were especially significant during Bob's time as deputy. The Resources Planning Act called for long term projections, which in turn required Research to provide specific information. The National Forest Management Act contained sections on topics like biological diversity, also generating need for scientific studies. The Forest and Rangeland Renewable Resources Research Act of

1978 provided specific authorization for current and planned activities. Congress affected more and more.

There were new initiatives in competitive grants, biotechnology, and research evaluation. Scientists were trying to determine the effect of acid rain on forests, studying endangered species, and coming up with ways to rehabilitate surface mining activities.

Bob retired from the Forest Service in 1986, but not from his profession. While with the agency, and fairly typical of deputy chiefs for research, he had been much involved with the International Union of Forestry Research Organizations. From 1976 to 1985 he had been a member of its Executive Board and also vice president. In 1987 he began a four-year term as IUFRO president; a major issue included reworking the administrative structure, especially the secretariat, for a far-flung organization that contained 650 participating institutions from 106 countries. Priority was also given to creating special programs for the Third World.

During this same time and to date, Bob has been a professor of forest management at Oregon State University, on a half-time basis. He guides graduate students interested in forest policy and international forestry.

His schedule remains full.

∾

Harold K. Steen
Durham, NC

VIEW FROM THE TOP

∽

Forest Service Research

R. Keith Arnold

INTRODUCTION

When Keith Arnold was director of forest protection research in Washington, D.C., during the early 1960s, I was an apprentice scientist in fire research at the

Pacific Northwest Forest and Range Experiment Station in Portland, Oregon. Keith convened a workshop for Forest Service fire researchers in Denver; there I met him only briefly. Our next meeting was not in person; I had written a book on Forest Service history, and Keith reviewed it for *American Forests*. He liked the book, even praised it, becoming one of my favorite people. We met face-to-face in 1978 while he was president of the Society of American Foresters. As a student of bad timing I approached him for a minor favor, just as he was rushing to the bedside of his stricken friend, Steve Spurr, whose name appears frequently in this interview. Although obviously agitated, Keith tended to my request. And that's just how well I knew Keith Arnold when we began working by mail to construct an outline for the interview that follows.

In August 1992, Keith met me at the back gate of Hot Springs Valley, Arkansas, a retirement community included in four thousand wooded acres. His fine home is literally at lake's edge; we taped the interview in a basement area with a full view, and I was at times distracted by pleasure boats cruising by just outside the window. His wife was away, and we had the house to ourselves, which means that Keith fixed my lunch—twice. One afternoon we toured the lake on his boat, and I could begin to understand why people look forward to retirement.

Keith is an organized person. Not only did we have an outline, but he had prepared twenty-eight hand-written pages of notes to guide his memory. He was always amenable to any spontaneous follow-up questions that I had, but we pretty much adhered to the outline and his notes. He reviewed the transcript carefully, making corrections of fact and shifts of nuance. As you will read, his Forest Service career was not a traditional one.

∾

FORMATIVE YEARS

FORESTRY EDUCATION

HAROLD K. STEEN: Why forestry? I don't know what other options you were thinking about when you were in high school, but somehow you wound up in forestry.

R. KEITH ARNOLD: I wasn't thinking of any career in high school as far as I know. Along toward the end, I guess as a senior and then in junior college, I came upon forestry, probably because my Sunday school teacher was a forester for Los Angeles county.

HKS: Is that right?

RKA: Yes. He was a young fellow and had graduated from Forestry School at U.C. Berkeley. Paul Gruendyke was his name. Paul took me to the Los Angeles county seed orchards and plantings. I got interested in the urban phase of forestry more, because I lived in Glendale at the time. We did a lot of tree growth measurements, and he said "obviously you're interested. As a very good student you might as well go to the University of California at Berkeley." One other possibility—and I think this did have a lot to do with it—my folks had a cabin at Big Bear Lake, which is just north of San Bernardino. I spent every summer there from age one until I was old enough to work. I thought that was probably one of the best forests in the entire world until I found out later it was between site 4 and site 5 [laughter] ponderosa and Jeffrey pine. I guess the third reason, and this is not very important, is that my father owned a retail lumber yard—my grandfather also was in the lumber business. I worked there in the summers. That was when you unloaded box cars and flat cars one stick at a time. I did a lot of lumber bucking. I don't know whether that made me think about growing trees that were lighter weight or not, but it really was Paul Gruendyke as a forester for L.A. county who did the trick.

HKS: When were you at Berkeley in forestry?

RKA: I went there in 1935 and then graduated in 1937. I went to junior college in Glendale, and I think it was fortunate to have gone to junior college because they did a better job of teaching. I'm a strong advocate of junior colleges. At U.C. Berkeley I had no problem getting through with high grades while working most of the time.

HKS: I have a cousin who graduated from the University of Washington forestry school in 1937. He never worked as a forester because he couldn't get a job and he wound up teaching school or something. What were the jobs like then?

RKA: There weren't any. I think our class had thirty-five or thirty-six graduates, and not a single one of them got a job in forestry. The closest one was John Zivnuska who worked in the Border Patrol out of New Mexico. John worked down there for a year or two before he finally came back to California and forestry. When the job situation opened up, most of my classmates went into the California Division of Forestry. I probably would have done the same, but I was at Yale when state jobs became available.

We'll talk about Jim Mace later on, but in forestry camp, which came after the junior year at Cal, he and I ran for camp manager and he won by one vote. But the students also voted that I would get my way paid, and so I worked for Jim Mace as assistant manager of camp, which meant that I organized all the work details. Jim made this statement the first day in camp, he took me out in the woods and said "this is going to be the best run camp that has ever occurred up here. Even if you and I flunk, we're still going to do a good job." Jim pursued that career goal all through his work with the California Division of Forestry.

HKS: Were you at all surprised that John Zivnuska went on and became the eminent economist and all that?

RKA: No.

HKS: He was good in school?

RKA: John was very smart.

HKS: What were the options at Cal? Forest management, logging engineering, or what.

RKA: There was just one curriculum, but you could specialize in watershed management, forest production, engineering—Emanuel Fritz taught engineering. Principally it was a fairly conventional curriculum in forest management.

HKS: When you were an undergraduate did you assume you were going on to graduate school?

RKA: No, I didn't consider it at all. I guess I might as well admit that I had a straight A average at Cal, and several of the profs said that I should go right on to graduate school, but I was ready to get out and go to work, hopefully in the woods.

HKS: Sure.

RKA: But when no jobs came up, I applied to Yale and received a scholarship. I was the teaching assistant at Cal's summer camp that year (1937) after graduation, and then went on that fall to Yale toward a master's degree.

HKS: You have on the outline that Walter Mulford was a major influence in your life. Was it because he was dean or was it because he was an influential guy?

RKA: It was for both reasons. I guess principally because he was dean and because he gave me several opportunities through my career that I'll have a chance to mention. Walter was involved almost entirely in recruiting top professors, building a strong school in a large university, and being accessible to every student all the time. He was not heavily involved in forest policy outside the school, but his history and forest policy courses were tops.

Yale was a little different. A professor at Cal who had graduated from Yale said that Dean Henry Graves would take on two students each year and give them each an individual course with him for three units. I decided I wanted that, so I went to Yale two weeks early and met with Dean Graves. You had to pick a subject of his interest. I'd been advised that he was chairman of the Parks Authority in Connecticut. I told him I would like to take a seminar in forest recreation. It resulted in a paper on forest recreation, its planning, development, and projections. He was the best editor I've ever worked with. He edited the *Journal of Forestry*, years and years before, and it took me three weeks, I remember, to get my first page through him. [laughter] This obviously was among the few best courses I ever took.

HKS: Forest recreation. That was not a typical subject in those days.

RKA: No, it wasn't. He was ahead of his time. He could see it coming and he was interested in balancing recreation in forests with other uses; so that's why he headed up the Connecticut parks and recreation program.

HKS: Tell me a couple of anecdotes about Graves. The ones that you read from his contemporaries are about his coal black eyes. His eyes, I guess, dominated the situation. What it is about Graves that made him so great?

RKA: That's a difficult question. He wasn't at all "stiff" when you were one-on-one with him. His laugh would echo throughout Sage Hall. I don't think there is any question that you described his actions and appearance perfectly. He was extremely friendly, and it was easy to exchange information with him and to argue with him, but he appeared in public to be quite formal. He really gave two students a year a special education. It was worth the whole Yale experience. I believe Henry Graves' greatness stemmed from his forward-looking capacity— some thirty to forty years.

HKS: This was a guy who was Pinchot's assistant. Did he have any sense of that history?

RKA: I am sure that he did but he was looking at the present and the future—at least with me. Gifford Pinchot came up a couple of times during the year and gave a seminar for the students, but other than that I didn't get any special sense of history. H. H. Chapman emphasized history in his management class. Of course there are endless stories about Chappy: how he blew his auto's horn at every intersection; how he walked out of a hospital with only a gown on; his weekly battles with the secretary of the interior. At one of his 8:00 A.M. lectures he kept his heavy overcoat on. We learned later that he had forgotten to put on his suit coat. No Yale professor would lecture in his shirt sleeves.

HKS: I suppose some of them are even true. Pinchot was rather elderly by then. Did you go to those seminars or presentations that Pinchot made at Yale?

RKA: Of course.

HKS: Did he have a strong personality? I'm curious about people who are in leadership roles.

RKA: Yes. Pinchot was plain tall and extremely forceful is my recollection. He obviously was old at that time. Now I would not consider the seventies as old. I guess we could figure out how old. He projected a sense of very strong persuasions and rather antagonistic approaches to problems.

HKS: So you had that short course from Graves. What other sorts of things did you study at Yale? The basic master of forestry program?

RKA: Other than Graves I didn't gain a lot with the year that I was at Yale. It was not the program's fault. It was quite duplicative of the curriculum at California.

HKS: Is that right?

RKA: It should have been obvious, because it was the general forest management course, and some of the people in it were in the two-year program. If you had an undergraduate forestry degree, you got your masters in one year as you may know. The two-year program produced the same masters degree for those who had a degree in some major other than forestry. As I look back, I didn't get anything new except for contacts with professors such as Graves and Chapman. I enjoyed the Yale summer camp in the southern woods at Arkansas more than any other part of the program at Yale.

HKS: So now you have a master's degree. What were your thoughts, by now you're sort of getting in the swing of things, you're looking ahead at more academic situations, right?

RKA: No, I was looking for a job at that time.

HKS: Still looking for a job?

RKA: I finally received two very different job offers just before the end of the spring camp in Arkansas. One was to stay on as an area forester at Crossett Lumber Company. We were on their lands with the Yale camp. The other offer was from the California Forest and Range Experiment Station—a field assistant's position at Pinecrest, about seven thousand feet in the High Sierra. As a native Californian, it was an easy decision for me to avoid the chigger-infested woods of the South compared to the beauty and challenges of the Sierras. So I did accept that job as a field assistant, and it lasted for about a year and a half.

The work at Pinecrest was under Duncan Denning, who was probably the finest ecologist and silviculturist to be found in the United States. I know that's a strong statement, but I spent a lot of time in the woods with him. He was the division chief for Forest Management Research. He was a poor administrator. He didn't like administration. He and Ed Kotok, the station director, did not get along. The Forest Service lost not only a good scientist, but experienced poor administration. He retired early out of frustration. I had a chance to help eliminate that kind of a problem twenty years later when I was station director in Berkeley.

The second year at Pinecrest was even better. Along with fishing and hunting, it appeared to be an ideal job. Director Ed Kotok used the Stanislaus field station for weekend rest and recreation. He was there quite often. One evening after a rather lengthy happy hour, he confided in me that I just wasn't the kind of person for Forest Service research. He advised me to move on. Sure, he says, you can do this field work all right, but I don't think that you will go far in the Forest Service. At that point I had never considered an advanced degree or more academic work. Guess I have to give him the benefit of the doubt and thank him for the right advice at the right time.

HKS: Did you have a sense of what he was referring to? Your lack of training or just your attitude or what. Why do you think he said that?

RKA: I don't know. I thought about it at the time and never have determined whether he was firing me or directing me back to academics.

HKS: Because research in those days wasn't very rigorous. I mean to not have a Ph.D. wouldn't have been a major deficiency would it?

RKA: No, no there were rather few Ph.D.s in forest research, and maybe he sensed that I should have a Ph.D., I don't know. But anyway, this was in April or May. In August I enrolled at the University of California economics graduate study and became a teaching assistant in forest mensuration during the school year and at the forestry camp in the summer. Thus began the work toward the Ph.D. received some ten years later.

HKS: So your Ph.D.'s from Berkeley?

RKA: No. I had about 2/3ds time available for Ph.D. work. The rest of the time I was a teaching assistant as well as working outside some for the Forest Experiment Station. In January 1942, Percy Barr, who taught forest mensuration, was a colonel in the army reserve. He was called up suddenly and left with no one on the faculty to teach his courses. Dean Walter Mulford asked me to teach mensuration. Here's an example of one of those many unique breaks I have had by being at the right place at the right time. So I taught mensuration in the spring of 1942 and then filled in for Percy Barr's time in operating and teaching the summer camp. The Navy caught up with me in July. The Ph.D. was on hold obviously for the four years of military service.

THE WAR YEARS

HKS: Did some of the training you had in the military apply to your later life?

RKA: One day I was in summer camp checking out student field work, the next day at eight o'clock in the morning, having driven all night, I was a naval officer and by two in the afternoon I was in full uniform. My orders directed me to Dartmouth for a two-month indoctrination course. There was no one to teach navigation so we had no contact with that critically important part of naval officer's training. My first orders were to go to Cornell. I think they looked at my teaching experience and the fact that I had taught surveying and so here I was ordered to Cornell to teach navigation.

HKS: Did you feel qualified?

RKA: I didn't know anything about the textbooks or the subject. When I went to the commanding officer I told him I had never had any contact with navigation in any way. He said well we have a top man that used to be at the Naval Academy. He'll teach and you just go in and go through one session. We have one-month courses here, and you can start teaching in a month. So the next day I went into the classroom; the teacher didn't show up. So after about fifteen minutes I got up and started talking about the world is round and there's latitude and longitude and a few things that I knew. The man who was supposed to teach it had a very serious emergency appendectomy. He was in the hospital for three or four weeks, so I taught navigation. I could handle everything by heavy reading the night before. Celestial navigation was different, and I needed help. I went to the hospital every day, got briefed and then went home and studied the books and gave the lecture the next morning.

We had two men who were yachtsman with much navigation experience. After my first lecture on celestial navigation, they both came up and said, "Lieutenant Arnold we'd like to congratulate you, that's the simplest explanation of celestial navigation we've ever heard." I admitted that the lecture included the sum total of my knowledge on the subject.

I applied for sea duty after about a year and was sent progressively to anti-submarine warfare school in Miami, San Diego, and finally Boston at the anti-submarine warfare instructor's school. I taught there for a while, but was transferred to the staff of the commander of the Atlantic fleet where we did sea-going tests of new equipment, wrote training and operating manuals, and worked on special assignments. I was one of the few who worked on the methods we would use to capture the U-505 now in the Museum of Science and Technology in Chicago. You may have seen it.

HKS: No, I haven't.

RKA: That advanced German submarine was captured by a plan that was developed in advance. I had some good experience in the Navy in writing briefs and training manuals. I got to sea half a dozen times, usually with new equipment to test. We did not simulate; we worked in the North Atlantic in real-life situations.

HKS: There's something about either the luck of your life or some temperament you had that you tend to wind up doing research and teaching rather than this other sort of thing. They could have sent you out there with depth charges, but instead you were doing research and writing manuals.

Graduate School

RKA: But it wasn't because I'd asked for research and training. Sometimes we had the depth charges also. After the war I knew I wanted to go ahead and finish work on my Ph.D. But when I returned to Berkeley, I found that I'd forgotten a tremendous amount, I almost had to start over in economics. Again Walter Mulford gave me an appointment as associate in forestry. I taught fire control and was responsible for operating and teaching summer camp. This was supposed to be done in half of the time with the rest available for study toward the Ph.D.

HKS: I suppose to the extent that there is an advantage to war, during the four years you were in the service there wasn't a whole lot of new knowledge being generated in the civilian ranks, so you didn't fall behind that much.

RKA: No, but never having given economics a thought for four years, I had to take graduate seminars over again. It became obvious after three years that I probably would not be able to get my Ph.D. at Cal. There is no such thing, as you probably know, as a half-time appointment, there's always more that comes up. I had a family at that time of two boys and a wife. We decided to go to Michigan for a Ph.D. Sam Dana was on my Ph.D. committee, Shirley Allen was its chairman. My Ph.D. thesis was related to fire policy with emphasis on economics. I think the most valuable organized course I ever took in any university was in group dynamics.

I needed units outside of forestry and economics to provide a rounded base for Ph.D. This was an excellent course for future application—how to work with small committees, how to work with large groups, how to recognize all kinds of leadership and the roles that needed to be played. I think I was able to apply those techniques in every day work from then on. Even though my Ph.D. work was in the economics of forestry and fire control, almost all of the research and work in the early years were in the field of meteorology and the physics of fire behavior.

HKS: How did you get involved in fire?

RKA: I grew up in the middle of fire. In southern California, I remember as a boy we had the hose up on the roof several times keeping the house from burning when nearby brush fields were burning. Also my teaching assignment was in fire control. That left no choice. I had to learn a lot about fire in a short time.

Probably the most important part of the Ph.D. thesis was the title. It was called the "Economic and Social Determinants of an Adequate of Forest Fire Control." Now that even sounds like a good thesis.

HKS: That sounds like a dissertation to me.

RKA: I believe that was the first Ph.D. thesis related to fire. That doesn't mean anything except that there was nothing to go on. Show and Kotok did some pioneering studies in fire control policy in the 1920s. Like flood control and earthquake damage, there is no way financially or organizationally that we can be prepared for the ultimate disaster. You can plan fire control in an economic and financial planning situation up to the level of what you might call a conflagration. The Yellowstone fires are good examples of that sort of thing, there's just no way that you can put together a program to have any effect on those major fires, it just can't be done.

HKS: Was this controversial? Did people disagree with that, people in fire control?

RKA: No, I haven't found anyone to disagree with it. You can't get that kind of money and even if you did you're not sure. The flood is a perfect example. The Army engineers' flood control programs in the country are designed to take up to, let's say, fifty- or a hundred-year floods, I don't know their criteria. So over long periods of time they have adequate flood control, people move into flood plains, businesses move into flood plains, then comes a five hundred- or thousand-year flood and wipes out everything. It's identical in fire, and we can't build the buildings to withstand the worst earthquakes. They're designed to stand up to some, but at some point in time an earthquake will be strong enough, it will destroy them. The only answer in fire control is a vigorous program of prescribed fire. If applied in wilderness areas and national parks, conflict is obvious.

As an aside, the Yellowstone fires would undoubtedly have occurred, unless there had been an aggressive prescribed burn program for the past fifty years. The existing "let burn policy" was inadequate: "if lightning started a fire in certain areas of the park, it was allowed to burn." There was no tie to fire danger rating or to long-term drought effects. I have a long-standing position (certainly controversial) that if conditions for burning allow a lightning fire to be allowed to burn, we could set a prescribed burn at the same time and place.

After I finished my Ph.D., unbeknownst to me, Sam Dana told the chief of the Forest Service and others that if the Forest Service ever needed a research director in the future, they ought to look at me. That came home to roost about six years later I guess.

HKS: When you're doing a Ph.D. in a relatively new field as you were then, what's the faculty role? Just somebody to react to? Because they didn't have any expertise, there wasn't any expertise at that time.

RKA: No, but the basic economic analysis was standard, and that often happens in the Ph.D. field. The candidate may or should know a great deal more than the people on his committee or her committee.

HKS: Were the data hard to come by?

RKA: No, the data in national fire reports and reports of the chief were adequate. After finishing the Ph.D., again Walter Mulford hired me back at Cal as assistant professor.

Early Years Teaching and Research

HKS: You and Mulford got along pretty well it sounds like.

RKA: Yes.

HKS: I can't quite sort out the line between teaching at Berkeley and work at the experiment station in Berkeley. Maybe as we go through your narrative here, we can pull that apart.

RKA: I can't pull that apart either. It varied so much and by so many different amounts. Sometimes I was full-time at the experiment station on leave from the university, sometimes I was full-time at the university but still doing research in cooperation with others at the experiment station; sometimes half and half. I have no records of dates when I was doing one thing or the other.

HKS: When you were teaching at that time, were the majority of the students returning GIs? You had large classes and they were dedicated, they worked hard and were ready to learn a profession, no goofing off.

RKA: No, it was the other way around.

HKS: Is that right?

RKA: Particularly at summer camp. They'd had two, three, or four years of discipline and they were ready to relax and have a good time. They were not uniformly good students, only by their own choice. In fact I had the first forestry summer camp in

1946 after the war, and we had two revolutions up there where they tried to get rid of me as a prof because I was insisting that they do some work. But most of those people got through and became top-notch foresters. They've been my good friends ever since.

HKS: Was the link you had with the experiment station relatively typical for Berkeley faculty? In that various professors had sort of a joint relationship?

ARMED FORCES SPECIAL WEAPONS PROJECT

RKA: Many of the younger ones did. I think John Zivnuska did, several others had cooperative projects with the station because the station had experimental areas and experimental forests and some of the profs would go up and do their field work at experiment areas of the Forest Service.

After the Ph.D. work, I was an assistant professor at Cal. My work with the experiment station became, as you mentioned, confused. I worked both with them as an individual employee and cooperatively. The first project that came on was the Armed Forces Special Weapons Project. This was a contract that the Forest Service took to determine the effects of nuclear weapons on forests. The military wanted to do it to know what the trafficability problem was. If nuclear weapons were detonated in forests, could tanks go through, could automobiles go through. What kind of effort would be needed to clear roads and that sort of thing. They were also concerned with the fires that would ensue after attacks.

Again the timing for me was fortunate. I didn't know anything about this subject. Apparently the Forest Service couldn't find anyone in-service that they wanted to release to head this project. It had some top engineers and physicists. They asked me if I would take leave from the university and run this project, and I decided to do so.

We had two areas of study. One was blast effects. We organized and did wind tunnel tests on small trees to know how much wind it took to break them or blow them over. We outfitted trucks to drive down highways at fifty miles an hour with a small tree maybe twenty feet tall, fully instrumented, standing upright on it. We developed a standard metal tree that we could put out at the nuclear test site in Nevada. It was supposed to react like trees and served to see what the difference was between the atomic blast and a steady stream of wind tunnels and trucks. We finally ended up moving a three-acre stand into the Nevada proving grounds. Every ponderosa and Jeffrey pine was instrumented to measure deflection. We recorded how many of these trees would break and how many would not. Our results were satisfactory, our predictions were quite accurate.

There were many tests that require recorders to be turned on. You might want camera lenses opened up or whatever. We had a lot of beds of various kinds of fuels that we had covered up to keep the dew from forming on them. We waited a minute or two before the bomb went off to uncover the fuels and start recorders. DOD called them blue boxes and charged three hundred dollars for each one. We needed fifty to a hundred of these blue boxes at three hundred dollars apiece, and to us dirt foresters that was big money.

HKS: Big money.

RKA: We determined that we didn't need anything more accurate than five minutes. I bought two hundred Big Ben alarm clocks at one dollar eighty-nine apiece. We put them all in a room in the Berkeley Experiment Station and wound them every day. When the alarm goes off, the winding mechanism unwinds, and it's a very strong pull. We hooked wires onto the stem and had them pull pins. Then we put two on each tache so that we had two chances for success. We had window blinds over our fuel beds. A pulled pin released the blind and it would go back and expose them.

HKS: It was a very high tech operation, it sounds like.

RKA: It was. The funny thing was that on every one of the first three bomb tests we had 100 percent success. The $300 blue boxes worked as needed about 75 percent of the time. [laughter]

HKS: That would be a good testimonial for the guy who makes these clocks. How about security clearances?

RKA: They were routine. In the Armed Forces Special Weapons Project we started to employ a fellow who'd been in the Forest Service for some time. His security clearance noted that he enjoyed cocktail hours and usually talked a lot. We couldn't get him security clearance, and we weren't allowed to tell him that that was the reason he couldn't have the job.

HKS: I remember seeing photographs of the same time period. Houses were built to test how far away from ground zero the house would be obliterated or just scorched or whatever. The same kind of testing but for forests.

RKA: Those houses were provided and that contract work done by the Forest Products Laboratory in Madison, Wisconsin. We studied ignition points in forests and in buildings. The fire effects gave us a chance to again study conflagration fire and ignition patterns in fire storms which went back to my thesis work. There was no question that the kinds of things that were ignited by the small atom bombs were the fine fuels such as grasses. The most common forest ignition point was in rotten woods. If you have a forest with a lot of rotten wood in it you'd get many ignitions in the rotted part of the wood. In houses you get ignition on curtains. You can have a whole side of a house burst into flames, but as soon as the bomb heat dies that would go out, it wouldn't hold.

As a follow-up to these field tests, we did a survey of Detroit and several other cities in the United States to determine fire patterns. We looked for ignition points in houses and developed some prediction systems on how large areas fires would be started. We studied the fires in firestorms in World War II, which gave us some good insights on how they would behave after they started. We also developed a rather simple world-wide fire danger rating system. This work was interesting to me; it gave me a practical insight into very large fires.

There must have been a thousand people at one time out at the Nevada proving grounds, doing different kinds of weapons effects tests. I think foresters stood out, because we were accustomed to working in uncontrolled environments in the woods. Most everyone there was a laboratory scientist. They had trouble working with wind, rain, and humidity. We went with our own tools, our own automobiles, we were completely self-contained and looked very good as our field tests had few failures.

I can think of one story that comes from that experience. We had nothing to do with the bombs, but after every detonation based on what happened to our materials we had to indicate what we thought that the bomb yield was. The first time they detonated a bomb about eleven o'clock in the morning was the first time it hadn't been at the highest dew point. The fine fuels and some of the materials exposed from other units had been always moist. When they detonated this bomb at eleven o'clock in the morning, the entire desert vegetation and some experiments for half a mile around went up in flames. As we were giving our reports everyone said this had to be a bomb with a greater heat intensity than had ever been before. We looked at our ignition and spread table related to humidity and said the bomb tested was the same. We were correct.

This leads to one story which may be a little meaningless. One night I was asked by Colonel Diller, our program manager, to come to meet with five or six people. They talked around a little bit about that there was going to be a test of a larger weapon in the Pacific area. They wouldn't tell me what it was, but they said they needed a test director and was I interested in being considered as a test director for this site. They said it'd take me six or eight months out in the Pacific, and I said I couldn't stay away from my family for that long. It was just not possible. One of them said Dr. Arnold, is your doctorate degree in physics or in electrical engineering? I said it's in forest economics. There was a dead silence in the room, and pretty soon they started walking out. [laughter]

I didn't hear anything more about what is now known as the first hydrogen bomb test. I did learn quite a bit about research administration and it gave me a chance to study conflagrations and fire storms.

OPERATION FIRESTOP

RKA: Jim Mace, who is probably my closest long-time friend, was the deputy state forester for southern California. Jim was probably the best forest fire general in the business. Of course in southern California you have plenty of opportunity to demonstrate that. He had been invited to Palmdale north of Los Angeles where they now land the spaceships. He had been invited by Donald Douglas to look at flight tests of the DC-7. They were flight testing it with tanks of water in it to simulate loading and distribution of passengers. When they finished they wanted to get rid of the water, so they just flew around the desert dropping water out of the plane.

This was Christmas Eve of 1950 by the way. Jim stopped on his way home, called me long distance and said that we're going to have air attack on forest fires. I said that's a good idea. He said and you're going to organize it. He says it's going to take a Ph.D. During that year we made plans to fight fire with chemicals dropped from airplanes. That had been tried with water in the early 1930s, but they did it with bombs. They filled bombs with water, and the bomb explosions would cause the fire to spread so that there was not enough water to put out the fire. The net effect was more serious fire situations.

HKS: And you used crop dusters, too, very small quantities of water.

RKA: That came during Operation Firestop, which was the name for the one-year research and development program.

HKS: Okay.

RKA: Conventional wisdom said that air attack on forest fires would not work. We had fifty thousand dollars granted by federal Civil Defense. That was the only direct funding. Everything else was contributed; my salary was contributed by the university. The entire fire research unit of the California Experiment Station was donated with Charles Buck as the scientific and testing officer. I was responsible for the organization and direction. We had help from Los Angeles County, the state moved an entire Conservation Corps camp into Camp Pendleton where we did the field testing. Region 5 of the Forest Service provided aircraft or personnel. The Marine Corps provided helicopters. We had twenty miles of bulldozed fire lines in there and we burned about two hundred test fires.

We started out looking for retardants that could be dropped. We found from laboratory testing that just about any common fire retardant could be used. Most were more effective than water. We had an overhead sprinkling system on tracks to run tests on small fires. We learned that you had to get four to ten gallons of water on one hundred square feet of brush to do any good. If you didn't get that much on there would be little effect. The problem in southern California is that water dropped during hot and windy days evaporates before it has any effect.

We had tests of water drops to measure distribution. We found that with the World War II torpedo bomber we could treat an area fifty feet wide and two hundred and fifty feet long with two hundred gallons. We also tried it with helicopters. Water and dissolved chemicals would just evaporate too fast. If you delivered it on the very edge of the fire, it would knock it down for a few minutes, but almost immediately the fire would flair up again. At the last test we were using sodium borate as a fire retardant. We didn't have enough to complete the test, so they mixed it with calcium chloride. That made it kind of a thick gravy-like slurry. When that was dropped in front of a fire, it clung to the branches and the test fire just up and stopped. Talk about luck.

HKS: Sure.

RKA: The Monsanto Chemical Company and the Pacific Borax Company said that sodium calcium borate was readily available. They dug this up in California to use as a slurry for weed killing along railroads. They delivered the mix, we ran more tests and found we had an effective agent for air attack on forest fires. Only crop duster planes were available. They could drop one hundred gallons. They used this sodium calcium borate, and that's why they were called borate bombers. We found out later that we could thicken almost any fire retardant with a chemical that came out of seaweed, it's a common thickener, I don't recall what it is. So from that time on air attack developed rapidly. Forest Service equipment development centers designed the boxes or bomb bays that would give the proper four to ten gallons per hundred square feet and so forth. Many other common chemicals could be used as long as they were thickened.

The Canadians about the same time used our free-fall liquid drop from float planes. Their fires do not usually burn as hot as those in southern California. It's cool enough for water effectiveness. They would scoop up the water from lakes without stopping the planes, drop it, and come back for more water. Although they'd use chemicals, they've had success with just plain water drops. Of course now it's completely operational, there are very few big fires or fast moving fires that don't have water or chemical attack from the air.

HKS: This is not Operation Fire Stop you're talking about is it?

RKA: Yes, this is Fire Stop. And it came because Jim Mace saw the drop from the DC-7 flight test.

HKS: In my mind's eye I have a recollection of you someplace around Bakersfield or somewhere piling branches and stuff to try to create a nuclear attack or explosion in terms of the amount of heat. That goes back to this special weapons project?

RKA: Yes, that was the study of area ignition—the kind of fire pattern that results from nuclear detonations. I think I know where you saw it. One of my early cooperative projects with the experiment station was the use of fire in land clearing. In other words how to develop prescribed burns that you could make safe enough for California conditions. We took areas that wouldn't burn normally, because the fuel was either so sparse or damp that it would kind of smolder along. There must be a clean burn if you want to clear the land or reduce the hazard. We made multiple simultaneous ignitions and called it area ignition. There is a report that indicates that area ignition can be used in some cases. I do not remember piling fuels for the tests.

HKS: I got you off the track on Operation Fire Stop.

RKA: No, I think that just about takes care of Operation Fire Stop. It was more trial and error than research per se. We tried a lot of things. We tried laying hose lines across mountains with helicopters. We even tried wind machines. They were a disaster, because the wind machine would create eddies and you'd have fire around behind the wind machine fairly soon. But the air attack using chemicals did succeed. I think the amazing thing is that if we had had financing for the entire period, the cost would be in the order of a million dollars. We did it with fifty thousand and contributed personnel and operations.

HKS: I suppose the availability of surplus military aircraft made a lot of things feasible then. They're relatively expensive to acquire if not to operate.

RKA: Yes. At that point the crop dusters were doing chemical drops in a different way. By just changing the figuration of their outlet valve they could work pretty well. They did most of air attack for seven years, and it was amazing how successful they were.

HKS: What happens back at school? You're always off doing these things. How about your students?

RKA: The fire course was only one semester. A teaching assistant took some classes. I flew back to Berkeley for most of them. It must have been a lousy job of teaching, I'm sure of that.

But more about teaching fire. I was inexperienced except for my economics study. Charles Buck, who was division chief in fire for the experiment station, was probably the best fire behavior expert in the country. He knew more about it on a scientific basis. He spent ten to twenty hours a week with me when I first started teaching fire. I went on as many fires as I could and attached myself to the most experienced crew boss I could find. Many of these old timers would have fought fire on the same area several times. Between Charles Buck and the work with practical experience, I became a fair predictor of fire behavior quickly.

HKS: I've never participated in any fires in California, but I fought some in the Pacific Northwest, very heavy fuel loads. The fire didn't move very fast but it sure did a lot of damage as it went along. But there everything was the weather. You sort of played with the fire until the humidity went up high enough. Is it basically the same everywhere that the weather is the controlling factor? Or can you really put a fire out?

RKA: Well, when you control a fire, you control a perimeter and let it burn out. The weather, particularly wind, is extremely important on an hour by hour basis. Fuel moisture has slower effects but is usually the determinant of how hot a fire will burn. A good example comes from the Yellowstone fires. The weather was bad, but it was the fuel moisture of three years drought that dried out all the heavier fuels. Normally in an ordinary fire, the tree and branches up to the size of your thumb may be the largest that will burn, or maybe even the size of your wrist, but they had trees up there twelve, fourteen inches in diameter that were so dry and even dead that they almost exploded.

HKS: The current five-year drought in California. Something is going to happen out there pretty soon.

RKA: That's why we developed a sense of conflagration potential. You're looking at the amount of the fuel and how dry it is. The very fine fuels, the grasses and herbs, they change fuel moisture on the hour. It's the larger fuels with moisture change over years that are critical. If the larger fuels are damp, the fire is easier to control. When the larger fuels are subject to long-term drought, you get conflagrations.

HKS: So Fire Stop was certainly successful. I mean you developed a body of very practical knowledge?

RKA: We got air attack started, and from then on it exploded and became operational almost overnight on small aircraft.

HKS: What's the date of this?

RKA: Early fifties, 1953 was the year. We had four months of field tests in the marine base of Camp Pendleton. Ten families moved to San Clemente. The state of California moved in the Conservation Camp. We had bulldozers from Los Angeles County as well as fire trucks and other equipment. We had two military Marine Corps helicopters. How we got all that together was largely Jim Mace's doing. Then we had to organize them all and keep in touch. Los Angeles county was a very strong contributor.

HKS: So by now you're becoming rather an authority on fire.

RKA: No one becomes an instant expert on fire.

HKS: Is this about the time then you actually moved into the experiment station or are you still teaching and doing research?

CHIEF OF FIRE RESEARCH

RKA: No, right at the end of Fire Stop, I was in my university office. George Jemison dropped in one day and said that Charlie Buck was leaving and would I like to become chief of the Division of Fire Research in the station. I told George no. I said I think I'd kind of embarked on this career of teaching and research and I'd better stay here. And he said well think it over and let me know in a day or so. I called my wife and told her. Thirty minutes later I called my wife back and said if it's alright with you I think I'd like to accept it. She said I've just been waiting for your call, I knew you'd accept it. [laughter] Fire Stop was just finished, and we had very strong cooperative relations with all fire control people in the state. There was a need to apply a lot of the information that was then in research, and as division chief, I did a lot of advanced training related to fire behavior, not fire control technology. We still had no major funding for fire research. We only had a scientific staff of four people, which was a little ridiculous for California.

HKS: Did any other stations have fire research?

RKA: The Northern Rocky Mountain Station had a strong fire research program, and there was fire research in the South, particularly dealing with prescribed fire. We put out all kinds of information and propaganda trying to get increases in fire research. But we can thank the Rose Bowl for getting us our major increase.

Several days before New Years Day there was a fire in southern California that looked like it was going to destroy the TV cable to Mt. Lowe. This was before they had electronic means of shooting signals around. If that cable were destroyed there wouldn't be any TV for the Rose Bowl. That brought the attention to enough California legislators and congressmen and maybe some others. We were able to get a million dollars for a forest fire laboratory and a sizable increase in fire research.

HKS: That's the lab at Riverside.

RKA: Yes, that's the lab at Riverside, that's how that got in.

HKS: I would have thought that the need for fire research would have been obvious.

RKA: It was in terms of losses. I'll bring up the problem of financing in California when I became a station director.

HKS: Now you've left the university. Was the experiment station on the campus?

RKA: Yes. The experiment station was on the campus in the forestry building. Part of my job as director a few years later was to move it off campus. We met with university people every day at coffee hour, it was a very close working relationship. Experiment station scientists were used in teaching and gave many individual lectures. It was an excellent arrangement for both the school and the California Forest Experiment Station.

HKS: You're trying to get money. How do you pick the topics to study? Is it easy, or is it launched in debates, committee meetings?

RKA: In every research project you have a project analysis which analyzes the problem itself, the whole thing. It tries to identify the places, the missing links in knowledge and the effort that might make a difference in whatever the subject may be, whether it's regeneration or genetics or fire control, etc. Then with some priorities studies are listed in order of importance, the administrator adds some dollar signs and hopes to receive additional financing. Some project analyses took as long as two years to complete, and they weren't just haphazard guesses. Some of them were excellent scientific papers in their own right.

HKS: A project of any significance would, in effect, be approved by Congress during the budget session, right?

RKA: Yes, they would approve an increase related to that area and for that purpose.

HKS: When the deputy chief is defending the budget, explaining the good work that's going to come, they talk about the specifics of research at the field level in Congress.

RKA: That's right. I guess we have to be honest. You'll have maybe five or six research alternatives, and if the number two or three approach is much more saleable (the number one approach may be basic research), you're going to push number two or three rather than number one. That's rather obvious I think. I was never bothered by that process. Normally, and quite often, if you do get increases for number two or three, then number one comes more easily. If it's true research, you don't solve the problem. I mean it's rare that you identify a problem that you can solve in one year or two or this sort of thing. It's an improvement of knowledge for later application.

HKS: At that time would you be recruiting a physicist to study the physics of fire or were they foresters that somehow were interested in fire?

RKA: Physicists, meteorologists.

HKS: So it was already pretty sophisticated?

RKA: Yes. In fact some of the early people in fire research, even before we got the Ph.D. level, were physicists and engineers. George Byram was probably the most intelligent man in fire research. He was a physicist. Wally Fons of the Cal station was an engineer. George Byram developed the concept and application of fire danger rating very early. Project Sky Fire in Missoula was replete with meteorologists.

HKS: Did you draw upon the manpower available from the fire control folk at the regional office for some kinds of studies or data collection?

RKA: Oh yes, we got quite a bit of help there and from the State Division of Forestry. Carl Wilson, who became division chief when I left, came directly out of fire control on the Angeles National Forest. But Carl's job was principally application and development.

HKS: I met Carl once. I worked in fire research at PNW. I remember Carl coming up from California, we went out and looked at some slash burning with him, he was intrigued by it. Region 6 was burning its slash better than Region 5 was and he wanted to know how come. Anyhow, you were there for four years. That's a pretty long tenure for you so far to stick at the same job.

RKA: A lot of my best friends say I never could hold a job for longer than three years. Fire research continued, and we developed one other idea. The CCC during the depression had put in fire breaks all over southern California. Fire breaks were common, but they were extremely expensive to maintain.

HKS: Sure.

RKA: Because in California in eleven years the brush grows back. Whether you burn or bulldoze the land it's normal chapparal in about eleven years. In fact in three to four years it can burn again. So we developed the concept of fuel break, which was to not leave the area barren. Of course it also eroded when it was barren, and that caused difficulties in the streams. We tried planting either a fire resistant vegetation or just a fine fuel, just plant oats or rye or something like that, plant them on the fire break, and then you could create an instant fire break by back firing. It's been used successfully in California ever since. It's been a fair contribution toward improving fire control in California.

HKS: Were you intrigued by prescribed burning as a way of reducing fuel load at that time?

RKA: Yes.

HKS: Did you have projects going?

RKA: Well, one of the first research projects I ever had was the use of fire in fuel reduction. It was cooperative with the state of California and with the experiment station.

HKS: I talked to Bob Buckman just two weeks ago. We were talking about fire, they've got some severe fire problems in eastern Oregon. Fire's been excluded for so long and they really have a tinder box ready to go off. He was saying the fundamental problem he sees in the use of fire is that the penalties for bad judgment are far greater than the rewards for good judgment so a prudent manager will stay away from fire, because statistically it just doesn't make much sense. Without trying to make you debate Bob Buckman, is that essentially a true statement? Is that part of the problem with fire?

RKA: It is in the West. The penalty Bob mentions is the conflagration fire concept in my thesis. In the South you've got a fairly wide range of conditions between the fire that won't burn and the fire that will escape. Prescribed burning in that area is largely related to weather and moisture. In California and in eastern Oregon, the difference between the extreme where you can't burn at all and where the fire will take off and you can't control it is a narrow range of burning conditions. I've got a couple of escaped fires to my credit that we won't talk about. [laughter] But that's the difference, and you have to be exceptionally precise in the West. There's another angle to it, if you have a longtime operation, you can create fire breaks. Your first half dozen fires may be critical, if you get those out of the way, then you can burn other fires up against them. So if you have a long-range plan and burn critical areas, then you eliminate or you greatly reduce the risk that Bob talks about. I would agree with him.

HKS: I suppose a large part of the problem then was that the weather forecasting was still too primitive, and a front comes through and that's the ball game.

RKA: I was on a prescribed burn with a man who was one of the best fire control experts and used prescribed fire a lot. We were observing a group of ranchers trying to start a burn. They spent up until one or two o'clock in the afternoon lighting fires and trying to get them going, and this fellow turned around to me and said you know, he says, I'll stake my reputation that that fire's not going to burn today. A half hour later that fire chased him and me and his reputation out of there. [laughter] It just took off and exploded. It is difficult. I think the important thing is to have a long-range plan. Not one fire by itself, but a whole series which makes a pattern. The more you burn, the easier they get.

HKS: Was Biswell active in his fire work at Berkeley when you were at the station?

RKA: Yes. He came from the South to Berkeley, and he had not experienced the differences in fire potential. Twice I made myself quite a nuisance to him by saying that no way should he burn. If he had burned a test fire, it would have burned up the town of Santa Rosa. But Harold did a lot of good work, particularly in simplified fuel areas of ponderosa pine with some open grasses or shrubs mixed in.

HKS: What was the primary interest in fire research then?

RKA: Fire behavior.

HKS: Fire behavior.

RKA: Because on every fire you have to predict where's this fire going to be in a hour from now and how hot will it be burning. Then where's it going to be ten hours from now. Can you put men into this canyon or on this ridge. So fire behavior was the most important element in California. In Missoula they concentrated on lightning research, which was most appropriate because lightning started 75 percent of the fires in that northern Rocky Mountain area.

HKS: You were dealing with a lot of arson in California.

RKA: A fair amount, but not as much as in the South. But most of the California fire problem was carelessness by campers and picnickers and people going through.

STATION DIRECTOR

HKS: Do you want to go into the station directorship?

RKA: I think we're ready for that.

HKS: Okay, so it's 1957 and you're living in the same house, driving to the same office, and my goodness, now you're director of the station.

RKA: That is an interesting story. George Jemison and I had spent a week traveling California to review fire research. He said he wanted an in-depth picture of it. We got back in Saturday night, we covered all kinds of subjects during that week. Monday morning when I went into the office, there was a note to go in and see the director. I couldn't for the likes of me see why I needed to see him after spending a week with him. He said that he was leaving the station and going to Washington as associate deputy chief. So I congratulated him, and he said McArdle would like to talk to you about being station director here. I thought a minute and I said, gee, maybe I could get there. I said, "Now I've got this and this, maybe I could go in Thursday or Friday." George said that McArdle wants to see you tomorrow morning. So I caught the red-eye special.

Mac gave me only one piece of advice. (Apparently Sam Dana had mentioned that I might make a good director sometime.) Mac advised that now you've had experience with probably the best station director that we have in the Forest Service, but you can't be a second George Jemison, you just be Keith Arnold.

This is another example of my being in the right place at the right time. Because of a death and then George's move, California had had three experiment station directors in about two years. They could not bring in somebody else from outside and go through major change. George had gotten on top and knew the key people and the key cooperators and all of that sort of thing. There had to be someone local who knew the ropes. I was the only local with a Ph.D. that they could think of at that time. I hadn't been on a forest lookout, I'd only been in the Forest Service a few years and all of that in California. I didn't think that I was qualified for director but the chief and staff did, and so I became the director.

HKS: McArdle came out of research.

RKA: Yes.

HKS: Did you find that good, that he understood research?

RKA: I never had a lot to do with McArdle. I can't tell you when he shifted the reins over to Ed Cliff.

HKS: About '62.

RKA: About '62, and that was about half way through my directorship.

HKS: But he had the RF&D meetings, so officially you got together as a group several times a year.

RKA: Through the years if a chief comes out of research, he pays a little more attention to administration, and if he comes out of administration, he wants to be sure that research knows that he's giving it appropriate attention.

HKS: I suppose.

RKA: John McGuire did not fit that mold, but I think McArdle and Cliff did. It was never serious, that was my impression. McArdle did pay a little bit more attention to the regional foresters and to the National Forest System. Not to any detriment, and Ed Cliff was always responsive to research. John I think hit a balance in between the two.

HKS: Tell me a little bit about Harper, everyone seems to be impressed with the man. You're working for him now, in a sense, or maybe Jemison is who you're really reporting to.

RKA: No, no you report to Harper. You reported to the deputy chief. Les played everything close to his vest. I had contact with him as a director and then in Washington as division director for forest protection research. Les had a keen mind, he was highly politicized; he and Senator Stennis had breakfast every Saturday morning. He used this effectively to keep Stennis up to date, and Stennis is one of the strong supporters of research. You never thought of him as ever having a good time. He was almost always business. We had him on an inspection trip to Hawaii, and he wasn't sure that there should be any research in Hawaii, there were too many distractions.

HKS: Yes. But this was the time of big growth for experiment stations, and new facilities in general.

RKA: Yes, and Les probably was largely responsible for directing and coordinating it—a time when research was certainly growing rapidly.

HKS: You must have had some assignments to deal with the California delegation.

RKA: Yes, although in a normal way—not actually assignment.

HKS: I don't know what facilities there were. You said when you were director you moved off campus so you had to get some money to do that. You must have been involved with talking to the local congressmen about the benefits, right?

RKA: Not at that move, that move was made in-house. We'll talk about that in a minute.

HKS: Alright.

RKA: For example, there was no question that George Jemison would follow Les. George was exceptionally able and he was in the right place. But when I was in there Les was in effect training me as someone who might move into associate deputy chief and on up. Maybe I wasn't perceptive, but it never dawned on me that that was what was going on. He was very strong in the South. I don't know that Les understood the West as much as he did the South. I think the same thing applies to me, I don't think I understood the South as much as I understood the West. You never thought of Les in a social sense of any kind. He was just all business. That's the way that I saw him.

To get back to the station directorship, there were two administrative things that bothered me. One was the station name. It was the California Forest and Range Experiment Station. You asked me a while back if fire was so important, why we didn't get more fire money. There were a lot of people nationally antagonistic toward California. It's a big state, it's growing, and you don't want more for California. California at that point was the only station with just one state. E. I. Kotok did quite well and I don't know how he did it except to spend a lot of time with California congressmen. But we received the normal growth that the Forest Service in general got. If it got an increase for forest management and research nationwide, we received our part of that. But it was obvious that the name was a handicap, and we were beginning research in Hawaii. We started thinking about new names. We thought about South Pacific where we'd have our own theme song; that didn't work too well. [laughter] We came out with the Pacific Southwest Forest Range and Experiment Station to make it kind of parallel to the Pacific Northwest Station, and it's had that name ever since. But in addition to recognizing Hawaii, the major reason for the name change was the problem of financing things like the fire lab in southern California.

HKS: Was range a very significant part of the research?

RKA: Yes. We had a fairly strong range management unit, it had two experimental ranges and then did a lot of cooperative work with the national forests. We had an experimental range in northeastern California, Black's mountain, and we had the San Joaquin range in the central Sierra. It was small but it was a very good program.

The other administrative matter was more substantive, and I thought more serious. I referred to Duncan Denning as a top scientist and a rather poor administrator. He didn't like administration and was quite honest about it. When I became director there were nine division chiefs. Every area of the forest had its own division in the station. Forest management, genetics, range, watershed, fire, insects, disease, economics, and products. Denning retired at age fifty, and we lost a great scientist. That would prey on my mind every now and then. So I tried to figure out some way to reduce the number of second level administrators. I'd say two thirds of those division chiefs were top scientists, and most of them were not cut out for administration. Largely, they were only interested in their own area— not concerned with the station as a whole.

John McGuire had moved out to the station as chief of economics research. John and I together did a lot of thinking on these kinds of things. On an inspection trip we came up with a concept. We decided to change to four assistant directors. One of those would be for application and planning, and each of the others would handle several areas; that is, forest management and genetics and insects and disease could be together.

It took quite a bit of selling. I spent about a year selling the idea to the people in the station. Then it had to be approved by Harper and the chief. As you probably know it's now the organization with some minor changes in all of the stations. Quite a few of the regions and the Washington office have gone the same way. I think it was one of the things that was important to get done at that time. We gained some scientists, and then we had people in research administration looking at the station as a whole, not as one narrow part of it.

HKS: Given the way his career unfolded, when you worked with John McGuire, did you sense that he was going to go far beyond where he was then, or did you think of him mainly as a good economist?

RKA: No, John was a very broad-gauged individual, seemingly very quiet. I enjoyed working with him when he came to Berkeley. He did not stay long, but if he'd stayed he'd have been one of the assistant directors, obviously.

HKS: Okay, they wanted John in economics.

RKA: They wanted him to have some experience. John had worked at the Northeastern Station and had worked for Ed Crafts in planning for a long time. Obviously they wanted John to have administrative experience, which meant he'd be a research director. He needed to have some other experience, so they sent him to California, and we became good friends as well as professional colleagues. Then he went back to Washington and returned when I left for Berkeley. In answer to your question, I didn't know whether he'd be chief or not, but I knew he would be in some leadership role. In fact, I guess at that time I wasn't worried about the chief. Actually, the California station director is probably the best job in forestry in the country. Your boss is two thousand miles away, and in the Forest Service you had a lot of delegated responsibility.

It was a challenging job. In that period of 1957 to '63 as director, we developed a full research program for Hawaii. George Jemison had gotten the idea started, had made preliminary contacts, and had sent Bob Nelson over to take a direct look at their research needs. George grumbled about Hawaii. He said here I do all the work and you get to go to Hawaii every three or four months.

HKS: What was the basic rationale for Hawaii. A foot in tropical research?

RKA: In the very early 1900s there was a Division of Forestry in Hawaii, and those foresters were botanists. In the early days, Hawaii had a lot of sandalwood. The whaling ships would come in and leave oil and whatever and would take loads of sandalwood to the Orient. They fairly well denuded some of the forests. Hawaii's water is in fresh water lenses, under each island. If those lenses dry up, you've got nothing but saltwater. The mountains are so steep that percolation is not good. For a while, they thought they had lost some of the percolating capacity and might even lose the lens. So the forestry department was designed to reforest about two-thirds of the islands. All of the higher elevation parts were in state forests with fences and gates around them. You couldn't even get into them. The policy was to go world wide and find species of trees that could not have any commercial value, which they did.

HKS: So this was research aimed literally at Hawaii's problems, and not some broader application.

RKA: Oh no, we're not there yet. This is the program that the state foresters of Hawaii developed in the early 1900s. But then as sugar cane and pineapple moved elsewhere, the islands of Maui, Hawaii, and Kauai all began to have severe economic problems. It looked like maybe they might have to look at timber production as one way to help the economies of the outer islands. This was before the big tourist inroad, they all went to Oahu instead of the outer islands.

So there was concern that there needed to be economic development. The research that we were over there to do was to look at commercial possibilities. There were some beautiful eucalypts over there, and there were some recent pine plantations on Kauai. We had watershed research to be sure that we knew what could be done without damaging the water-percolating capacity. Then just plain silvicultural research. Russ LeBarron, who'd been in the northern Rocky Mountains for a long time and then was forest management chief in California, helped start the work. When he retired, he went over there. It was the kind of research that was done thirty-five or forty years ago in Forest Service in this country, but it was aimed at how can we help the economy of the outer islands. Now the tourist industry has taken care of that economic problem and so now they have broadened the base of research over there to include more fundamental research in tropical forestry, semi-tropical forestry, and that sort of thing.

HKS: I guess it's a wonderful research ground for studying the impact of imported species.

RKA: They're all exotic, right.

HKS: It strikes me a little strange. At budget time, yes Mr. Congressman we really need some money so we can put a research station in Hawaii. I am curious why this was approved.

RKA: Because the Hawaiian delegation put up a very strong front and their economic condition was critical.

HKS: The information gained from that. Can that be generalized and applied to other islands?

RKA: Yes. I've lost touch with it, but I think it's now the Pacific Islands Forestry Research or something like that. We look at applications for Guam and other areas.

Another problem was the move of the station from the campus. We were the single largest non-university user of university space, and as space became tighter on campus, we were asked to move. We had a building through General Services Administration built about two blocks from the campus. A location near the campus was important. We had a lot of our employees working toward a Ph.D. on the side, either on a part time basis or on leave. Then the farther we were away we would lose the direct cooperative contact with School of Forestry and other scientists. We lost that anyway because no longer did we share the same coffee room, which was the most productive place for individual contact.

We did strengthen the cooperation with University of California and the California Division of Forestry, that was easy for me to do. George Jemison had started a cooperative study of the wild land research for California which involved the state of California, the forestry school at Cal, and the experiment station. We finished that up in the long-range wild land research plan for California, with priorities for all research going on and for the list of things that needed to be done. That paid off for me. A couple of fellows from a congressional committee came into Berkeley one day and said they were reviewing Forest Service research and could they review us. I of course said yes and they said well now we want a list of all your low priority research projects. I said I don't have any low priority research projects. They didn't like that approach and got fairly antagonistic. There's always high and medium and low. The wild land research plan which had been in effect about a month listed all of our projects with a priority rating which happened to be high. I handed them the report. They walked out of the office and did not return while I was there. There was just one other thing that I did as director. I was always concerned that up until that time a scientist, to get ahead grade-wise, had to go into administration. I tried to make certain that as a division chief in fire and then as director that I had a scientist with no administrative duties of equal or higher grade. We kept one or two senior scientists at the station, and that was the finest recruiting for top young scientists that we had. Here we could point to this man. He's the same grade that I am, and he's a pure scientist.

HKS: Did you have a pioneering unit?

RKA: Yes, we had one in fire physics. We picked up a navy physicist, a researcher. That covers what we did. At that time genetics was coming. That was a good time because research was increasing in terms of funding, and we had good support in the stations from the Washington office.

HKS: You were the director until 1963. In 1962 Rachel Carson published *Silent Spring*. Certainly the sort of general public interest in what we now call biological diversity and the balance of nature became household words. That must have had an impact on relationships and interests in Congress and certain kinds of research. Did you feel that immediately or was that a longer term thing.

RKA: It was longer term. We got quite an impact on it because the lake one hundred miles north of Berkeley was her observation area. It gave more weight to ecological research certainly, but I can't recall that it caused much immediate change. And I don't think we got any congressional support because of that.

HKS: My recollection at the station in Portland is that it was considered a bug problem, DDT and the spraying. The entomologists were rather upset by the publicity.

RKA: I hadn't thought about it, but our entomologists were most concerned at the same time. That's true.

DIRECTOR, FOREST PROTECTION RESEARCH

HKS: Okay. You finally get to leave Berkeley.

RKA: I figured being a native son, I was ready to stay in Berkeley the rest of my life, I thought it was great. But there was a new position in the Washington office. Les Harper had nine division directors in Washington reporting to him, and that's just too many. Most of them had their eyes focused on their narrow unit and it was pretty hard to use them as a force for either planning or direction in a general sense. So I was asked to come in as a director of forest protection research. I was there for three years. It reduced the number of people reporting to the deputy chief from nine division directors to six. I had fire, insects, and disease, it was called the Division of Forest Protection.

HKS: Does that really make sense from a scientific point of view? Where you have two biological things and a physical thing.

RKA: No, and now it's changed. Forest fire is separate and insect and disease are together. Because of my background I guess that it was all right. I think I ought to mention that in fire, insects, and disease we had three gentlemen who were top scientists in their field. A. A. Brown in fire, Jim Beal in insects, and Ray Hansborough in disease. Now you can bet that they were against this 100 percent. They no longer had direct access to Les Harper, they had to go through me.

HKS: Les created a layer then, where everyone else stayed in place as it were.

RKA: The final judgment on budget came from me rather than their working with Les. But they were gentlemen and they not only lived with it they supported it. There was no backbiting. It was a difficult situation for me, they could have made it impossible. But here again it was a good move, and you're right that fire and insect and disease don't go on together. But with my fire background and then the station background, because in the station I think I spent more time on insects and disease than I did on fire, just for the same reason. But I would certainly like to have it on record that those three men, taking in effect a demotion, didn't let it bother them either personally or professionally. They did their job and lived with it.

HKS: I suppose it would have been difficult if Harper had picked one of those three to head it up.

RKA: Those three were more scientists than administrators. I don't mean that in a negative sense, I mean that they were science oriented. They saw the picture as related to their specialty, but not really interested in Forest Service research as a whole. Again it worked out for the best. They had plenty to do, and the growing research in the stations required more of their attention at the scientific level related to the scientists in the stations, and to university relations in their field. I had a very large number of special assignments from Les Harper—to prepare position papers and letters that normally he or George would do easily. I thought it was part of the job, I wasn't aware that he was trying to give me experience that might lead to that job of deputy chief sometime in the future.

HKS: Were there many, how can I say this, good scientists in the Washington office, as opposed to people who were scientists that were in management? You characterized the three gentlemen as people who were really better off as scientists. What do scientists do in Washington?

RKA: Scientists in Washington are responsible to evaluate and inspect the scientific programs in their area in experiment stations and the cooperative work with universities. They do a lot of work in evaluating the station's staff, looking for moves that would improve research and bring on future leaders. They write position papers and answer letters that require scientific expertise dealing with their specialty, and it would take a top scientist to do that. They were largely responsible for the international scientific cooperation in the early days, in IUFRO and FAO. If it came to insects and disease they organized and participated in the meetings. Jim Beal and Ray Hansborough were very effective in international forestry in the area of insects and disease as was Carl Ostrom in forest management research.

HKS: Were their colleagues in the other specialties of equal stature as scientists, and the jobs were all basically the same, analyzing the research?

RKA: Yes, and working with the deputy chief to program priorities in various stations. Carl Ostrom, for example, was a top scientist but also an exceptionally able administrator, as was Herb Storey, and they both became associate deputy chiefs later on. Usually the Washington office division directors were able in the administrative area. But with nine of them you couldn't really develop progress or programs that you might want.

HKS: Okay. During this time period, Ashley Schiff wrote a book called *Fire and Water: Scientific Heresy in the Forest Service* that took on your specialty. What was the reaction within fire research? What was your reaction to this book?

RKA: It was completely naive. I read it, and he obviously had not been on many fires and had undoubtedly reached the conclusions in the book long before he wrote it. I can't recall the details now but I know I read every word in it.

HKS: On the fire side basically it was that the fire protection people were so dominant that the benefits of prescribed burning were supposedly suppressed so not to interfere with the larger mission of putting fires out.

RKA: That was just not so. In the South prescribed burning was pioneered by H. H. Chapman and Forest Service people and carried out. I didn't know of any reluctance to do prescribed burning except in the areas where as I mentioned the available moisture and weather conditions between no burn and conflagration was very narrow. And Schiff did not really recognize that potential.

HKS: McIntire-Stennis. You refer to the cooperative agreements with the universities. Obviously that's very significant, so comment on how this expands the Forest Service mission, or what did it do really when the universities had access to funding?

RKA: There was careful coordination between M-S and the Forest Service grant program. It allowed university scientists funds for better cooperative research projects. M-S could only be used by state universities in the land grant colleges. The Forest Service largely supported non land grant college research at schools such as Yale. I think at first the Forest Service was a little bit concerned that it might reduce Forest Service research budgets, but I don't think it had that effect.

HKS: We have pretty well covered your time in Washington as director, except that I have to put on the record that that's when I met you. About '64 there was a fire research workshop in Denver, and Dave Bruce, Bill Morris, Owen Cramer, and I came. Everyone went around the room and explained what they were up to, and so we shared information on what was going on at the various stations. You came in and asked if we had all the money in the world, what research would we want to do. Of course I was very junior and everything to me was a learning experience. What impressed me was that more than a few people were really intimidated by your scenario. Is that a technique you've used on other occasions to get people to use their imagination?

RKA: Yes. That came out of the group dynamics that I mentioned before. I used it from time to time to try to get priorities in focus. But really you hit it, it's to see who has imagination and who doesn't.

HKS: I remember some of the presentations by people who appeared to be relatively senior. They were very nervous.

DEAN OF MICHIGAN SCHOOL OF NATURAL RESOURCES

HKS: Alright, it's too good to be true, you were staying with the same kind of job for a while, but now you're going to leave the Forest Service and go to Michigan as dean. How did that transpire? Was there a job announcement and you applied for it?

RKA: No. Steve Spurr had been dean of the School of Natural Resources there. He was moved to vice president for research and head of the Rackham Foundation at the University of Michigan.

The University of Michigan had four departments in Natural Resources at the time: Forestry, Landscape Architecture, Wildlife and Fisheries, and Conservation. They brought in several foresters as a possible dean of natural resources and none of them were acceptable to the other departments other than forestry. Ken Davis was acting dean. Ken wanted the job very much and openly tried to get it. Ken just could not relate well to conservation. He was economic minded and I guess rather bull headed on relationships. I don't mean to be unkind to him, we'll talk about that in a minute. But anyway they'd been at it a year and had not found anyone.

Steve Spurr and I were on the SAF Council and were meeting in Oregon. He had to leave the meeting early, because he was going up to see the ice break up on the Yukon River, a typical Spurr trip. I took him into Portland airport. We got to talking, and I said I'm always interested in schools. I said that sometime I want to come back to a school situation. He said well why don't you come out and visit us. I said well, maybe. Two weeks later he had arranged for me to go out and visit the school. Ken Davis was the first person that I talked to out there. Ken was very strong, he said he wanted the deanship. He came there fourteen years before with the idea that he would follow Sam Dana. He said Spurr got in the way. He said now I'm going to keep on working for the job but I don't think I'll get it. He said if you're selected as dean, you have my 100 percent support, but meanwhile until that happens I'm working for it. A true gentleman. Anyway, I accepted the job, and it was an interesting one. I was interested in the broader aspects of forestry which it offered.

HKS: Michigan was one of the leading institutions.

RKA: Yes, in forestry education. It was Sam Dana's school. It was a school that was always ahead in education. It was formed as a Department of Forestry in 1903. I don't know exactly when Sam came into the picture, but it became a School of Forestry and Conservation in 1927, there broadening the forestry education. Now that was Sam Dana. And in 1950 it became the School of Natural Resources, and Sam Dana brought in and developed the conservation department along with forestry and fisheries and wildlife. Landscape architecture moved over from the architecture school at the request of the department head, Walt Chambers. He was feeling stymied, he couldn't do what he wanted in architecture and he thought he'd have a better chance to go on his own and pay no attention to a dean if he got into the School of Natural Resources.

It turned out there were many areas of common interest, as you would know, and they became more important. Besides the original departments, there were new programs in resource economics, urban planning, and resource administration. A recreation curriculum was just being developed. We had four departments, forty faculty, and four hundred students in round numbers—it makes it kind of easy to remember. In my job interview with President Robin Fleming, he pointed out that there was going to be a reduced university budget, because the state of Michigan was in difficulty. He said small independent schools are more apt to be eliminated than just across the board reduction. He said I'll admit to you that the School of Natural Resources is one of them. Per student, it's the most expensive school in the university. He said that the dental school was another possibility that he had in the back of his mind that if push comes to shove they might have to eliminate.

HKS: In the '60s with the environmental movement and so forth there was burgeoning interest on the part of undergraduates in environmental education. That hadn't quite happened yet?

RKA: No, but it was right on the threshold. Some of the best students in the school were in conservation.

HKS: Okay.

RKA: It was starting. But Fleming made the further statement that there's no way that I can close this dental school in the university. Every state legislator has one of our graduates boring on his teeth. [laughter] He said I just want you to know the risk, and I want to support the school and will do it. Well that led to the need for several actions on my part. One was to make the school more important to the university with more cross campus courses, more service courses, more attention to the planning, and more integration within our departments. Forest products as a program had been eliminated several years before. It used to be a major activity, but it had been eliminated. There were still two full-time tenured products professors on the forestry payroll who did nothing. They weren't even doing research. They were just occupying their offices. It took two years to have them move elsewhere before we could recapture that money.

The new programs were in the right direction for keeping the school. It became my goal to support those more, so priorities went to improve these new programs. We brought in an urban planning group from Michigan State. They helped develop this broadening. But the changes did impact the traditional departments. This is the first time I've publicly ever mentioned this particular problem. In three years the president told me the school was safe, that he didn't have any problems with it. That is one of the reasons I was able to leave to go back to the Forest Service.

Let's talk about Sam Dana. I was dean of his school, and he was retired, but he was professor emeritus. He was there every day, he taught some policy seminars. He never attended faculty meetings. He would not come in and sit down in my office. If I had occasion to ask something, he'd stand and walk out. It was unbelievable that he and Steve Spurr both left the job to me. Now if I called Sam and said are you going to have happy hour at your home this afternoon? He always said yes. I could go out there and we'd talk about anything. I could ask him and he'd give me advice, but never once did he come up with any suggestion or direction in the school unless I asked him for it.

HKS: Both Buckman and Dickerman are fascinated by the history of Forest Service research. Between the two of them they've been collecting odd bits of memos and so forth. And they sent me copies of all the stuff they had collected. One was written by, I think, Earle Clapp about 1935 when he was about to be kicked up to associate chief. He was looking for his replacement, and he was analyzing the candidates. One was Sam Dana. There's three or four others, I don't know who they are now. But the critique of Dana is pretty harsh, and I'd like to have you react to it. That Dana was overly ambitious and self-serving. In the '30s he would have been middle-aged I guess, does that ring true at all? I was amazed at that critique.

RKA: That surprises me. He was on my Ph.D. committee. The fact that he stayed at Michigan when he could have been a university president or a high ranking government official in a number of places had he so wanted. If anyone was ever quiet and unassuming, it was Sam.

HKS: It's intriguing that his superior critiqued him that way. In some way he got crossways with Clapp.

RKA: That might have been. You don't know what might have happened in one instance but that was not Sam Dana. I probably have more than one hundred hours with him just one on one talking about forestry, forestry education, and anything that I wanted to talk about. That is amazing.

When I did leave Michigan, there was forestry alumni criticism that I used the University of Michigan as a stepping stone to the deputy chief's job. I could not answer them. I wasn't about to talk about the possibility that there might not have been a school. After my initial talk with the president, I had developed a five-year plan for the school. After three years it was on target, some by my efforts and some by just natural development. Lyle Crane in the Department of Conservation had a very broad look at the entire picture of ecological and social systems impinging on forestry, and that led to a lot of things.

HKS: Do you want to talk about Steve Spurr at this point?

RKA: Yes. Steve is probably the most intelligent individual ever related to forestry. He wrote that whole series of textbooks. Wherever he was he was ahead of the game. He knew about the school in Michigan, yet he never talked about it unless I would bring it up. He and I had prepared a paper for the World Forest Congress. I think we were both on the same wavelength. He was an excellent administrator and a top scientist, he had both of those skills. I was never a strong scientist in the narrow sense. I did have most of my interest in the administrative parts of it.

Steve was physically oriented. He ran almost every day. Even as vice president of the university he played water polo with the water polo team. He had to be active all of the time. In the middle of a Society of American Foresters council meeting, he would get up and walk over to the corner and do fifty push ups and get himself awake again. I think his greatest contribution was *Forest Science*. He came up with the idea for *Forest Science* and then was the editor.

HKS: Everyone who talks about Steve Spurr is almost in awe, certainly admiration, at the diversity and the talent and the goodness of the man.

RKA: Steve had Parkinson's disease and had a slight stutter when I first met him, even before Michigan. I always thought that his brain was going faster than his mouth. But he'd had Parkinson's for a long time, and it was under control. He had had open heart surgery at the University of Texas. It worked very well, but in treating for the by-pass they lost control of the Parkinson's. The medication that was needed to preserve and make sure that the tissues were not rejected in the open heart surgery caused the loss of the Parkinson's. He lost a lot of control physically, but not mentally.

I never knew about Parkinson's until he was giving a paper at Albuquerque at the national SAF meeting. Halfway through he said "And I thank you ladies and gentlemen" and sat down. I was sitting in the front row. He took a couple of pills out of his pocket, he thought he was having a heart attack. It turned out that he wasn't, but when he was in the hospital, the doctor said that he was taking medication for Parkinson's. And from then on he went down physically pretty fast, but he still kept an office at the university. He still typed. He swam half a mile every day that he could. He walked when he couldn't run. He was a fighter beyond anything that anybody could imagine.

HKS: At a AFA meeting about five, six years ago, he received some award, one of their several awards for outstanding service to American forests. He was at the head table. Everyone was aware of his frailty at that time. It wasn't at all clear if he was going to be able to stand up to make the acceptance speech. There was a very long pause. I heard afterwards that his wife was ready to take the script from him and read it in his behalf. But he got up and he walked three or four steps to the lectern and he read in an absolutely flat, toneless voice, he got through that. It probably was one of the most dramatic moments any of us had ever experienced. The house went crazy with applause for his ability to pull himself together for the few moments that it took to make that speech.

RKA: From then until he died a couple of years ago, I'd been in his home over night many times, and he had a talking board. There's a board about a foot long with one inch squares. He had to touch a square with each syllable to talk. Without the board he couldn't talk. He was just amazing.

HKS: So the school at Michigan is in good shape.

RKA: At least the school at Michigan was secure within the university and it had been given a push in the direction that Sam Dana started it in the 1920s of broadening the base. I think now most of the larger forestry schools have gone that way to natural resources training more broadly or to emphasize specific areas of research and teaching. In the 1920s, '30s, and '40s there were three outstanding forestry schools: Yale, Michigan, and California. Today, we can only look at specialities within a school.

HKS: Let me back track one question. When you left the Forest Service to go to Michigan as dean, was any part of your decision because of being director of forest protection research wasn't very exciting or you looked ahead and you've done about all you're going to do in the Forest Service and that's a chance to try something different?

RKA: No, no there was no dissatisfaction of any kind with the Forest Service or my work. It was just opportunity offered. I don't think I would have gone to very many forestry schools, but I liked the School of Natural Resources. It was a professional decision that appeared to me to be the one to make at that time. I might say that at that time Ed Cliff talked to me and, for the first time, he said you know, we have you with one or two others who obviously are candidates for Harper's job at some point in the future. That was the first time that I had ever known it. He didn't argue about my going but he tried to give me a little encouragement to keep me in the Forest Service. I've never left any of the positions I had because of being unhappy with either the situation or the promise of the future.

Deputy Chief for Research

HKS: So now you're about to go back to the agency. How did that happen?

RKA: That is an interesting one. The Western Forestry and Conservation Association had a meeting in San Francisco in December of 1968. John McGuire and I were having breakfast, and he pointed out that George Jemison was leaving shortly. They were having difficulty in finding someone professionally qualified for the job in civil service terms. I don't know who it was but there was someone in the Forest Service qualified, and he had passed away. Dick Dickerman was obviously qualified. He had been George Jemison's associate deputy, and Dick had all the qualifications. His wife was ill and he was unable to travel. That job required travel, both national and international. So Dick had removed himself from consideration. John told me at breakfast that Cliff was worried because this might be a way of making the first political appointment in the Forest Service of a nonprofessional nature. John said that I was the only one now eligible. I just made the offhand comment if that's the case I'll come back.

HKS: What's John's position at this moment? He wasn't chief, Ed Cliff was still chief, right?

RKA: Ed Cliff was chief. John was special assistant to the chief. Again, Christmas Eve. Here is another example of my being at the right place at the right time. Ed Cliff called about 4 o'clock in the afternoon in 1965, December 24th, and he said he'd talked to John and he said I wanted to call you right away. He said you know, since you talked to John, Secretary Freeman has been talking to Secretary of Defense McNamara, they play handball every week. Freeman was saying that they didn't have anyone in-house who could administer the Forest Service's research program. Under the civil service rules there was nobody eligible. McNamara told Freeman, I've got forty or fifty research directors over here, they can direct any research in the country, he said that he would be glad to let Freeman have one. And of course that upset Ed Cliff because that would mean the first nonprofessional appointment in the Forest Service. He said would you consider accepting the job if offered, and I said yes. I said in fact I'm coming into Washington next week. He answered that he needed the application on the 26th of December. So I hurried down to the post office before it closed and got some forms and spent all day Christmas Day filling them out.

HKS: The standard civil service forms?

RKA: The standard civil service forms. About a month later they went through all of the reviews and so forth and so that's how that came about.

HKS: Was it because of the high rank that the civil service was so involved? I was going to ask the question earlier when you were out at Berkeley, it seemed rather casual hiring. I mean the people weren't off the civil service roster apparently, from the way you narrated the story. Did it become more difficult at higher levels, or was civil service tightening up the rules as time progressed, or have I missed something here?

RKA: No, there is no short-cutting of the civil service procedure.

HKS: Okay, that was the problem.

RKA: Then when you apply it goes through the normal civil service. Even in Berkeley, George couldn't guarantee me the job. He said I'd like you to apply, and he says that based on what I know I don't think there's any question that you would be selected.

HKS: Alright.

RKA: It was a matter that there was no one qualified on the basis of experience and education who could qualify for the position. As a director of forest protection research with university experience, apparently I was eligible. But anyway the official offer came and I moved down there about in May.

There's one major difference between universities and the Forest Service as research was carried out. In universities you can have a lot of argument or discussion or various viewpoints on whatever you're going to want to do and then you can have a vote and decide that this will be done, this is the direction we are going. But most tenured professors continue to go whatever direction they want to go. I don't mean this in a derogatory sense, but very few of them are interested in the "school as a whole" or the "university as a whole," they are interested in their area. As tenured professors they don't have to have more than an interest there and some of them can continue to be antagonistic. In the Forest Service, you'd have maybe violent discussions and so forth, but once you decided everybody goes in that direction. It's just a pleasure. I learned that the hard way at Michigan, of the independence of professors. It was repeated again at Texas. It makes the leadership job much more difficult.

HKS: Oh sure.

RKA: Anyway, it was a tremendous joy to return. First with Dickerman as associate deputy. I mean here he was fully qualified and probably better qualified than I. Anytime I'd go on a trip my desk would be empty when I came back. Usually it was full when I left because I'm a dirty desk person as you can see, I usually spread things out. Amy King was secretary. She'd been secretary for Harper and for Jemison. She knew everyone, knew the congress people. She could probably have handled the deputy's job. Most key positions were well filled as station directors and the division directors. We were in an area of increasing budgets; Les Harper I think probably generated a lot of that, and George carried on.

It's always easy to kind of look good if you have funds to improve programs rather than to do it the other way around. It was the continuation of environmental concerns started in Michigan.

I would have put in my vote to come back in order to participate in the beginnings of the environmental and social concerns related to forestry. The National Environmental Policy Act was passed in 1970, and in fact President Nixon had labeled the '70s as the decade of the environment. This had very strong support from Ed Cliff. Ed was just great, just really going out of his way to make me feel not only at home, but to assist in the directions that needed to be changed or whatever decisions were made.

HKS: Let me follow up on that. One of the back benchers in the Washington office, he used to say he sat in the second row at chief and staff meetings, characterized Ed as overly tough. He didn't ask for questions, he came in to meetings, read the agenda, and the only person with courage to challenge Ed would be John McGuire. Do you share any of that view. I mean is this a matter of temperament on the part of Ed that he seemed gruff but wasn't or what was that. Why would that person have said that?

RKA: I don't know. That does surprise me.

HKS: You weren't talking about the same man this other guy was.

RKA: No, I don't know how many times he was in meetings with Ed. I never felt intimidated by Ed. I travelled to international meetings with him as his principal staff man and of course was in his staff meetings.

HKS: Ed may have been rough with National Forest Management.

RKA: But I never felt the least bit intimidated by him. In fact if something came up at the staff meeting that I didn't want to talk about in public, I felt really comfortable with going in and talking with him after the meeting, privately. Maybe because of the way that I came back he treated me specially. I knew Red Nelson very well, who was the deputy chief of national forest administration, and I never sensed that he had problems with Ed. So I don't know where that came from.

What Does the Deputy Do?

HKS: You are answering this in pieces here, but what I'd like to be able to come up with is, "What is it like to be deputy chief?" What does a deputy chief do? I mean, you get up in the morning and brush your teeth and eat breakfast, you drive into work. I know there's no typical day. There's so much going on that you have to delegate a huge amount, what's left for the deputy?

RKA: As deputy chief there was a heavy overload, and I soon considered it an issue. Most of it grew during Harper and Jemison's administrations, but it was a sixty million dollar program. Now it's over a hundred million. You're responsible for the work of a thousand scientists in eighty scattered locations. As deputy chief at least the general thought was you would try to visit every location every three years in terms of a review of the work and inspection of the program. Some of that the associate deputy did, but the deputy chief needed to keep in touch.

You're a member of the chief's staff, and there's a chief and staff meeting almost every morning. And you're considering all aspects of national forests, broad budget decisions, letters from congressmen. You're acting chief one month of the year, and the chief is in and out all the time. You cover as necessary. I always picked the month of December. I figured out usually if you try to take December off there's always some emergency that came up and they'd call emergency staff meeting or something, so why not just work in December, which I did. There's no small workload attached to being prepared, to analyzing things, to go from a chief and staff meeting to your key people.

There was the Nixon White House, and you'd get a phone call, "this is the White House calling," I need an answer to this question in three minutes or seven minutes or something like that. Most of the time you knew what problem had entered the White House. Amy King would immediately get the right person on the telephone if it were something that I didn't know about. When I was deputy chief, there were one hundred and five members in the research staff in Washington, that includes secretaries and key people, and it included a fair group, twenty or twenty-five, in international forestry work. Keeping in the loop of all relevant activities was no small job.

The cooperation with McIntire-Stennis took some time in program development. We had joint meetings with McIntire-Stennis leaders regularly to look at the research programs. McIntire-Stennis only applied to land grant colleges, so you still have Yales, Dukes, and other places where there were forestry schools. We tended to direct a little more of our grant money to those institutions if they had the right people. You asked about McIntire-Stennis before. I think it was helpful to have universities with their own funds coming directly, because they had to learn how to get the funds. It took some of the pressure off of us. Details of budget preparation and congressional hearings take time. Then there were requests from Congress, whatever it is. I want a request of what you need or why are you doing this research or whatever. There could be as many as ten of those a day, sometimes there would be a week without any.

HKS: In response to an inquiry from constituents?

RKA: It could be that, sometimes some constituent would ask his congressman why are these people doing this, or I think we need more research here. The congressman would call us and we'd provide appropriate information. That takes time. We had several people who could do it. Besides the associate deputy chief, Carl Ostrom, Herb Storey, and others were helpful. International forestry took lots of time. The State Department did not handle international forestry matters. Forest Service Research had the full responsibility. The published papers and national meetings were important. I counted up, in two years I gave thirty-three professional papers scattered around the country. I didn't write all of those, but I'd say probably that I prepared a quarter of them myself. I just wrote the first draft and then had somebody smooth them out. In other cases somebody wrote me a draft and we'd kick it back and forth.

HKS: That's a lot.

RKA: Many of those related to the environment and forestry, the new changes. Maybe I accepted more than I should, I don't know. Then there was other related work. I had the Department of Agriculture assignment for leadership with NASA. NASA had an inter-agency committee, and I represented agriculture and forestry. That resulted in a trip to Russia for the Earth Resources Satellite Program. These programs mapped the waters and mountains, detected insect outbreaks, looked at erosion, etc. This was an antagonistic meeting. The work of that meeting took over one month. I was the agricultural representative of the inter-agency work group on meteorology. We had ties, of course, with the Park Service. We had research going on in Yosemite in California on insects and disease in the parks. So you were asking what does a deputy chief do, that's kind of a listing of the sorts of things.

HKS: Did you have much control over your agenda?

RKA: No.

HKS: Was every day a surprise?

RKA: Every day is a surprise. I think we were working from nine to five-thirty with a half hour off for lunch. I made the habit of getting into the office about seven in the morning, and my prepared agenda lasted for two hours. From then on it was rather haphazard except for scheduled meetings.

HKS: Much social obligation? Meeting with congressmen for luncheons and that sort of thing.

RKA: I didn't do much of that, no. The chief did quite a bit. I think Les Harper did. I did little of that. My social activities related rather heavily to visiting scientists from other countries. Any time that we in the U.S. would visit other countries, people were given a blank check to entertain us. In Russia they just liked to entertain because that got them real alcoholic beverages instead of just plain vodka. In terms of my level at the Forest Service, there were no entertainment funds. So the entertainment in the States came out of my pocket.

HKS: Is that right?

RKA: So I wrote off between a thousand and two thousand dollars a year in entertaining international visitors in research. I'll give you another example. At the World Forestry Congress in Argentina, all the large countries had receptions. Russia, France, Germany, Scandinavia, and so forth. Of course the U.S. has to have a reception. Sometimes at World Forest Congresses the State Department would provide some funding. They didn't in Argentina. Our reception was at the American Embassy and everything was fine, except that we paid for it. We asked each member of the U.S. delegation to contribute twenty-five dollars, and then John McGuire and I gave three or four hundred dollars apiece to cover the balance of the costs.

HKS: Is that typical throughout the government from your observations, that there will be no money for entertainment?

RKA: At the secretary's level there is an obvious need for entertainment funds. The State Department and Defense Department have those kinds of funds. We occasionally would request funds from the State Department if the meeting was at a high level, but for the most part it was on us. We used the Cosmos Club a great deal. I didn't mind. I thoroughly enjoyed that part of the work when we were in foreign countries. We invited many to our home. They would prefer that to a restaurant or club.

HKS: I can understand that.

RKA: We did a lot of entertaining. Another social obligation was to entertain in one form or another station directors when they were in Washington.

I mentioned earlier that there was a heavy overload, and so I requested a second associate deputy chief's position. There was really too much for two of us to handle. Herb Storey, director of watershed management, was the first incumbent. Research still has two associate deputies. I think an interesting change is that International Forestry has been moved from a responsibility of the deputy chief for research to the office of a new deputy chief.

Administrative Issues

HKS: Right. Interesting to see how that evolves. How about administrative issues?

RKA: The first administrative issue was the overload on the deputy chief for research and the addition of a second deputy. The application of research was always critical. Quite often new research findings, particularly if they affect the National Forest System way of doing things, are a little slow to be applied. That was why we had in each station an assistant director for planning and application. His responsibility was to work with regions, with industry, and with state and local governments to improve the application of research findings.

HKS: Generally was industry supportive of the Forest Service research program?

RKA: Generally, without fail, yes. Industry was, I'd say, most supportive of Forest Service research.

HKS: They could help you a lot in Congress.

RKA: Yes, and they did. In general terms industry applied specific research findings faster than the national forests. It's something that they wanted and it meant dollars and cents to them, and they really moved. There are many exceptions to that statement. We tried to think of ways to improve that. One was the deputy, or assistant station director for planning and application. Another way was development of research and development units. In other words an R&D unit.

We started R&D units in insect research with gypsy moth and southern pine beetle. Units were organized with scientists and administrators operating as a team. Funding was mostly from research, but also from the regions. They quite often were pretty good-sized programs, I think the southern pine beetle got up to about a million dollars in one year. There are still some research and development units, but the concept has not gone as far as I thought it might go.

Then another issue was the general push to encourage more university research, hold down federal employment, and do things of that sort. The McIntire-Stennis program was run through the Cooperative State Research Service, as you know. The Forest Service gave grants to individual professors for specific research projects where success could be achieved at a lower cost. We had planning sessions with McIntire-Stennis where we reviewed priorities so that we could avoid overlapping and duplication. It's my impression that it worked reasonably well. The McIntire-Stennis program had its own organization with an advisory board and a chairman. They could lobby Congress hard, because they represented universities. I think all in all it might have been more efficiently operated had the Forest Service been responsible. I can understand why the department didn't want the Forest Service involved, and I think I would have made the same decision to put it under the Cooperative State Research Service.

HKS: By and large is university research of acceptable quality?

RKA: Yes, but it is difficult to generalize.

HKS: There's so many graduate students involved, a lot of apprentice scientists working in university research.

RKA: But it's still a responsibility of some professor, and it depends on him. I've heard John Zivnuska several times say the university is the worst place to get research done. He said it semi-facetiously, but semi-truly because teaching should be the primary responsibility. When a prof supervises graduate students, he loses teaching. It's almost like our original division chiefs in the experiment stations. If you have five to eight graduate students, you have a full-time job keeping them occupied. You're not able to do the work yourself. But that's a matter of selection and particularly performance. And if it turns out that some research is not satisfactory, it's easier to change university grants than it is to change Forest Service projects. You have to take time to plan and direct and move people where they're most efficient.

HKS: Were competitive grants in existence at this time, or is that later?

RKA: That was later.

HKS: Okay.

RKA: We had Public Law 480 grants which were international too and were monitored by our Washington office staff. The Northeastern Station pioneered an idea of a consortium. At the time that I left we had two, we had one in the Rocky Mountain Station and one in the Northeast. The concept was to fund a given research problem or program and organize a consortium with five to ten universities and the Forest Service. Together they would plan and then parcel out the funding to individuals in universities and the Forest Service to do the research. It worked very well. I understand that the PNW station has four or five consortiums now in various problem areas. That was a new way to get universities involved in forestry research. Those are two issues, the one on application and the one on more work with universities. Another administrative problem comes up periodically. Some regional foresters suggest that stations be administered under regions.

HKS: Yes.

RKA: Regional foresters, if administering research, could set priorities. Usually those priorities would be on immediate problems. The Park Service places its scientists under park superintendents, and the result is no science in research. There was a clear separation of research and forest administration in the early 1900s. The chief of the Forest Service at that time outlined why you need two separate units. The Forest Service has always recognized the value of basic research, and under the regions you're not going to have any. Also you're not going to recruit top scientists.

You're going to have lots of administrative studies and you won't keep your best scientists in research organizations. Nor do you get problem analyses that are related to science. You get problem analyses that are related to problems on the regional foresters desk today. The idea of research under National Forest Administration came up now and then. We would go through the arguments that science had to be separate..

HKS: It was 1915 when, under Graves, that research was given official status. Obviously research had been done before that, but that's when the Branch of Research was created.

RKA: Your memory is far better than mine.

HKS: Do you have any observations you'd like to make about the budget process, or about the hearings in Congress?

Budget Process

RKA: The budget process that we went through each year started with division directors in Washington, D.C. They accumulated needs and wishes from the experiment stations and came up with options, needs, the timing for expansion, and so forth. But also they produced what we called the budget book. The deputy chief was involved in setting priorities and achieving a balance with other Forest Service programs within guidelines set by the secretary of agriculture's office. The final document for research was a book about two inches thick; individual pages with plenty of side margin indicators that described both the stations and the programs within the stations. Those books were the source of information for hearings. When I first went to Washington as deputy chief, I'd spend about two weeks on "the book." I'd actually stay home away from the office memorizing facts and having key people come in for briefing and practice. When I was there we, the Forest Service, tended to put a lot of weight on environment: wildlife management, range management, basic ecology, and so forth.

After the budget was reviewed in the secretary of agriculture's office, it went to the Bureau of the Budget and came back to us as the president's budget. Almost invariably the emphasis was changed from environment and broader concerns to production, which is timber and grazing. The Forest Service has taken a lot of flack through the years that it was product oriented rather than environment or resource oriented. There's nothing that it could do about it, because the president's budget was the only public budget document. There always were more changes toward the production end in the secretary and the Bureau of the Budget offices.

HKS: I've read someplace that Congress behaved the same. The Forest Service would usually get the bulk of its request for timber management but only a small fraction for recreation, for example.

RKA: That's right. Let me tell you about Ralph Nader. Have you run into his work on the Forest Service?

HKS: I know generally what he has done.

RKA: He came into Ed Cliff's office one time and said that it was obvious that the Forest Service was in bed with the lumber industry, and he would have a team come in to document that fact. And sure enough, several young lawyers and secretaries showed up. They were given office space and provided any information they wanted. About a year later, Nader in reviewing this work found that the Forest Service was presenting a balanced budget. Nader removed the whole lot and put a new crew in.

HKS: I remember the book now that you're talking about it. He puts out a lot of stuff where he writes the foreword or introduction, but the book is by one of these special crews that he assembles.

RKA: It came out that he was not able to document the emphasis on production.

HKS: The budget process in testimony. Is the chief always there. Officially you're there to help the chief. Do deputies go over by themselves to testify on budget?

RKA: No, but they testify on the part of the budget with which they are concerned. You go over as a team. You have the chief, the deputies, and the associate deputies.

HKS: How about assistant secretary.

RKA: Yes, someone from his office is there. He doesn't testify, at least my experience was that he didn't. The assistant secretary or a staff member was there to be sure that we didn't push a little too hard on certain aspects beyond the president's budget. After one meeting with the appropriations committee, Ed Cliff called me in and said assistant secretary so-and-so thought that you were selling a little too much. I've forgotten now even what it was. I had waxed a little enthusiastic about something.

But you have the budget books, and Dick Dickerman always sat right behind me. Anytime a subject or question came up, maybe three-quarters of the time I knew about it. Otherwise, I could turn to the book, and Dick would hand me the appropriate pages out of his book so that I could respond. You can say "I don't have the answer to that question but I'll provide it for the record." We tried to keep it at a minimum. In fact, almost uniformly, the Forest Service was complimented that its people seemed to know their subject much better than the staff from other agencies.

HKS: Did you ever rehearse before you went over? If he says this, who's going to say what, or did you know your roles well enough that you just responded to the situation?

RKA: The chief would normally respond to that subject, and he would often ask one of us to respond. We did not need to rehearse. The chief presented the overall budget, and then State and Private Forestry, the National Forest System, and Research deputies would present more detail about their budgets. Even then the chief might join in.

HKS: Were there times when you testified to Congress other than budget?

RKA: Yes, you testify on bills. They were trying to limit clearcutting once, and that was...

HKS: Monongahela was during your time.

RKA: Yes, one time when I was acting chief, Julia Butler Hansen called. She had a new congressman who wanted to eliminate from the budget anything that would allow clearcutting. Three of us went over and presented the Forest Service view about the problem. She said "we're going to lunch now," she says you come up with language for the budget bill. She was quite direct as you know. As she walked out, she turned and said, "Get some language that I can sell to that son of a bitch, would you please." [laughter]

When I was with a NASA team in Russia, I flew back the Saturday night before Monday budget hearings, and I didn't get my two weeks preparation. I remember Bob Buckman came over on Sunday and briefed me all day, but I was still on jet lag. We went to hearings Monday, and after the hearings were over she said Dr. Arnold would you stay for just a minute? I said, of course. When the others filed out she said, what's the matter, you didn't seem to be on top of your subject today. [laughter] I said I guess it's because yesterday morning I was in Russia. In my book she was fair, she was hard, she understood, she knew her subject area.

SHIFTS AT THE TOP

HKS: You worked with two chiefs, both Ed and John. Would you like to characterize in some way when one chief leaves and another comes in? Is there any change in the operation really? A different name on the door, but does the Forest Service have enough momentum?

RKA: I think that's right, it has enough momentum and there's enough stability that things don't change overnight. I think more it's a matter of management style than anything else. The Forest Service at that time had exceptionally able people.

One thing that the Forest Service has done is maintain that a professional be in the chief's position. Now in the Bureau of Land Management and the Park Service that is not the case, and I've heard from Boyd Rasmussen who was over there for a while that there when a new head comes in you may have a major shift, and quite often do. Because he may or may not have any background in your area, he may be some congressman who didn't get elected.

HKS: The Park Service under Nixon, I can't remember who that person was, but he was from North American Van Lines or Pepsi-Cola or something. He certainly wasn't out of the Park Service.

RKA: There's one answer to your question. The chief is a professional, he's been in the right chairs. He can come from Research, State and Private, or National Forests. I'm not aware of any political activity within the Forest Service that had factions trying to put a certain person in as chief. I know of some individuals who wanted to be chief, and they made it known. The chief has been a professional, been appointed in a professional manner, and acted as such. It would be very difficult for some major change to take place.

AGRICULTURAL RESEARCH SERVICE

HKS: I don't know if this is an administrative issue or not. ARS, were you liaison officially to the department for matters of research?

RKA: Yes, we didn't have a lot to do with ARS. There was conflict earlier on when range management was moved to ARS. We rarely had any difficulties with ARS at the time that I was there. But there was a time, I think it was while Harper was deputy chief, that they were trying to take over more of the research in the Forest Service.

HKS: During the other kinds of reorganization, where the Forest Service and the BLM and so forth might be merged, research was one of the issues. What would happen to Forest Service research. I think it was McGuire who was explaining that under the Nixon plan, which you may have been involved with, that State and Private Forestry would go to HUD or something and Research would go to the... I mean really split it up. So you'd wind up with forestry not together, again, even after the reorganization. But that kind of reorganization wasn't something that you were concerned about when you were deputy.

RKA: No. President Carter tried the same thing while I was at Texas. Steve Spurr and I were consultants informally to the White House and to the secretary of the interior. They wanted us to be in support of moving the Forest Service to the Department of the Interior. We emphasized that the Forest Service had to be an integrated, self-contained unit. What happened was that Carter got involved himself, and he kind of eliminated any possibility of anything happening after task forces had been doing a lot of work. I don't have my records on that. He finally did not try to move the Forest Service to Interior. Interior made another push to have the Park Service handle all recreation on all federal lands. That again got bogged down in the White House.

Forest Products Laboratory

HKS: One more subject, and you may not see this as administrative issue. The Forest Products Lab. The way it's presented in the chief's reports and other Forest Service reports, it's always AND the Forest Products Lab. It's like everything else AND the lab. Is that an accurate characterization? It's not a team player, and somehow it's so different that it doesn't fit in with the rest of the team.

RKA: Yes.

HKS: Was that an issue or a problem for the deputy?

RKA: Yes. It was both an issue and a problem and it still may be. The Forest Products Lab was an efficient and effective research and development unit. They were ingrown. They didn't want anyone from the outside to come in and they didn't particularly like the system of the research problem analyses that went on. They were most productive but they did it their way. Finally we got key lab people to come into Washington in the products division director's job. That helped a lot.

    The Forest Products Lab wouldn't build on research done at the Southern Forest Products Lab or at universities. They had to do it themselves and use their own data. We ran into that in the Armed Forces Special Weapons project. We had worked through maybe four atom bomb drops and had certain basic information that would be of assistance to them. They started right over in working with effects on structures. That sounds like criticism but it was just their way of working. You described it very well, it was a different unit. It was under research, but again it kind of went its own way.

HKS: Does it make sense for the government to do forest products research? Does the government do oil research? Is this unique, this particular subject? Is it an accident of history, or is there real logic in terms of a complete research program that there be a Forest Products Lab?

RKA: There's real logic there. First thing is that forest products can describe the kind of tree that should be grown, length of cells or vessels or specific gravity or all of that sort of thing. The Forest Products Lab through work in computerized sawing can make a given log more efficient. If you get more lumber from a given log you reduce the impact on the forest, less timber to be cut. No, I think it was most appropriate that it was not outside of the realm of the support of the Forest Service. ARS does all kinds of poultry research and research on almost every crop in the United States, genetic and otherwise. And I can't...

HKS: So, if there wasn't a Lab, the experiment stations would almost have to invent a replacement of some kind in order to know how to round out the research?

RKA: That is correct. In Berkeley we only had two people in products. They applied Lab results to local conditions. It's something that has to be done because it impacts the kinds of demands that are put on the forests for cellulose.

HKS: The forest industry is certainly engaged in products development.

RKA: Yes.

HKS: This to me would be the only rationale I could think of to question the existence of the Lab, because you have the number six industry in the U.S., in terms of sales, engaged in products development. Why does the government have to do it too, why can't the industry do it? I'm not advocating this, I'm just asking a question.

RKA: The Lab does do basic research. The Lab, I'm trying to think of the right words here, the Lab has to take research up to the development phase, but the application and development of it should be industry. Up to World War II there were many small units in forest industry. Application and development were probably appropriate. Today with most industrial units very large, they should and can do their own development.

HKS: In your experience was there ever a time that industry was nervous that the Lab might develop something and patent it for the benefit of the public that somehow cut into the special interests of the company that was about to corner the market on a process and patent it.

RKA: I did not run into it. We were finally able to make the Lab more an integrated part of research. We attracted some of the younger people in the Lab out to positions in the stations or the Washington office. In fact, one of those that we attracted out is the director of the Pacific Northwest Experiment Station. We were able to integrate the Lab more and more into the Forest Service. Dickerman and I, when I was there, worked on that, and he continued. Les Harper had given up on Lab integration.

E. I. KOTOK

HKS: It just took too much time to do that. That's all I have on administration. You said you'd like to talk a little bit about Ed Kotok.

RKA: Yes, I hadn't brought him in except when he invited me to leave the Forest Service early on. When George Jemison came to the California Station as director, Ed Kotok had retired and was living in the Bay area. Ed decided that he could help George run the station. He got to spending an hour or two or three hours a week with George. Finally, out of desperation, George asked me to find a job in fire that Ed could do. We developed a study outline of fire control and fire research that Ed could work on, and it was appropriate because he had run the very early fire study in California back in, I guess 1925 or '30 with Show. His health broke down several months after that, so it was never completed.

HKS: The early work on fire, was that the controlled burning?

RKA: No, that was fire control in California. It actually led to more rapid initial attack. They studied the fire damage and size of fires in relation to the nature and timing of the initial attack. It was used as a basis for spiking up fire control and fire control forces in the West and probably all over the country.

CHARLES CONNAUGHTON

RKA: I did mention briefly the difference between the Forest Service as an organization and universities as organizations. I think I had a view of the Forest Service probably different than anyone else. I came into the Forest Service in a middle management position, as division chief of fire. Most key Forest Service people had paid all their dues and moved through appropriate chairs. I had not. Then after going to the Washington office, which I never thought that I would ever do, I moved out to Michigan and then came back again.

There was only once in that entire period that anyone in the Service referred to me as a late comer or someone from the outside. He was one station director that I didn't get along with too well. We just didn't work on the same wavelength. Other than that, I had I think a chance to view the Forest Service as an outstanding organization. It was highly professional. It was a team playing outfit from start to finish. I'd mentioned that once you decided on a direction or made a decision, everybody pitched in. It was more than that. I know that many Forest Service people were asked to move or to take particular jobs that personally were not attractive to them, and it was rare that they ever refused to move.

One exception was Charlie Connaughton, an example of someone who had said early on he would never go to the Washington office, and he never did. Yet I'm sure had Charlie followed a path of normal moving from the field to the Washington office that at some point at time he would have been the chief. I don't think there's any question about that. He was strong enough and had all of the abilities. But Charlie was a field man, and by the calendar he kept very close tabs. He spent 50 percent of his working time in the field, out on the forest, and 50 percent in the regional office.

HKS: Wow.

RKA: I've seen Charlie in December when he looked at his calendar and found he was seven or eight days short in the field. He took off for the forest at that point.

HKS: That must have been almost a record among regional foresters.

RKA: I don't know how other regional foresters were, but he insisted that he was a dirt forester from the very beginning and he was proud of it. He was going to remain that way no matter what. He also had been a station director early on, but he never tried to run research from his regional office.

HKS: Bill Towell talked so highly of Charlie, who was very active in AFA, about his broad vision. The anecdote was dealing primarily with the '70s when so much forestry legislation was enacted. But Charlie saw that was what was missing in the '60s, there was no organic plan for the Forest Service in terms of what Congress had said, lots of bits and pieces and traditions.

RKA: There were many forest supervisors who thought that they had the best job in the world and would never leave it to be promoted to a regional office or on to the Washington office. They made excellent forest supervisors, but they didn't have the opportunity to contribute at higher levels. The Forest Service, as far as I could see, was unique in the area of government organizations.

HKS: I think everyone acknowledges that, whether they're advocates or adversaries of the agency.

RKA: Yes.

HKS: In my interview with Max Peterson, of course he worked for Charlie for a long time, most of his anecdotes were how hard-nosed Charlie was, or at least everyone was kind of afraid of Charlie. Your dealings would have been much different.

RKA: I have only respect for him, and I went to Charlie for advice at times. He came to me for advice. Charlie had some problems when he was president of the Society of American Foresters. He spent a lot of time as any president of SAF did on SAF affairs. Some of his forest supervisors got upset and were about to complain to the chief that they had a regional forester who wasn't available to them. I was able to meet with three or four of the key people who were most unhappy and get them to see the big picture. And he certainly had a mind and a capacity to look at the very broad picture.

HKS: When I interviewed Max, he didn't say it this way, but obviously it showed his broader interest. He was an engineer but he poked around to find the background of issues and he talked about being involved with the research people, because he wanted to know more about how things happened. And you had mentioned briefly earlier that you had wanted to hire Max in research. Can you elaborate on that?

RKA: Well, not much. I was director and even as division chief in fire research, Max obviously had amazing talent. We needed engineers in research. I had several positions that I talked with Max about from time to time. He just wasn't interested in being in research. He was supportive of research and was involved in cooperative studies which required engineering work out of the region. But other than that, there was no direct tie to Max.

Minority Hiring

HKS: It may not have been significant during your tenure as deputy chief—minority hiring, which is such a major issue today, diversity in the work force and all of that.

RKA: It was not a strong effort. No I shouldn't say that, it was there and we were looking. We had difficulty finding scientists, and the one thing we did do was to work with Tuskegee Institute. We put a research scientist down there, Brian Payne.

HKS: I know Brian.

RKA: He lived there and helped develop a pre-forestry program that would feed black people interested in forestry to the Yales and Michigans and Dukes and other forestry schools. It had some success and is still operating. But yes, I'd forgotten about that until you mentioned it. Chief and staff asked research to handle the project. We had several minorities in the Washington office staff, but it was difficult to have, at that point in time, blacks in the field. I remember one black in forest insect research in Missoula in the 1960s. He was working at the northern Rocky Mountain. Many restaurants would not allow him to eat with other scientists.

HKS: In the Rocky Mountain region.

RKA: In the Rocky Mountain region.

HKS: I didn't realize that.

RKA: So we moved him to California where he had more freedom, and later on he was in the Washington office.

RESEARCH ISSUES

HKS: Okay. Those are all of the issues that we have under administration, so let's turn to research itself. What were the issues? This was the late '60s and early '70s, we had NEPA, we had Earth Day, a lot was going on we can see in retrospect. It's not always that clear at the time. What was the impact on research?

ENVIRONMENTAL DECADE

RKA: Environment is the principal issue that I can see from this vantage point. I didn't realize how much I'd been sensitized to the environment and to the changing roles of foresters while at Michigan with Lyle Crane in the Conservation Department. There's no question that Nixon hailed in the '70s the environmental decade as he opened it up with the National Environmental Policy Act, NEPA. But as I saw it, we had the ecological, physical, social, and political environment all beginning to intertwine.

My goal, I really didn't state it as such, really was to move forestry research to the cutting edge of environmental policy. To emphasize the need for recreation research, to emphasize the need for basic studies of ecosystems, and particularly in recruiting young scientists with broad interests. To use research as kind of the leader to help move the Forest Service into this environmental problem area.

I went back, just because of this review, through all my publications. Up to 1968 they all dealt with forest fire or research needs. But then from 1970 to '72, of thirty-three formal papers, eighteen dealt with conservation, environment, and the changing role of foresters. I was amazed; well over half of them. I want to feed into our conversation some of those papers because I think they describe where we are today. I had no way of feeling that the forestry movement in this environmental era and the political impacts would be anywhere close to where it is now. But in a first presentation to the RF&D meeting after I became the deputy chief, I stressed the increasing concern about the environment and how it impacted forests.

WILDERNESS

RKA: We had to look very carefully at the whole balance of nature. Our population, culture, and technology were having a much greater impact on forestry and forests than it ever had before. The interesting thing at that time, and even today, the Forest Service is still wearing a black hat and is not getting credit for what it really has done. You know that the Forest Service started the wilderness movement and had many millions of acres set aside even before the Wilderness Act.

HKS: Sure.

RKA: Yet the Forest Service is kind of seen as anti-wilderness. Research was studying then thirty of the thirty-three recognized ecosystems in the United States, and we were encouraging greater detailed studies. The Forest Service was the largest single employer of landscape architects, even at that time.

HKS: Those numbers always amaze me. Diversity of work force, I mean there's a different kind of diversity.

RKA: Of course we were the largest single employer of ecologists, pure ecologists, along with everything else. I suggested that the Forest Service needed an environmental analysis group comparable to forest economics. Steve Spurr and I, earlier on, actually in 1971, prepared a paper and presented it on "The Forester's Role in Social and Economic Changes" at a world conference on forestry and forest education. The highlights of that paper could probably be written today. Steve and I were not looking that far ahead, but we talked about forests offering a tree-based environment for production, psychic well being and social well being. I hope you don't ask me what psychic well being is.

HKS: Intuitively I know what it means.

RKA: Foresters were looking to develop more simple ecosystems. Foresters needed to produce more critical and complex forest ecosystems, and cities need to be surrounded by forests and have forests threaded through them. Foresters have to learn to manage for environmental beauty.

I think it's very interesting that Steve Spurr, in about 1960, gave a paper at the National Wilderness Conference in San Francisco. Steve at that time pointed out, (why he wasn't ostracized I don't know) that we have to manage wilderness ecosystems the same as any other ecosystem. Now he did indicate that there were some scientific needs to keep some part of wilderness completely inviolate, maybe not even allow anyone in. He said that wilderness lovers enjoy the wilderness as it is today and that foresters know enough to keep them just as they are. But it requires active management in insect and disease control, fire control, fire use, even some cutting to keep the appearance as it is today. Now that, of course, has never happened.

Charlie Connaughton was at that meeting, and he gave the closing comments. There were several hundred people there. He said all of you will go home and you've had a good say on wilderness and we've dissected the problem and he said and you'll play golf or read the Sunday paper. But we in the Forest Service will make sure no water is polluted. He said somebody this week will get hurt in the wilderness and we'll run a rescue to get them out. He said wilderness is on my mind almost every waking moment along with other forest problems. It was an interesting comment.

NATURAL AREAS

HKS: How about natural areas in wilderness. Was this an issue or are natural areas set aside in some administrative fashion to not be manipulated, to have those within the wilderness area.

RKA: They were called research natural areas and they could be anywhere. Some were in wilderness areas, also there were plenty of research natural areas outside of wilderness. I guess when I was deputy chief we added some fifty or sixty research natural areas. We wanted research natural areas in every major ecosystem. They were selected with the help of regional offices to be sure that we didn't locate them where a road might have to go ten or twenty years hence. But natural areas were not related to the wilderness concept. The concept was related to the opportunity to study ecosystems undisturbed in contrast to the managed ecosystems around them. They weren't large areas. They didn't compare in area to wilderness areas.

HKS: All right.

RKA: Then Steve and I in that paper made this statement: "Foresters will be working in the strong glare of conflicting public opinion and often in conflict with the inputs from other professionals." That certainly predicted what's happening today. But it was clear enough that you could see what was coming. I never would have imagined that we would have had the impact of lawsuits and other purely delaying tactics that go on today.

We said that foresters will have to reverse the trend from more efficient man-simplified systems to more complex systems. That foresters are going to have to be responsible to people and involve people in their management decisions. And we have to obviously have a holistic approach in all forest practices. I'm bringing these up because you mentioned before the major issue of emerging environmentalists. I always thought of myself as an environmentalist, and I think most foresters do. We have our basis in ecology.

In a paper for an FAO meeting in Rome, I talked about the complexity of the forest environment. Globally, forests produce over one half of all photosynthesis and all transpiration, actually the air conditioning effect on the world. Regionally our forests regulate floods, they reduce soil erosion, they could be even used for sludge disposal and sewage waste disposal as well as being a basis for the wood and paper products. Then you get down to the microclimate of forests where they ameliorate weather. They have places for kids to climb trees and so forth. We could go into a lot of detail but this was apparently what I arrived as deputy chief with. I was not then aware of how much emphasis I was giving environment. But all of those attributes of forests affect people and often provide antagonistic choices. I think probably the big problem awaiting resolution is that the people of the United States have not decided what they want to do with forests. We have factions, and I guess we're going to keep on having factions.

HKS: It looks that way.

RKA: Obviously we used all of these things I've been talking about in determining program priorities. We tried for a lot more emphasis on wildlife, soils, air, and recreation research as well as basic research in the ecosystems. I think we mentioned before that even though we gave initial priority to those things for increases in budget preparation, increases were made in timber and wood production related activities. One comment is important. The Washington office, and particularly Ed Cliff, changed rapidly, but the field part of the Forest Service was slow to move.

HKS: Let's follow up on that. When I worked in an experiment station, we would hear complaints from the field that research never did anything that they could use. It was too technical. The field forester didn't know how to apply this stuff. It seemed to me that it wasn't a real argument. Those people were adequately trained, they could in fact read this stuff but apparently the rewards weren't there, there wasn't an incentive. What was your gut feeling of why the field was slow to respond to changes in technology?

RKA: Part of it was, and I've given it quite a bit of thought, that they had a job to do. Most of them were overworked, they had very heavy loads in the field with things that had to be done. Those loads were always being disturbed by major fires that might take a man away from his job for several weeks at a time, so that they really didn't have the time to look for ways of applying new information.

We had in California, two or three instances that I recall. We asked the regional office for comments on a major paper. Every now and then a comment would come back from somebody in middle management that well we really can't publish this because this indicates that our current policy, whatever it is, is not good or should be changed. All I had to do was mention it to Charlie Connaughton, and we would go ahead and publish based on the research results.

I've been trying to think why the field was so slow in looking at the environmental movement. We had those people that were excellent field foresters doing their job, but they didn't have time to sit back and look at the entire picture. They had so much timber to mark. They had trails to build. They had people to manage. They had all kinds of emergencies: lost people, fire, insect and disease attacks, these kinds of things. They really couldn't back off and say what went wrong. A lot of the clearcutting problem came the same way. If you've got so much timber that you're supposed to get out, and you don't have the time to mark all the timber, it's pretty easy to put a line around it and clearcut the area. In the next year you clearcut some area next to it. All of a sudden you have a very large area that's clearcut. I think the only time I ever saw Ed Cliff completely nonplussed was right after the Monongahela. He made a quick trip to the field. And he came into the staff meeting on Monday morning still just shaking his head, "I can't believe it." He said clearcutting is good but we can't just clearcut those tremendous areas.

HKS: So he went on the Monongahela to look at that?

RKA: Oh, you bet. And then also he looked at some of the areas in the northern Rocky Mountains.

HKS: Max was telling me that he and Ed toured some Region 1 clearcuts just before Monongahela, or at that time. They both were appalled at what they saw, and by the rationale and the justification from the local people. That's sort of an amazing comment to me, that in this reward system the guy on the ground is getting rewarded for being the cutout type, he's not being rewarded for changing with the times.

RKA: Yes.

HKS: Research isn't the issue, it's somehow in the layers of bureaucracy. The reward system is not responsive to change.

RKA: We never had a period of such rapid change. It all hit in a very short period of time. Foresters, those who had been out for ten or fifteen years, had been taught production forestry.

It's just the sheer pressure of doing the daily job, in the national forests. We just didn't have any dead wood scattered around. The field people in the field season were completely overwhelmed. Working fifty, sixty, seventy hours a week, even more. And you don't build into a regular work schedule large fire control needs where you may be out for several weeks.

HKS: This is your interview, but I've two anecdotes that maybe you can react to. When I was on a ranger district cruising timber I ran across some silver fir and the needles were chewed off pretty bad. I snapped off a branch and sent it in to the entomologist at the experiment station in Portland and asked him to identify it. The answer came back through channels, with a really angry note from the supervisor's office, that they have the answer to all the questions and it was absolutely inappropriate for me to go to the scientists. Well that's a problem. I could have gone up through channels but I didn't even realize there was a channel to go through. Does this surprise you?

RKA: No, many field foresters didn't like to have any feeling that they didn't know what was going on on their forest. There are plenty of exceptions to that. Many used research at every opportunity.

HKS: The bug, by the way, was not a serious problem. The other one was in reverse. When I was at the experiment station examining a series of field plots throughout the Cascades, measuring the effect of slash burning on the vegetation that would come in afterward, the first step in the process was to go on the ranger district, introduce myself to the ranger, explain that we had the plots there and I'd be working on this district for a day or two and just say howdy, maybe get a map of the district. I visited about twenty districts during that summer. Only one ranger said sit down, what are you working on, I want to hear more about this. Most were fine thanks, good luck, but several were really hostile. One wouldn't allow me to stay in the crew house, because that was for Forest Service employees. Does that surprise you that research is not viewed with great romance by some of the people in the field?

RKA: No, I would wager that he was probably fifty years old and was still at the ranger level.

HKS: Could be, I don't recall.

RKA: We had forest areas in California where we just didn't bother to go because it was too much effort to get work done. We had other areas where everything on the forest would just stop to help in any way. For the most part, Research and National Forest Administration worked well together.

HKS: I don't want to interrupt your thought train here but I've got two other general questions about the environmental movement, as it were. You hire a scientist, really competent, has a certain expertise, field technique and whatever. And the world changes. Now we're going to study more environmental things. Given the long term nature that research tends to have, how do you have a course correction when you're dealing with people, not with changing the subject, but how do you get people to change. I mean they've just written a famous paper, everyone is congratulating them and suddenly you say well don't do that anymore, we've got something of a different nature we're working on.

RKA: I couldn't say that to him, and would not. But we would ask him in the light of current problem areas that he review, revise, or develop a completely new problem analysis. You can't tell an able scientist what to do. But in the problem area to which he's assigned you can ask him for a problem analysis. That was the way that you made changes with an existing staff. Maybe the problem analysis would show that they didn't have the right mix of people in that, so slowly you could look for a place where this individual would fit and then bring somebody else in that would do the job. Those are the approaches that you had to, had kind of a problem. Creation of new research teams and recruiting new scientists would create change.

HKS: Relating research to the so-called real world, in Washington, the political world, did you or your staff work directly with the White House or the science advisor to the president when NEPA was on the drawing boards, Earth Day was around? Did the White House call the Forest Service and SCS and others and say hey, what's this all about and get feedback? Does research get involved in policy itself or is that somebody else?

RKA: Rarely did research get involved in policy except in economic studies. The chief or Ag. forestry staff would use the material. The examples you mention were pretty well cut and dried before I was in the Washington office. But no, there was no call from the White House to research. It would go through the department, then the department would ask the Forest Service either to provide two or three people to work on the task force or just prepare draft material.

HKS: So you were just another member of the public in terms of what the president might have in his State of the Union address on an issue. You weren't involved in that sort of political aspect of forestry.

RKA: When he was making a State of the Union address, I would ask all research division directors to make statements that they thought important in their area. In chief and staff each deputy would have three or four items for possible use in the State of the Union address. Chief and staff would make up a list of four or five items that they thought were important to mankind and the country and send them to the secretary. Occasionally we would get one or two sentences in the speech.

HKS: I just heard an anecdote, and probably there's some truth to it, but it's so fun to burlesque it, about President Bush's activities at the Rio earth summit. Two days before he went down he announced one hundred fifty million dollars for international forestry. The next day in International Forestry, so I've been told, they got a call from the White House saying how are you guys going to use this money? [laughter] There were some follow up questions I guess from the press and they didn't have the answers, so they wanted some answers to tell the press. I'm sure there's some cynicism in that anecdote, but one could see that happening.

RKA: I could believe it, sure.

HKS: Okay, I stopped you midcourse.

RKA: No, you didn't, because we came through the major issue, which obviously was the environment. I thought that I might be almost out of order by pushing on the environment. Yet Ed Cliff and his staff, working as a whole, encouraged me to do all that I could. I wasn't out of line nor did anybody say slow down. I can't take any credit for any changes or improvements, because the policies were clearly stated and changes were made out of the Washington office. It was maybe ten or fifteen years later before there were major impacts in the field.

HKS: Ed took a lot of heat, even abuse, for a few sentences in a statement he made at an RF&D meeting, probably his last one as chief, saying that the Forest Service was not prepared for the environmental decade of the '60s. There were those that likened that to Eisenhower warning of an industrial military complex; like alright, chief, if it wasn't ready, then maybe you ought to speak out as to why it wasn't ready. You've spoken about Ed in kindly ways, and I'm not asking you to be unkindly, but were those shots at Ed at all fair or was everyone overwhelmed and surprised by what happened? At the rate of change of attitude toward the environment? Maybe the Forest Service was better prepared than any other agency.

RKA: I think that was what I tried to point out, that the Forest Service did not get credit for a lot of things that it had done. Ed Cliff was not at fault in my view. The population of Forest Service employees in the field were not ready. And they weren't ready five years or ten years later. The Forest Service policy of delegation down to the ranger district made such major concept changes slow to implement.

HKS: I went to forestry school in the '50s. When *Silent Spring* came out, I said yeah that's right, that's what I learned in forestry school, that biological control is better than other kinds of control. I don't know how typical I am, I don't know how typical my education was, but most of the environmentalists' criticisms of forestry are compatible with what we learned was the best theoretical practice in the '50s in school. You can't always afford to do it, maybe we don't know how to do it yet. I'm trying to follow up on your statement that the foresters weren't ready in the field for the change. And yet if they went to school in the late '50s, early '60s, to the extent that I'm typical, they should have been better prepared.

RKA: How many people were in field positions that had any real say on what was done? In the '50s and '60s those people were educated in the late '30s or '40s. I think you were right at the change point. I would hazard a guess that in the '50s about one-half of foresters were prepared to listen.

HKS: So the people in the field, the ones out driving the pick-up trucks around, they actually were ready, but middle management was another generation of education and priorities.

RKA: Your statement might apply to the '70s and '80s, but not the earlier times. The now retired supervisor of the Ouachita National Forest as late as five years ago saw the Ouachita Forest as the best timber producing forest in the South. I'm maybe doing him an injustice, but he treated the forest as a supervisor in the '50s would treat the Ouachita Forest. Again, the Forest Service in the East gets no credit for taking abused farm lands and severely eroded watersheds and converting them to beautiful productive forests. Today, the Ouachita is one of the greatest forests in the country, both in beauty and recreation as well as timber. The Forest Service has changed millions of acres of land into productive forests. One very important point. For the most part, timber sales are the only way to treat forests. They are largely the only source of funds.

HKS: I read a comment by some Forest Service guy, I think he's in the Washington office. He was talking about the eastern forests, essentially all of which were acquired as cutover and abandoned lands, pretty much abused in the environmental sense, and now they're back looking like the forests we have around here in Arkansas. The Forest Service sees it as an enormous accomplishment. And the environmentalists see the Forest Service as the one that's causing all the environmental damage. It's a conflict of cultures. I guess there is no way of actually bringing these groups together.

RKA: Old growth forests uncut are generally not the places that you stay in for recreation. Almost all campgrounds are in areas that have been cut and managed for years. A typical western old growth forest, as you well know, is kind of dark and damp and really it's interesting to look at, but it isn't a place where you actually stay for some time.

RESEARCH PROGRESS

HKS: Let's get to something more specific about research.

RKA: We had to react to emergencies when the clearcutting and the Monongahela issues appeared. There were many task forces formed to study the situation and to determine how critical it was and what could be done. They always had one or more research foresters or research scientists with regional or with national forest people. Carl Ostrom prepared that excellent publication, *Methods of Cutting Appropriate for Forty-Five Ecosystems in the Country.* It was not the subject of new research, it was just a compilation of what had gone on before. But in looking at those issues, I think all we could do would be to mention some of the kinds of things that research was accomplishing at that time. The Forest Products Laboratory, we mentioned before, was into low-cost housing very heavily. They still had basic research on fiber and cutting methods, they were computerizing sawmills at that point to actually determine the most efficient pattern of sawing. I don't think there's any question that the Forest Products Lab was moving very rapidly at that time. Skyline logging came in, again in response to environmental concerns. There were some slopes particularly in the Northwest I guess where that was used more. Skyline logging, helicopter logging, and developing methods of treating logging waste.

Project Skyfire had been emphasized for many years in the northern Rocky Mountains out of the Missoula forest fire laboratory. They had determined that cloud seeding could reduce the incidence of lightning fires. The only problem with that is that most of the major storm will have thousands of strikes. You can work on one cumulus cloud or maybe ten or twenty, but you can't seed several thousand over large areas.

HKS: Weren't there some lawsuits, or fear of lawsuits, from farmers that if you seeded you might take their rain away. A state farther east doesn't get as much rainfall and they blame Project Skyfire for doing this?

RKA: Yes, we had that occur in California a couple of times. You had always that possibility. In the Yosemite National Park, we had a very strong crew of forest insect researchers studying insects, impinging on the park, and they had put out the word once that in several areas of the park they were going to do some aerial control with pesticides. Bark beetles attacking lodgepole pine in higher parts of the park were causing some serious problems. It happened that the weather wasn't right and the spraying was postponed. You should have seen the articles that hit the newspapers. People were counting dead fish and birds that had been killed by non-existent spraying.

I mentioned Jim Mace a couple of times, he brought us into that Operation Fire Stop. Jim had a very serious fire near Riverside in southern California. The crew was getting nowhere. Jim on his own very quietly went out and out of his own pocket paid for a cloud seeder to go up. There were big clouds right over the fire. A gully washer thunderstorm followed. They couldn't even get the fire fighters out of the fire camp. [laughter] It was not publicized at that time. And I don't know whether it's been used very much anymore or not. It's the only time that Mace tried it. [laughter]

HKS: I would imagine. So there's no question that the seeding caused the rain, it wasn't a coincidence?

RKA: I wouldn't think so. Again, I wasn't there but the cloudburst came at the proper time right after he seeded...

HKS: Pretty dramatic...

RKA: Over a big fire you can have smoke columns and clouds go to fifty thousand feet, so they're prime for it. A national fire danger rating system was developed at that time, which allowed you to compare the potentials for fires in different parts of the country. George Byram many years before had a fire system for the Southeast. Every fire research unit, I think, built its own fire danger rating system. We put together a task force of some regional people, but mostly meteorologists and foresters, who worked out of Fort Collins in Colorado. We moved one man from Berkeley there, and they developed a national fire danger rating system that is, as far as I know, with some modification still in use today. I think that was a rather important contribution. With computers there are mathematical models of mass fire behavior. I can remember a series of charts that I'd pull out of my pocket to look at to predict what might happen to fires depending on changes in humidity, wind, temperature, or inversions.

HKS: One of the tasks I had when I was in fire research was to analyze all fire reports for Region 6. It was amazing to me that the primary cause of the fire getting out of control, a lot of these were escaped slash fires, was an unexpected wind shift of 90 degrees. So the front goes through and these guys don't know about it. They could get a weather forecast for three thousand feet elevation or four thousand but they didn't know. I mean the people on the ground, the ranger, apparently was not aware of the technology available. I don't know how accurate these weather forecasts were but you get weather forecasts. Those fronts came right over the regional office on their way to the Cascades. They could have had about a forty-five minute warning, or so. I'm talking about fifteen or twenty or thirty fires a year got away because a front went through and the guy on the ground didn't know it. And that's a pretty simple kind of technology. It's not exotic, it's not hard to understand. But that doesn't surprise you either, that the ranger...

RKA: We in research were heavily involved in fire training in California, and ours was related to behavior. We didn't try to get into techniques of control. Groups of fire bosses and fire control planners would have a two-week session almost every year. We taught how fires spread and such things as what were the signs that showed that the inversion layers were going to be penetrated. Look for little whirlwinds. Many signs wouldn't give you as much as an hour, but most were in the ten to fifteen minutes category.

HKS: Time to move some people from one side to the other.

RKA: I've mentioned Jim Mace and we're talking about fire. I have four slides that show this. There was a critical fire in southern California near Camp Pendleton. Jim was driving toward the fire camp and was five to ten miles away. On the east side was a bulldozed line and maybe fifty men with fire hoses that had been strung from the top down. I spent many hours trying to find out what he saw. He used his radio and said take the men off the fire line on the east side of this fire. He heard the radio going to the crew bosses saying roll up your hoses. Jim radioed again, drop everything and run, either up or down, which ever is closer. About ten minutes later that fire blew up and became a big fire storm in this whole canyon. Those were the things he was very sensitive to but in no way could he explain what he saw. The important thing is to have somebody on the fire with no control responsibility, always to look at what's this fire going to do and feed that word to the planner or the fire boss.

HKS: You mentioned the clearcutting of the Monongahela and the fine paper that Ostrom pulled together. Anything more on clearcutting from the research perspective that you'd like to talk about?

RKA: The use of clearcutting on the Monongahela came as a result of research findings. I don't recall all the details, but this was a low quality hardwood area generally, and in order to make it productive you did need to clearcut the areas and then either wait for natural regeneration to come in more slowly or plant. The details of that particular technology came out of research at the Northeast Station. What went astray was the combination of small clearcuts to form a large clearcut area. I don't think there was any question that there was too much clearcutting.

HKS: Even today?

RKA: Even today many environmentalists think that the Forest Service pioneered in the use and developed clearcutting as a method of timber harvesting and replanting. It's been used in the Black Forest of Germany for several hundred years and in Japan for three or four hundred years. I think there's a place for clearcutting if you need to improve the wood quality of an area. In the redwoods and in parts of the northwest, if you do a partial cutting the wind will take care of the stand that's left.

HKS: Le Tourneau makes these huge machines, great big mesh wheels in the front, where they just smash down I don't know how many acres a day of hardwood. You probably don't need a scientist to talk about that, you just get rid of the hardwood and plant pine.

RKA: If your goal is timber production, that certainly is alright. I'm sure that research has looked at that in terms of the impact on the soils and the productivity of the area. Without question I know that they have, but I'm not aware that they're involved in the actual application.

There was an aerial spray job in Arizona in some of the timber areas. Lawsuits were threatened. We sent Jim Beal and Ray Hansborough as part of a Forest Service team to get the true picture. It was a routine aerial spraying to control some insect, and I don't recall now what it was. In the little town they were telling how many of the gardens had been damaged and so forth and they said why there's a lady up here, up the road that has the prettiest Colorado blue spruce you ever saw in your life and for the first time in twenty-five years she wasn't even able to decorate it at Christmas time. The spraying ruined the tree. Ray Hansborough thought that that was a good case to look at and he went up and the lady came out and was talking to him and he said I understand you've got problems with your Colorado blue spruce, he said it looks alright to me. She said I didn't get to decorate it last Christmas because I was seeing my daughter in Los Angeles.

FIRE IN WILDERNESS

HKS: Fire in wilderness. Something you want to talk about?

RKA: Yes. You win some battles and you lose some. I don't have the exact date. The Park Service started a little prescribed burning in Sequoia National Park. Underbrush and ground fuels were building to the point that the giant sequoias could be damaged by fire. It was an excellent program done in small fashion, very carefully and just at the time when they knew it was completely safe. Other parks did some.

Until that time the Forest Service had put out fires in wilderness areas. We didn't use mechanical means; we walked in and walked out. The general thought was that fire should be extinguished principally because it can spread out of the wilderness areas and become major fires. The policy became that in wilderness areas the Forest Service could, again under appropriate circumstances, measured by fire danger rating, let lightning fires burn. There's no question that that was a good policy as far as it went.

I argued quite a bit in chief and staff and in other areas that if the conditions are just right so that if a lightning fire strikes we let it burn, why not drop the fusee and do it on purpose. Then you can start the fire where you want it, you can start it when you want it, and you know when it is going. Once you have a few fires that have been allowed to burn, then you can have even safer burns. There's no question that technically prescribed burning is a very sound policy, but it violates the wilderness area concept. It's alright if the lightning starts a fire, but it's not alright if man starts a fire.

HKS: Maybe if you had an Indian going with flint and steel and started it, it would have been alright. [laughter]

RKA: The interesting thing was the Yellowstone fires that drew national attention a while back, and the Park Service was criticized for allowing some of the early lightning fires there to start. It was in the *American Forests* description of the fire situation and the Park Service policy. Park Service paid no attention to fire danger rating. They had areas where if a fire started they would let it burn. They of course completely ignored the conflagration potential of three or four years of drought plus insect-killed lodgepole pines.

I was amazed that at the time of the Yellowstone fires, they still had just a strategy of allowing lightning fires to burn without any review or override by fire danger rating or conflagration potential. But I think, for the record, that the fire conditions were so bad at that time that allowing the lightning fires to burn didn't affect the final fire damage.

HKS: I was intrigued that one of the more frequent official spokesmen on nightly news during the Yellowstone fire was Dick Rothermel, who you probably remember from the Missoula lab. I thought that's interesting, I wonder how many people watching it realize he's a researcher rather than a dirt forester who really knows what it's about. As a matter of fact he's an aeronautical engineer, he's not even a forester, and yet the Forest Service trotted him out to talk to the media. He's an affable guy and articulate.

RKA: And Dick had been studying fire for many years. Using only nature-caused fires is inefficient and sometimes dangerous.

HKS: Given the affection that the environmentalists seem to have toward fire, is it ludicrous to consider asking Congress to amend the Wilderness Act to allow this kind of management in the wilderness areas? Who would oppose it?

RKA: There are many people who would oppose it. A wilderness must be completely natural and you don't do anything in there deliberately. It would just never fly with the true wilderness lovers, who love the idea of the wilderness. I feel that people should be barred from many wilderness areas.

HKS: But it's alright to rescue someone if they're injured.

RKA: A young lady up in the wilderness area in the state of Washington fell and broke her leg, and her partner hiked out. I believe it was the local Forest Service ranger who instead of hiring a helicopter to pull her out in a few minutes took a couple or three days before she was out by sending in a crew after her. He wasn't a ranger too long after that as I recall.

It is amazing, it's the same fervor that you get with some endangered species, and those all came about after I was there so maybe we don't need to talk about them.

HKS: The act was there during your time, but they didn't realize how significant it was going to be. You got out just in time, I think.

RKA: If I looked at changes through the period of time I was there, and again from the standpoint of enjoyment of the work and opportunity to do things, I was there at the right time.

HKS: Got a couple of letters here from people at the experiment stations. I said I was going to be interviewing you and Dickerman and Buckman and what should I ask these guys. Many wanted to ask each of you to come up with the three most important or five most important scientific issues of your administration. Fire and wilderness was one.

RKA: It was quite important, even letting some lightning fires burn was critical. Yes, it was an important issue.

HKS: Wasn't there a fire in wilderness in Minnesota that got away and caused a lot of damage and during fire review the ranger, or whoever had charge of the fire situation, was reprimanded? Well you don't remember.

RKA: It could be. There was one in Michigan, I think, a while back that did the same thing and there was one in Arizona. It was where those who had made the decision did not pay close enough attention to the fire danger rating and fuel conditions.

HKS: It may have been because of the people I worked with in Portland, Dave Bruce and Owen Cramer, we spent an awful lot of time talking about the national fire danger rating system. We made some field trips, and there seemed to be a lot of opposition to it from the regions where fire wasn't so important—why have such a complex danger system when fire wasn't an issue. Obviously in California, the sky is the limit. You trot out whatever technology you have. Is there more to talk about on that system? You've referred to it.

RKA: No, I don't think so. It was difficult to get the same examination of potential for serving fires in different areas of the country with different rating systems. With different systems you might have difficulty in determining the priorities for adding manpower or equipment where you're stretched pretty close to the limit.

INTERNATIONAL FORESTRY

RKA: I'm down to "international."

HKS: Okay, let's go for it. There's quite a bit to talk about there. Does it make sense to you that international forestry was in Research as opposed to National Forest Administration or State and Private Forestry?

RKA: That makes a great deal of sense because the International Union of Forestry Research Organizations was the first worldwide research organization. Now there are thousands of them in every field that you can have, but that was number one. It came about because the Swiss and the Germans wanted to look at each other's sample plots way back in the 1800s. The bulk of the international exchange to date deals with research.

We did send foresters to developing countries and other places. The principal impact and the principal advantage to the U.S. was in research. I may have a little bias on that. One in international forestry was the balance of participation among international programs. You had FAO programs, we had the North American Forestry Commission under FAO. We had IUFRO which is the International Union of Forestry Research Organizations. And then because it was assigned to the deputy chief for research, you had to have the balance between U.S. research and international work.

We had a few scientists who wanted and tried to spend all of their time in areas of international interests. Of course you can overdo a good thing. It did revolve largely around research, which is the direct communication between parties. Public law 480, which utilized local currencies, particularly in developing countries and elsewhere in Scandinavia and Germany and Italy. They couldn't pay us for whatever services might have been done in U.S. dollars but they would build up in our embassies foreign currency funds. Those were used to finance congressional travel in countries, and used rather widely for that. They also were earmarked, some of them, for research by foreign scientists. That meant that we had to know the people that were in forestry research in other countries and know what their capabilities were.

HKS: So when you got to India then you would switch to their payroll as it were, and they would finance salary and everything.

RKA: No, no they wouldn't finance anything of ours. I can't recall that we financed any travel out of PL 480 funds. Let's say that we had two hundred thousand dollars in Germany, and we were interested in several research areas. We would ask for grant applications.

HKS: So German scientists could apply to do the work, okay.

RKA: They applied. We got a lot of research work done at a low cost to us.

HKS: I could see that.

RKA: I guess two-thirds of it was handled directly by the Washington office staff. But in many areas the key person who knew most about that problem was a research scientist in the field. He would be put in touch with the appropriate person and would evaluate the grant proposal and so forth.

HKS: Was there much difficulty dealing with different standards of research in different countries, different cultures?

RKA: Yes.

HKS: The quality of the product, was that generally acceptable?

RKA: Yes. That was our job to see that it was. Most research was in universities. I would say there wasn't a real problem. When we started cooperating with Russia we had a problem with research quality—some very good and some poor. In Russia some research results were ordered to conform to communist thinking.

HKS: I would assume for some countries where there is large amounts of foreign aid, there was almost a problem to find enough things to spend the money on.

RKA: No. I guess maybe eight or ten million dollars was as large as it got with us, and when I was in Washington it went down to two or three million. It was a good program, it was certainly cost efficient. Probably the major contribution in research was the actual scientist to scientist interaction. You develop close professional relationships that really paid off in exchange of information, both by letter and at scientific meetings.

Soviet Union

RKA: One other thing that international research did was to improve the relations between countries. When Nixon started with Russia, he authorized cooperative research interchanges and we were part of that. We had to use our own money, we did not get extra funding. We had three different missions that I know of to Russia. George Jemison took one and John McGuire took one. That was kind of interesting, I was the organizer and John was the head of the U.S. delegation. The head of the Russian delegation was not a forester. He was an ag economist and took John off with him. I had a title something like program manager. Their program manager and I did all the negotiations while John got to see a lot of Russia.

HKS: The photographs you've given me show you and John at the dinner table, and John and you are laughing, someone standing up says something funny, probably the translator, but...

RKA: That's true, and that was not a dinner, that was just a part of the working sessions. I got in personally on one other one—one of NASA's.

If you recall, we had pretty well severed diplomatic relations with India back in the 1960s. As a start towards normal relations, a Pugwash group of scientists went over to negotiate cooperative research.

HKS: Is that an acronym, Pugwash?

RKA: No, Pugwash is a town somewhere up in Nova Scotia. I don't know very much about Pugwash. A group of scientists who developed the atom bomb, part of the nuclear weapons, got together to try to make sure that they were never used. They were a policy-forming scientific group. Roger Ravelle was one of those, and he was requested to form a Pugwash group and go to India as a start at reestablishing relations. It was in agriculture, forestry, and in other industrial applications. Although we were briefed by the State Department, the State Department could have nothing to do with this, it had to be a private venture. I went while I was at the University of Texas. It was another example of using scientists to begin to open diplomatic relationships with other countries.

HKS: Is there any more to say about the work you did in the Soviet Union specifically? You talked about the quality of research, was it because of lack of training?

RKA: No, part of it was lack of equipment, part of it was the size of their research program. They wanted a lot of what we have, their people obviously had translated almost all of many publications from the U.S. in forestry. They had some fairly good work on insect and disease, biological aspects of that. Their forest genetics people were told what their results were going to be before they did the research.

HKS: Is that the aftermath of the so-called Lysenko business of the '40s?

RKA: Yes, and that was beginning to change.

HKS: That's amazing that was still there in the '70s, right?

RKA: Yes, it was changing then. But they had to be very careful. Their forest economists were quite able people.

HKS: That's interesting. All the problem the East Block is having now privatizing because of a lack of economic structure or experience.

RKA: I think it was a matter of individuals. The head of their delegation was an ag economist and my counterpart at the negotiating table was an economist.

HKS: What were you negotiating?

RKA: We were negotiating exchanges of scientists and agreeing to exchange plant materials.

HKS: Is it supposedly a one on one, was that the goal? You did something for the U.S., and the U.S. would do something equal for you?

RKA: About every two or three years, we went over there and they came over here. I don't know since I left how many, at least one trip over here for them, and I'm sure we've been back, but we've had several scientists spend several months to a year over there and they've had some scientists here too.

HKS: Is most of their work in Siberia where most of their forests are, or where you in the European part?

RKA: We were in the European part, we didn't get into Siberia. When I was there with NASA, they were antagonistic—both sides were. We were supposed to exchange photographs from space. We were not allowed to admit that we had taken pictures of anything that could be discerned smaller than a football field, when we knew we could almost read the newspapers. But we would admit, only to that, and they initially said they had never taken a photograph in space.

HKS: This is all pretty silly when you look back at it.

RKA: We went back and forth for four or five days. The assistant secretary of NASA who was heading the project came into my room. Of course our rooms were all bugged as we knew. He said Keith, if those Russian SOBs don't lay photographs at 9 o'clock in front of me tomorrow morning, we're going to tell our ambassador that they're not cooperating and we failed. At five minutes till nine they placed a big batch of photographs on his desk. [laughter]

HKS: You stayed after the official group left.

RKA: Yes, I stayed and met with their forestry people through the agricultural attaché. We set up the general plan for the forestry exchange. In contrast to the NASA exchange, the forestry exchange was completely friendly. We had what they needed and wanted very much. We wanted to get acquainted with their work. We had many pleasant dinners and this sort of thing. They asked us where we wanted to go and we were allowed to go. We were a small busload, I think there were five of us and maybe eight or ten Russians. They even closed schools in some small towns to wave American flags as we went through. It was a completely different setting.

Of course we knew some of the Russians through IUFRO. There were several who were regular attendees at IUFRO, and out of it came a definite scientific exchange. Our orders were to be sure that we did not give them a big advantage over us, because we knew that they didn't have as much as we did. But we got quite a bit out of it and they got quite a bit more.

You've got to remember that some of the Russians were KGB people who made sure that they talked only about forestry. At the World Forestry Congresses their security people were quite evident. Actually, their scientists were as friendly as they were allowed to be.

It was international forestry that created a lot of the overload that led me to request and get a second assistant deputy. There's no question that the work exchange through IUFRO had grown a great deal with Harper and Jemison.

IUFRO

HKS: Did you feel any sense that you had inherited a commitment to IUFRO through Jemison?

RKA: It was a Forest Service commitment to IUFRO in research, and I'm not complaining.

HKS: Did you personally have to be high profile in IUFRO the way George and Les Harper were?

RKA: Yes. I was on the executive board for a number of years and led a protection interchange group before that while George and Les were still there. I didn't know I was being groomed for some future activity. There's no question that it took a lot of time. We had the World Forestry Congress in Argentina, and of course the U.S. participation with that fell directly on our shoulders. There were about thirty or forty Americans down there.

FAO at that time was taking a greater interest in forestry, and they elevated the forestry and forest industries division to a forestry department with two divisions, headed by an assistant director of FAO. The headquarters for agriculture and forestry were in Rome. We had a number of meetings there to bring FAO along and to hopefully increase FAO support for service foresters in developing countries. The North American Forestry Commission's activities in Mexico, Canada, and the U.S. also took time. I don't think we got as much out of that, nor did Canada, as the Mexicans, but I think really the North American Forestry Commission was a real help to productive forestry in Mexico.

HKS: At the SAF meeting in Albuquerque, you were president of SAF then. I heard comments. The American foresters were really surprised at the low level of technical background the Mexican foresters had, those who came to Albuquerque. Theoretically they would have been among the more advanced, they were bilingual they could read English and keep up and so forth. So maybe what you say is the issue, how to bring Mexico along.

RKA: It certainly was in forestry. They had a few able people, but politics was involved in every decision made in Mexico. All political decisions, very few professional decisions.

HKS: Is that right?

RKA: In terms of forestry.

HKS: Give me an example of that. A decision was made that was not good forestry but it was good politics, does something come to mind?

RKA: Not specific examples, but the location of industry, the nature of cutting, even their participation. They had to be very careful of what they said in the meetings.

HKS: The forests surrounding metropolitan areas must be really be harmed these days...

RKA: Oh, there aren't any. They were cut for firewood.

HKS: The concept of urban forestry doesn't really work in Mexico.

RKA: No. Of course Mexico City has lots of shade trees and that sort of thing, but not in the poorer parts of it. The southwest region of the Forest Service, and particularly out of Albuquerque, had very close ties with their Mexican counterparts and they had joint meetings every year.

HKS: I would think that State and Private Forestry would be emotionally oriented to a lot of this international work, but apparently it does not get involved.

RKA: No. Probably because of the nature of their funding.

HKS: The kinds of cooperative activities that they deal with with states would work very well with Third World and so forth.

RKA: Never even heard that brought up before this time. Research is the most important and by far the largest exchange. Anyway we had some two hundred nationals from other countries that would visit this country every year for professional forestry training in the forestry schools. We arranged trips for visitors and developed their itinerary. In one year visitors spent twenty-eight hundred man hours with Forest Service people. It had to be organized and managed.

CENTRAL INTELLIGENCE AGENCY

RKA: I guess it's not classified anymore, but we had a fairly sizable international forestry staff, maybe eight or ten people, who were studying purely the forestry potential, the forestry programs in various countries for the CIA. It had nothing to do with clandestine operations. Those reports were available to anyone.

HKS: The CIA would fund that?

RKA: Yes, they were funding it. But again, the reports were not classified, the source of the money was the only thing secret.

HKS: Oh, I see.

RKA: If some industry or someone wanted a report, we would send them a report on the country.

HKS: When I was a senior in forestry school, taking photogrammetry using Steve Spurr's book, the CIA would come on campus to recruit. They always wanted to recruit a few foresters because of our alleged skill with aerial photography interpretation. They wanted two foresters, young graduates, to go to Europe to do photo interpretation. I guess there was a lot of that going on. But it was open, I mean we were told it was the CIA who wanted foresters to go to Europe.

RKA: Yes, our work with the CIA was not classified except for the source of the grant money. Why, I don't know. It was there when I came in and was there when I left.

SECURITY CLEARANCE

HKS: You made an observation, I think it was you, at the Denver Fire Research workshop about Jack Barrows. He was being considered for some Washington position and the one thing that hadn't been completed was his security clearance. I wondered what these guys do in the Forest Service in Washington that requires a security clearance. This is the sort of routine thing you're talking about?

RKA: No. That security clearance was a different matter. Any chief or deputy chief or key person in Washington may need some security clearance. At the chief or deputy chief's level, it's secret or top secret, that sort of thing. Some space projects or advice requested from DOD may require clearance. A. A. Brown was retiring, and Jack was being considered to be the director of the Division of Fire Research in Washington.

INTERNATIONAL AFFAIRS

RKA: Tom Gill was interested in international affairs. He was concerned that most international cooperation came via research. The practicing professional forester had little opportunity to meet with his counterpart. That was important to developing countries. He was most concerned about developing countries. So he and Les Harper organized the Union of Societies of Forestry to help other countries develop professional societies of forestry comparable to the SAF. The SAF supported it in general. When I became a deputy chief I inherited that activity. It was the job that I didn't care to inherit, because it took a lot of time, it wasn't overly productive at that point, yet it was important enough to do and it looked like I was the only one to work with Tom Gill to continue it. In 1974, I organized and directed in Finland the Second World Congress of Foresters. It gave emphasis to societies and how they could support education and training.

FRANK WADSWORTH

HKS: I think this fits under international issues. We're in a process of completing an agreement with the Forest Service with international forestry for me to interview Frank Wadsworth, who's been in Puerto Rico since 1942. What would you ask Frank Wadsworth about?

RKA: The Puerto Rican National Forest was run by him through research for some time. Now it is administered by the Southern Region. Let's back up a little bit. Frank and I were graduate students at Michigan at the same time.

HKS: I didn't know that.

RKA: We shared an office, so I got very well acquainted with Frank. As foresters we were concerned, this was a forty or fifty year old wooden building with oiled floors. With each step oil would come out. We were writing our theses, but because of fear of fire we made a copy of everything and kept it out of the building. It didn't burn while we were there but two years later it burned to the ground and destroyed I don't know how many theses. Frank was interesting. He married the daughter of Gus Pearson of southwest fame. Frank worked with Gus for a while in Flagstaff. I think it was before his Ph.D. work, and at this point I don't know how Frank got to Puerto Rico.

HKS: I talked to him on the phone for five minutes. I asked him that and he said he married Gus's daughter, the law on nepotism came in, he had six months to change where he worked. This story I'm sure has been embellished over the years, but the first job opportunity that came through was Puerto Rico. He said we had to go look at a map to find where it was.

RKA: They were working on the first experimental forest of the United States at Flagstaff.

HKS: Fort Valley.

RKA: Fort Valley, yes. I think he saw the forest in the broad sense of its major contribution to mankind as well as to production of wood. He progressed in Puerto Rico till he was the forest supervisor. I don't recall whether that was his title or not, but he was responsible for the forests as well as research. He was brilliant, well trained, and so he has become Mr. Tropical Forestry over the years. I had one conversation with him about coming back and taking some place in the U.S., but he wanted to stay there all his career, which he did.

The Puerto Rican parrot came on the endangered species list, and Frank was greatly concerned. He worked with the department on endangered species. They had built at the ARS research center, just out of Washington, a special cage and had raised other parrots to be sure that they could handle the endangered ones from Puerto Rico. It came time for them to bring the parrots to the U.S. I got a phone call from Frank and he was just frantic. The parrots were sitting down at the airport and they were holding a DC-7 airplane flying into Miami, all introduced plants and animals had to go into Miami and be held there until they were declared to be healthful. They were worrying about the Newcastle's disease. Newcastle disease had wiped out millions of chickens in California, and Florida was concerned because Puerto Rican parrots had been known to carry Newcastle disease. They wouldn't allow them to land in Florida until someone would indemnify the state of Florida for damage if Newcastle's disease spread from these parrots. A few calls around to the secretary's office found that nobody over there had the slightest interest in indemnifying Florida. [laughter] Stupid me said that I would sign the indemnity, which would allow Florida to sue somebody. So they brought the parrots over and by golly about two months later one died. I can tell you I lost a little sleep over that.

HKS: Sure.

RKA: But it turned out they were taking so many blood samples, checking for Newcastle's disease, that that's what killed the parrot. Anyway the parrots got to the research center out of Washington, ARS. They've successfully bred and moved parrots back and apparently now, I don't know whether it's endangered or not, but there are many more parrots in Puerto Rico.

HKS: Frank sent me a copy of an extraordinarily detailed resume. Apparently the agency requires it every so often when you're up for a certain review or maybe a promotion at a certain level. We're talking about a thirty-eight page summary of his career. What impressed me about it was the number of trips he took. It looked as though whenever the word tropics would come up in Washington, D.C., somebody said let's get Frank to take care of that.

RKA: That's correct.

HKS: About two-thirds of the world was his beat. He couldn't do any work in Puerto Rico, he was going someplace, on a committee, making a study, filing a report. Who else, I mean Frank really was it initially, right?

RKA: Yep, and he was highly respected worldwide. He could represent the U.S. in excellent fashion.

HKS: Let's get back to going to Congress and asking for money. Was tropical forestry a tough one to sell at budget time? Or were you trying to increase the amount of money available, to have more Frank Wadsworths.

RKA: No, we did not push tropical forestry, because there are few tropical forests in the U.S. Hawaii is mostly semi-tropical and the tropical part of Hawaii is not important.

HKS: I was thinking more in terms of the Forest Service hiring a Brazilian scientist to work full time for the Forest Service, or some other tropical nation. But that wasn't on the agenda then.

RKA: No. We tried to strengthen forestry in Puerto Rico. Frank was assigned because he was there, knowledgeable, and spoke fluent Spanish. But no, there was no big attempt to get on with tropical forestry. Carl Ostrom was knowledgeable in it from his viewpoint as division director of forest management research. We had a few other people who had worked in the tropics. I don't recall who at this point. But Frank did most of the work. Ostrom went to quite a few meetings over the years where a broader viewpoint was needed.

HKS: Of course the Forest Service was acquiring through normal process a lot of Peace Corps returnees that had hands-on experience. They had been trained in forestry, maybe came back and went to grad school, so we're building up a potential infrastructure for tropical forestry in the U.S. That wasn't the plan but it happened that way.

RKA: Yes, Yale and Syracuse had excellent programs in tropical forestry.

HKS: I had on my suggested outline the translation services. I guess I got that out of the chief's reports. Was that a USDA function or were you guys involved directly with that?

RKA: You know, I don't know.

HKS: It's in the chief's report for some reason.

RKA: We tried to translate key materials. We didn't translate nearly as much as other countries did, for example, Russia. I can't tell you how it was done. Dickerman or

Buckman probably know. But it wasn't a big operation. It might have been under contract with the Library of Congress, I'm not sure. It was not adequate. We lost a lot, and our individual scientists had to work their way through papers if they knew a little bit of German or French or Spanish, or they paid out of their research funds to have translations.

HKS: Okay, let's shift gears. You're at the SAF meeting in Albuquerque.

RKA: I had promised the head of the Mexican delegation that I would welcome them in Spanish. I wrote out a speech and had a friend of mine who was fluent in Spanish put it into Spanish on a tape. I drove from Austin, Texas, to Albuquerque and played that tape all the way, and learned my speech in horrible Spanish. He was fluent in English. The comparison was not a good one. [laughter]

I guess we're about ready for the University of Texas.

UNIVERSITY OF TEXAS

HKS: What's the background? Had you been talking to Steve Spurr over the years about joining up with him, or was this a sudden thing?

RKA: This was sudden, as most of my changes were. I had told Ed Cliff when I returned to the Forest Service that I would come back until I was eligible for retirement, which was about five years. I'd always thought of going back to universities to teach a couple of semesters and enjoy part retirement in nine-month appointments. I'd looked around the United States several places, and John Gray at the University of Florida had offered me the job that Les Harper had had down there, which was a position I liked. I didn't care for Florida as a place to live as much as some other places. It was a little too far from the West. One night in the spring of 1973, Steve Spurr called me. I think we were even asleep, it was 10:30 or 11 o'clock. He said he had a position at the University of Texas that he had been trying to fill for several months but could not find the right person for it.

HKS: He was president at that time?

RKA: He was president.

HKS: Okay.

RKA: He said the position was in the Division of Natural Resources and the Environment, which was a catch-all for anything that they didn't know where to place administratively. The University of Texas had no vice president for research. This unit was responsible for the McDonald observatory in west Texas, for oceanography and the marine science institute, for the geological survey of Texas, and several other small units. It had one other professional, Ross Shipman. As usual, I was fortunate to have an excellent co-worker.

Ross knew the university inside and out and he knew the Texas legislature the same way. We enjoyed working together and had a great time. My academic appointment was in the LBJ School of Public Affairs in which I would give one seminar a year. The rest of the time was in the environmental program. I told Steve before saying yes that I had never been to Austin, and if he would move the university to someplace that was out of Texas, I'd consider it seriously. All he said was pay us a visit. We went to Austin and liked it. And we thought that it was an ideal place to work for a few years and retire then in Austin. Without telling Steve, we bought a house before he made a formal offer of the job. On the day that I was eligible for retirement from the Forest Service we moved to Texas and took up this job. As far as the administrative part of it goes, it was very much like that of a deputy chief or dean, working with university scientists.

There were some very able people in the McDonald observatory and marine sciences and the geological survey. It was really enjoyable to work with those key people. There were some problems that were unique to Texas. I found out that at the University of Michigan most problems were kind of buried. They were not publicized. In Texas any time you had a major problem you would run it up the flagpole so that everybody could see it. They're out in the open.

I enjoyed the LBJ school and became acting dean about a year after I was there. The dean at that time had two or three DWI problems and had some problem with drinking, so he was removed from office. There were two factions of the school, neither one wanted the other faction to have either an acting or a regular dean, and they both considered me neutral. I ended up as the acting dean. In fact I was acting dean twice after the dean left rather suddenly to head up President Carter's federal personnel office.

Elspeth Rostow, the wife of Walt Rostow who had been on LBJ's staff, was a most able person. She taught at the U.S. War College. They'd send a plane down for her, and after teaching she would be flown back to Austin. She was named dean. She accepted on the premise that I would be the associate dean. She did a lot for the school, both bringing the faculty together and increasing outside support remarkably. She and her husband taught public administration together. She would let nothing interfere with that one day a week. In fact it was taught live and on TV at the same time. She was most innovative.

HKS: Did you have a chance to get to know Johnson himself?

RKA: President Johnson had died. I got well acquainted with Lady Bird, however. The Johnson library of course was right next to the school. Lady Bird had some oaks dying on the Johnson ranch. We had that point of contact also. You'd be in your office in the LBJ school. A bright young man, probably thirty, well dressed, would come walking in or stick his head around the corner and disappear. About five or ten minutes later Lady Bird would walk in. And she usually had something related to the library that she wanted assistance on or quite often just invited me to a social occasion. One time she had all of the key staff (secretaries of departments) under Johnson there for several days. There was a dinner that happened (another one of my lucky times) when I was acting dean. So Lillian and I were invited. They exchanged stories about "the old man." She was remarkable. She knew everyone's name and used it. She'd go out of her way to introduce people, and made it difficult for the Secret Service young men to stay with her because she was just very friendly and knowledgeable. And she also knew how to get work done, too.

We might just talk a little bit about Steve's firing at the university. Steve was a very able university administrator. Faculty appreciated him and respected him and trusted him. Students, it was unbelievable how he managed to maintain rapport with students. He taught an introductory course in forestry just because he wanted to teach. He never wanted to be in a university without teaching. He taught just that one course, but it gave him a tie with students, some three to four hundred at a time.

The chancellor at that time, the head of the U. T. system, thought Steve was after his job. I don't recall his name, he was apparently quite able politically, but I'm showing a little bias. He was not a strong chancellor in my book. The chancellor did all of the contact work with the legislature on budgets. The presidents and others in the system deferred to him because he was the budget contact. He was afraid of Steve and he needn't have been. Legislators, anytime they had a question about the university they would call Steve, and Steve, immediately on every call, and I know this for a fact, would immediately apprise the chancellor that so-and-so had called and asked this. Steve suggested many times that they call the chancellor. But the chancellor took all of this as a threat.

Right in the middle of the budget session, when he was meeting with the legislature, he fired Steve. Steve had had open heart surgery several months before. He called Steve in and said you know you're really on kind of a slow bell and you've had this heart problem. Wouldn't you like to retire? We have a house out here that you can live in it as long as you wish and we'll provide you a car at the same time. Of course you'll have an office and a secretary. Steve said well let me think about it. Steve knew what was going on and he didn't want to be bought off as had others. This was not unusual at the University of Texas.

HKS: I understand that.

RKA: It had happened before. Anyway, Steve went in and told the chancellor the next day that he'd decided that he would not retire. The chancellor said well then I'm removing you today from the office of the president. He said of course you're a tenured professor in the LBJ school, but I want you out of the president's office within forty-eight hours. If it hadn't been in the middle of the budget session, my guess is that the Board of Regents would have removed the chancellor rather than Steve. But there was no way to do that without upsetting the budget process. Lady Bird refused to vote on the matter. They confirmed what the chancellor wanted. The chancellor said this is it. But if it had been done at a time when they had some time and could adjust, I'm fairly confident that Steve would have been there. Not in the chancellor's position, but in his president's position.

HKS: Was he unduly upset by being fired?

RKA: He was upset, obviously. He was my boss one day and the next day, since I was acting dean of the LBJ school and director of the Division of Natural Resources and the Environment, I was his boss. His academic appointment was one half in each unit. We had many laughs over that. Surely Steve was upset but not emotionally. I was at the national SAF meeting in New York when I got a phone call at 3 o'clock in the afternoon that this was going on. I left on the 6 o'clock plane and got back to Austin. All the faculties were protesting. In fact they were unable to hold a university faculty meeting for over a year because no quorum of professors would attend a faculty meeting.

HKS: That must have been rather devastating for a lot of the programs.

RKA: Well, they went on. The regents appointed a vice president of the university, Maureen Rogers, as acting president. It seemed that the chancellor thought that maybe she was weak enough that he could control her. It turned out that Maureen Rogers was a very strong person. She ran the university and was eventually made president in her own right. But Steve just went right along, he started some research and more writing on his own. He gave seminars in the LBJ school. It was at about that time that the Parkinson's disease caught up with him and he was a little bit on the slow side. He never gave up or evidenced anything but that he had a life to live all the time.

HKS: Would you characterize the university as being less rough and tumble than the Forest Service? I mean both in their own way could be rough and tumble.

RKA: The university, both Michigan and Texas particularly, was much more rough and tumble. I can give you one example. I was acting dean in the LBJ school. A phone call came to my secretary; a Mr. Head, who was a legislator, wanted to come up and see me. And she said well Dr. Arnold has so-and-so appointments and what not and whoever it was on the other side said Mr. Head will be there at 2 o'clock. I immediately called the president's office to see who it was. He was the chairman of a subcommittee on university budget.

HKS: Important person.

RKA: He was important, so obviously I had time for him. He came in and was very pleasant and said he had just been reelected and that his campaign manager had a Ph.D. in English. He had promised his manager if he were reelected that he'd have a tenured professorship in the LBJ school. He said, do it now. I stuttered around that we didn't have any positions or funding for such a position. He referred to my ignorance and said that those could be provided without any question. I said well I'll talk with the president's office. He said you call me tomorrow.

We obviously would not think of a tenured professorship. We did plan a research unit in the use of communications in the federal government. We got enough money out of the president's reserve fund to pay a good salary and a secretary and travel expenses. So I called Mr. Head the next morning and said that I was just pleased to be able to accommodate his needs and that we would look forward to so-and-so, whatever his name was, coming in. He uttered a few four-letter words when I described what we had done and hung up. When the LBJ School budget came out from his committee it had changed from about a million dollars a year to zero. It passed the House of Representatives at the zero level.

HKS: Oh God, rough and tumble as they say.

RKA: Three days later, I learned that Mr. Head (the legislator) had borrowed money for his campaign from a west Texas bank and from a bank in Austin. Both of those loans were called. Anyway, you were asking about rough and tumble, that's it.

The university was well run except in the research area. I enjoyed working there. I think it is important to know how Steve reacted to all of this turmoil. It did not affect him as an individual. He wasn't bitter about it at all, he knew the risk that he was taking and lived through it. Shortly after Maureen Rogers was named president, she established a vice president for research. I think it's something Steve would have done because it obviously was needed. We had about eighteen to twenty million dollars worth of research going on in the university and no oversight except on those few things that were in the Division of Natural Resources and the Environment. Eldon Sutton was brought in as the vice president for research, and I was the assistant vice president, which meant doing about the same things that I did in the Division of Natural Resources and the Environment, but on a broader scale. The job became almost identical with that of deputy chief.

HKS: Is there a medical school in Austin?

RKA: No, the medical school's at Houston, and there was a medical school in San Antonio. We had a nursing school but no medical school. As we began the work in research, we found a large number of research units, institutes, and so-called research programs. Someone would have received a grant of some kind. He or she would immediately print some stationary, hire a part-time secretary, and be a research institute.

We devised a system of a standard charter to describe the work of research units and funding, and of course listing the publications and so forth. It took about two years to review all of those with a research committee. The legislature didn't fund research at the University of Texas at Austin for one year. We had a two or three million dollar state budget. We used that to close down a lot of these small units who had been getting by. When research funds were reestablished, we were able to start financing key young scientists to start their research careers.

HKS: Was this before the oil boycott? I'm just trying to think what happened to the revenue in Texas.

RKA: This was well before that. I was long gone from there so I had no contact with those hard times. My position was also that of a handy man for the president. If there were a special recruiting job, I quite often went off and worked on it. In the case of several firings of key people in the university, I was the one that did the actual notification. The same thing went on with Maureen Rogers, I had a good working relationship with her.

On one Friday we had to fire the director of the Marine Science Institute. Marine science was big. We had a big biological station at Port Aransas, and at Galveston we had ocean-going research vessels with geologists and others involved. This individual is a renowned scientist, but used his sexual proclivity to keep young people working for him. We fired him, and he promised, because he had close ties to the governor's office, that that would be changed on Monday. On Saturday morning the president called a meeting to select me as acting marine science director.

HKS: You'd taught navigation after all, so...

RKA: But nobody knew that. It was quite disconcerting to the marine science people to have a dirt forester run marine sciences. Actually the biological research part was quite easy. Whether the medium is soil or whether it's air or water, the basis for apprising and helping scientists is the same. I didn't tell anybody of my four years in the navy. That experience was most helpful in directing the activities in marine geology.

We were talking about the rough and tumble stuff. The one who was fired didn't exactly go out peacefully. The rear window of my car shot out one time when I was driving to Port Aransas. He had my telephone tapped. He wasn't that smart and I never could figure out how he knew everything that was going on and reacted in advance to it. We were trying to eliminate him as a professor as well as an administrator. It takes a little time to remove a tenured professor.

HKS: Sure.

RKA: I retired then from the university on the planned date of retirement. They found later that the telephone had been tapped and it monitored all calls out of the marine science director's office.

HKS: The extraordinary importance of the oil industry to the health of the Texas economy. When you're studying natural resources policy there's a lot of opportunities to criticize the oil industry for this that and the other thing. Was that an issue? Was anything off limits?

RKA: No, never. I was not involved with oil. That teaching and work was of course in the school of engineering. It was a large school and well financed. You see the LBJ School of Public Affairs was principally to educate people in the public life and public administration. We had no one giving seminars or related to the petroleum industry.

　　We might bring SAF up here, it was occurring during all of this.

SOCIETY OF AMERICAN FORESTERS

HKS: Yes, that's fine.

RKA: Except for meetings and papers and incidental contact, most of my professional activities at that time were related to the Society of American Foresters. Early on in California I went through most of the chairs, committee assignments, and so forth.

HKS: Okay.

RKA: It was the northern California section that decided that they wanted to put me up for president. That's how they made the nomination, and since I was well known in the Forest Service and had strong ties to the West and some to the South, I was elected.

　　I was a member of the SAF council for six years, and of course the presidency is a longtime commitment. Two years as vice president and two years as president. I was to change that to one year at each level. This was the time when the SAF was flourishing. We had over twenty thousand members. Now I guess we're down to seventeen or eighteen thousand. Anyway, I was president in 1976 and '77, and I was vice president when John Beal was president in the two years prior to that. A vice president can have his committee assignments lined up. He doesn't have to wait till he's president, he can be ready to move. But if you organize in that year as vice president, you are ready to go. Several able people I know would not commit themselves for four years.

HKS: When you were a candidate, did you have a platform as such?

RKA: I don't think so. I can't remember. We all write out what we see was the need for SAF and right now I couldn't tell you what I said, except to be of service to the profession and to represent the profession in major policy issues.

HKS: If you had been still with the Forest Service would that have been a problem, in terms of the Forest Service concern about being too dominant in SAF.

RKA: No. Charlie Connaughton was president of the society.

HKS: That's true, he was.

RKA: I was nominated while I was still in the Forest Service. Even before Steve Spurr had called me about Texas I talked with John McGuire. If I accepted this opportunity to be nominated it certainly would be demanding. I knew of Connaughton's problems in California.

John agreed with me if we needed some extra help in the deputy chief's office. We thought we could handle it, because I had the two associate deputies. It would mean maybe 20, 25 percent of my working time to do it right. John just said well you work on what you need and we'll make it go. Even before I accepted at Texas I pointed out that if I were elected that was a major program that I would have to devote time to. That was no problem to Steve as long as I did the work that was at Texas. A lot of time comes out of nights and weekends. The Washington headquarters property developed through Tom Gill, and I think that that's something that you might want for the record.

HKS: Definitely.

RKA: I worked with Tom Gill on the International Union of Societies of Foresters. In 1969 Tom asked me to join him and Gifford Pinchot, Jr., for lunch at the International Club. At that lunch he propositioned Gifford Pinchot, Jr., to join him in financing a headquarters building for the society in Washington that would be named after its founder, Gifford Pinchot. Gifford, Jr.'s major interest was in oceanography. He said that the building was a great idea but he had recently provided a major gift in oceanography and fisheries biology. His interest and all of his spare funding at that time had gone into marine biology. He said that he couldn't do it and would not be able to participate. When he left I'd never seen anyone just more disappointed. Tom Gill sagged, and I thought maybe he was going to cry. He said, you know for many, many years I've thought of leaving most of my estate, he had no sons or daughters, to Yale University. He said it dawned on me about a year ago that I could make the greatest contribution to forestry by giving some $500,000 to the society for its own headquarters. As we sat there and talked I said, you know there's another way of getting matching funds. Why don't we see if the SAF membership would match your five hundred thousand (not from an individual, but from the membership). He just lit up, came alive again, "a tremendous idea," he said.

He and I talked and we agreed to look at a million dollars as about what was needed to do what we were going to do right. With his five hundred thousand, if we had three hundred and fifty thousand from the membership, we could borrow a hundred to a hundred and fifty thousand and not overburden the society. He named several conditions. The first was that I would present the opportunity to the SAF council and he would remain anonymous. He did not want them to have any knowledge that he was the anonymous donor. Secondly, he had hoped that the society would get a headquarters which would allow some of the other societies, similar societies, to locate with them. He wasn't specific but he said it might even, we might even be able to organize some kind of a consortium. Those were the three conditions that he wanted.

HKS: When did he die in all this?

RKA: Afterwards. Charlie Connaughton was asked to head the campaign and before they finished raising, it became obvious that we were going to do it. In fact they raised more than the three hundred and fifty thousand through pledges and gifts. It was at that time, just before the campaign ended that he died. Unbeknownst to me he had changed his will from Yale to the SAF. If the SAF accomplished its goal of raising $350,000, and if it could obtain the bank loan, then the $500,000 would go to SAF, otherwise it would go to Yale. That was how that came about. Then of course there was an excellent committee under the Forest Service deputy chief, Gordon Fox. Gordon Fox looked for sites. They found the Wildacres site, and all that's become history.

HKS: Talk a little bit more about Tom Gill, the person.

RKA: I didn't have a lot of contact with Tom Gill. We met probably once a month for lunch to discuss the International Union of Societies of Forestry.

HKS: So the tie with you, primarily, was international?

RKA: Yes, that was the complete tie. Tom was quiet and a little bit of a recluse, I think. He did attend some meetings. I only had three or four conversations with him about this affair. We pretty well settled it at that first meeting, and we talked by phone. I kept him apprised and I told him right off that he was going to have a hard time turning down requests for funding, for matching funds. It turned out right. Hardy and Charlie Connaughton and others just couldn't figure out why he wouldn't be one of the major contributors to the matching funds needed. Of course it didn't come out until later that he was the anonymous donor.

HKS: We have his diaries, we have some of his photo albums that show him in Honduras or wherever standing in front of big tropical trees. Was his interest in tropical forestry long lasting, I mean in later life other than he just was for it. He worked for the Pack Foundation? He did so many different things, I don't know what kind of a forester he was, but was he really only active in tropical forestry during the '20s and '30s.

RKA: I am not sure. His international interests were principally in tropical forestry. He wanted the practicing foresters to have more international ties, that was why he was behind this International Union of Societies of Foresters. He knew the work in IUFRO and was most pleased with it; that gave scientists a chance to exchange, but it was not for the practicing foresters. He thought that the counterpart of the SAF formed in a number of countries would help forestry.

HKS: I contributed something to Charlie's appeal. There were those who said it doesn't make sense to move the national headquarters out of downtown Washington, outside the beltway as it were, because if you're going to be effective you've got to have quick response time for requests from the Hill and attend all the social life in central Washington, D.C. Was that much of an issue? Was it difficult to pick an area as far out as Wildacres as opposed to trying to find something closer in.

RKA: That was an issue in developing the criteria for selection. There was some thought that headquarters might be in Richmond or someplace out of Washington. That was discarded fairly quickly just for the reasons you cited. But when it came down to Wildacres, the subway was planned to go there. It was appropriate to buy the Wildacres site. It was several years before there was a direct subway to almost anywhere in the city. That issue did not apply to Wildacres. It was easy to get from the airport to the headquarters and so forth.

One of Gill's suggestions as I mentioned was that he would like to see the SAF be kind of a focal point for a larger grouping of natural resource related agencies. Out of that wish and with the property, the Renewable Natural Resources Foundation came about. It was formed at the instigation of the SAF. Hardy Glascock was very strong for it. In fact, the original Wildacres deed was deeded to the foundation and early on all of its support came from SAF personnel. At some point things turned sour. In Albuquerque, I remember giving quite a eulogy of Hardy Glascock in my president's report, how ethically strong he was and keyed into the nature of what professional forestry needed and so forth. Sometime after that he completely discarded forestry and SAF as an interest. He was the original leader of the Renewable Natural Resources Foundation. As a member of that council, and as president, we made a number of serious errors. Ben Meadows was the only negative person that I heard. Ben said this was poor business, we'll regret it. Ben was right. But other than Ben I think the council was completely in favor of the concept of the Renewable Natural Resources Foundation and its initial control over the site.

HKS: That was to serve primarily as a clearinghouse for funding that affected the whole area.

RKA: And to invite other people, through the foundation, it could build buildings and invite other agencies in wildlife...

HKS: So officially SAF was only one of the many organizations in the foundation.

RKA: It was formed with ten. SAF and I believe ten others, as I recall. Some of the others pledged some funding, I think the largest was thirty thousand dollars. We had authorized incidental use of SAF people on this. It turned up later that Washington office employees of the SAF were doing foundation work, much more than we anticipated. Some correspondence from other societies about RNRF were never brought before the council. And the council was not informed as to the degree of participation of SAF people.

HKS: This is Hardy's management we're talking about.

RKA: Yes, and we're talking about Hardy's management directly. There was a fund drive, and they asked Charlie Connaughton to head it up for industrial funding for a headquarters building for RNRF. At that time it became obvious that there were some problems. RNRF tried to evict us from Wildacres. We had Wildacres turned back to us through a deed transaction wherein we owned 90 percent and RNRF 10 percent. The reason for the 10 is that in Maryland that's the lowest percent of ownership that can have a say in management or direction.

HKS: By that time, though, Hardy was now director of the foundation.

RKA: Yes, I don't recall the dates that Hardy resigned from the SAF and became the director of the foundation, but shortly after that SAF was voted out. They met in a session and didn't invite John Barber who was then executive vice president of the society. When he arrived at the meeting he was informed that SAF was no longer a member. I'm not aware of what was said or done at that meeting.

HKS: I can see that Hardy may have had ill-will against SAF, but why would the other members of the foundation have voted yes? Can you speculate on that?

RKA: I suppose that they acted on the information they were given.

HKS: Was Bill Towell your vice president? Because he was president during some of this turmoil.

RKA: Bernie Orell was vice president. Hardy knew where the money came from, where the support came from. The entire operation was SAF, again leading back to Tom Gill's original suggestion. Hardy's attitude and behavior is extremely difficult to understand.

HKS: I knew Hardy very slightly when he was still with the trade association in Portland. When he was selected to be Henry Clepper's successor, I just happened to know Ed Heacox, who was on the SAF council at that time. I asked him how you select somebody. He said Clepper ran his own show but the thing we know about Hardy is the council can control him. It's interesting how things turned out.

RKA: Yes. I was on that same council with Ed when Hardy was employed. Henry ran things but you knew everything that he was doing. Henry could be dissuaded. It took a little argument and a few tears. He could come up with a good tear jerker, but Henry was certainly Mr. SAF for many, many years.

There was one other controversial policy matter that took quite a bit of time. While I was president, in fact just before that, the Justice Department had taken off on several engineering societies, saying that there were canons in their code of ethics which, in effect, led to restraint of competition. The matter went to court through these societies. They were levied fairly heavy fines and paid court costs. We're talking about sizable sums of money. With that in mind, we found very similar canons in our codes of ethics.

HKS: Give me an example.

RKA: Such things as all professionals should be able to compete for consulting work by price level. Consulting foresters were in competition with foresters in State and Private Forestry and foresters paid by industry who had some latitude for doing consulting on the outside. They would underbid the consulting foresters because they were financed partially in some other way. The consulting foresters were very unhappy with any changes in the code, which would give more freedom to industry and State and Private Forestry.

HKS: We face that all the time. We're constantly underbid by professors who receive support from the university to do history projects. We just can't compete at all. So that's the issue you're talking about.

RKA: That was one of the issues. I can't remember the details. They asked me and I wrote a long article for their bulletins to use and we didn't put their material in the journal. It caused a lot of hard feelings, principally because we had no choice as far as I could see. The council backed me on it. We had no choice but to make the changes dictated by the Justice Department.

HKS: I would think that's the way life is. Were you ever involved in a code of conduct violation where a forester was charged?

RKA: Yes, there were one or two cases like that, I don't recall them but the council appoints a committee to review the matter and makes a recommendation to the council.

HKS: I can't remember who I was talking to at SAF, but he had a recent case where a forester, through some process, had become a convicted felon and was in jail but he hadn't violated the code of conduct. They thought there was something missing in the code of conduct, if you could be a gangster and still be a good honorable forester, that there was...

RKA: You've probably noticed I had a remarkable faculty for picking up things in a hurry and forgetting them just as fast. The division directors in Washington, D.C. said they could make me an instant expert in almost anything. I could take information and digest it and so forth and appear before committees or do whatever was necessary, but three days later I couldn't tell you what it was about. The same weakness applies in some of the things we've been talking about. That's why I am vague about some matters we have discussed; I'm not trying to get out of talking about them. Right now I wish I had kept a diary.

HKS: There was a canon that's always bothered me. As a matter of fact it may have been during this very time when the code was revised and put up for vote as a member. And I voted no. And when the results were published in the journal, the changes were adopted overwhelmingly, I think the vote was like nineteen hundred and forty-two "yes," three "no" votes. I mean it was on that scale. I was one of three weirdos out there. The reason I voted no is that one of the canons is loyalty to the employer. To me loyalty to the profession should have been higher than that, to be idealistic in the sense of a code of conduct. In the real world we waffle on certain things, but in terms of what we aspire to, that's always troubled me. Do you recall off hand, was that canon debated much?

RKA: Yes, that canon was debated in the council a great deal. It was a difficult thing. I remember that at one time we said that foresters should try to influence the policy of their employer, the landowner, or whatever. The forester didn't have to take the position that the landowner did. But once the landowner decided what he wanted to do with his property, the forester should do it in the best professional way. That's where I remember that one coming out. But it was a matter of what the landowner wanted to do and how it was done. The forester should put his profession first in trying to help the landowner decide what to do, but once he did it, then the professional had to do the best job he could for the landowner.

HKS: How important is a code of conduct? I realize every organization has one, but I'm not sure how often each working day a field forester thinks about that. Do you think it has much influence on the way people behave?

RKA: I doubt it. Yet every profession needs a code. There's much to do over any change in the canons. But I doubt if anyone often considers ethics before he or she decides to do something or not do something.

RETIREMENT

RKA: I was ready for retirement from the University of Texas by 1980, and at that time my university retirement systems came in. With eight years in Washington, D.C., and the teaching in the LBJ school, I had a good handle on forest policies. It occurred to me I might keep in touch with deans of forestry to see if they had any openings for sabbatical leaves or some position to fill temporarily; I'd be glad to teach a class in forest policy or policy seminars. It worked out quite well. Colorado State University was first, and I spent a semester up there thoroughly enjoying it. I looked kind of for parts of the country where I wanted to live for awhile. I hadn't spent a lot of time in Colorado so that was enjoyable. I went to Oregon State with Carl Stoltenberg for another semester. And the University of Maine after that. But then meanwhile we had moved from Austin to Hots Springs Village, where we are now, living in the middle of an oak-pine forest on the lake. The University of Arkansas, Monticello, which is south of here, needed someone to teach forest policy. They could not fill a full-time tenured appointment right away. So I taught one semester for three years. I was there when the University of Arkansas Forestry School got accredited.

HKS: You'd go down for the week and come home on weekends?

RKA: I taught a three hour course, three lectures, and I gave lectures Monday, Tuesday and Wednesday.

HKS: I see, accommodation for your travels.

RKA: I went down Monday morning and came back Wednesday afternoon.

HKS: Without getting too personal, is the compensation equivalent to the actual performance. Is this really a bargain for the school to get someone like you?

RKA: It was a bargain. I had my travel and living expenses covered and a little more. I wasn't in it for the money. But I didn't take a full professor's pay at any of them. In fact, in Maine Dean Fred Knight just covered my travel and living expenses. They had funds for that sort of thing. Kind of interesting, a couple of years later John McGuire did the same thing and they paid him an honorarium along with it. Fred explained to me they had more money available when John was there than when I was there.

Conclusion

HKS: The standard question is any regrets; also, what part are you most proud of.

RKA: I think in terms of the deputy chief position, that I probably was an average deputy chief. There was a lot of momentum going when I arrived, and I really was a little bit ill prepared for the job. I had relatively few congressional contacts, even in California it was done by other people there for the station. I enjoyed it thoroughly, and I don't have any regrets. I kept the thing going and supervised moderate growth, and particularly with Dickerman's very valuable work, we solved all problems. But I don't think things grew so rapidly as with Harper or Jemison. I did do a lot toward moving in environmental issues.

I think my forestry career just happened. You've asked several questions was I thinking about this job or that. From the standpoint of the division chief in fire in California, I think I could have stayed in that position the rest of my working life and thoroughly enjoyed it. I felt that about every position. I had never intended to leave Michigan. In fact, there was informal discussion that maybe I ought to look at the vice president for research job there sometime in the future, particularly when I left. I never had a career plan, but I certainly was in the right place at the right time many times.

I used the GI bill to work on a Ph.D. at Michigan. The Armed Forces Special Weapons Project gave me good experience in research administration. I learned a lot of what was good and what was bad from watching people. I would not have been the station director in Berkeley if it hadn't been for the fact that they needed someone in a big hurry. And Harper, I don't how he picked me except that Sam Dana had told him about me. The chance meeting with Steve Spurr made it possible to go back to Michigan, and then, of course, Freeman and McNamara made it possible for me to return as deputy chief.

HKS: Seems like you get a couple of your jobs because they're in a hurry. I'm not trying to say that you weren't really qualified but they needed somebody at the moment. You were qualified so, and available, and...

RKA: That's why I say that somebody had to have a plan, and I do believe in a higher authority than all of us. It just seems to me that there was a plan, but I wasn't knowledgeable of it until after it was over. But that takes care of anything I have.

HKS: I think it's an excellent interview. You were certainly well prepared.

∾

M. B. Dickerman

Dick Dickerman maintains two homes, one in Alexandria, Virginia, and the other in Cooperstown, New York. He was summering in Cooperstown when I caught up with him by phone in 1992 to schedule this interview. He was good-natured but firm that Cooperstown was for relaxation, and being interviewed sounded too much like work. We would have to wait until fall. But fall held other conflicts for both of us, and it was not until early January 1993 that we found ourselves sitting in the den of his Alexandria home, working through an interview outline that we had developed by mail.

He has lived in this home since 1966. It is comfortable, sitting at the top of a knoll looking toward the District of Columbia. When he was active in the Forest Service, it was a quick bus ride to work. His den holds much memorabilia, which is dominated by western objects—a mounted antelope head and Charlie Russell paintings. The head was a gift—he doesn't hunt—and the prints he collected early in his career. Our lunch each day was prepared in advance by Mei, Dick's China-born wife. The sandwiches and desserts were so elegantly assembled and arranged that we hesitated before plunging in. We were both well taken care of. Later, Dick reviewed the transcript, correcting errors of fact and clarifying statements.

∾

THE FORMATIVE YEARS

HAROLD K. STEEN: Somewhere along the line, maybe when you were in high school, you decided you wanted to be a forester. Describe that experience.

M. B. DICKERMAN: Basically it goes back to my family. They were early settlers in Connecticut. They tried farming, but I guess farming wasn't a part of their makeup, because they never did very well at it. My dad, being the youngest member of the family of nine, became interested in woodlots. He took on the job of supplying the brick yards in southern Connecticut with fuel wood for their kilns. Every year he would take several hundred cords of wood into the brick yards, and that meant cutting, skidding, and hauling timber.

That's where I became involved, helping my dad. He was an observing individual and had an interest in how trees grew. That's where I came into forestry. Also when I was a Boy Scout, I had a scout master who had studied plant physiology and was very much interested in plant sciences. He encouraged me to go on into forestry. So when I finished up high school in New Haven I decided to go to the University of Connecticut. In the 1930s, they had a very small forestry department. I spent four years there. I might just say as an aside that some people would say, well, you lived near New Haven, why didn't you go to Yale.

HKS: Sure.

MBD: The answer is my dad wasn't very enthusiastic about that. He had had some dealings with the people at the forestry school there and he didn't think he was treated too well.

HKS: If you'd wanted to go to Yale, it doesn't have an undergraduate curriculum in forestry. Is there a program at Yale that you could have gone through four years and then gone into the forestry program?

MBD: Yes. At that time there was. I don't know right now what their set up is. I could've gone there, but this was the mid thirties and the financial burden of going to Yale would have been substantial. Anyway, my dad was happy to have me go to Connecticut. As far as I'm concerned, it was a fine choice, at that time it was a small school—five hundred students.

HKS: The whole university or just the forestry program?

MBD: No, the whole university.

HKS: Is that right?

MBD: At that time it was called Connecticut State College. The forestry school had a couple of profs.

HKS: Anyone famous?

MBD: Albert Moss was a longtime member of the faculty at the University of Connecticut, a graduate of Yale, and a highly competent, well-trained forester. He established close personal relationships with all his students.

HKS: Was it an SAF accredited school?

MBD: No.

HKS: With just two faculty members, it couldn't have been.

MBD: No, it never was as far as I know. But they gave the basic forestry courses, so you had to schedule your courses in alternate years. I personally got a lot out of it, because I was sort of a shy individual and going to a small school helped me develop. I spent the four years there and enjoyed it tremendously. I became involved in lots of student and forestry activities. One of the highlights of my entire career was going to the University of Connecticut.

HKS: You took all the basic forestry courses.

MBD: Yes. Dendrology, mensuration, management, silviculture, and many others including a minor in economics. When I graduated in 1934, the country was in a deep depression, things were really rough going. I put in an application for a fellowship out at the University of California, Berkeley. I felt that I needed to go on and get some more training. I didn't hear anything from them, so I took a job on forest land acquisition with the Forest Service in Missouri. I was there for, oh, maybe two or three months, and all of a sudden, I received a reply from the University of California offering me a fellowship. I accepted and went out to Berkeley in the fall of 1934.

HKS: You would have been in school with Keith Arnold.

MBD: Keith was a couple years in back of me I believe.

HKS: Okay.

MBD: He may have been a sophomore when I was a graduate student or something like that. Hank Vaux was at California the same time I was in graduate school. I was interested in forest economics and marketing work, and so Emanuel Fritz became my adviser. He interested me in studying the marketing of redwoods. I spent much of my two years at Berkeley studying marketing of forest products. I thought very much about going on for a doctorate, but by the end of the second year, finances were beginning to get pretty tight, so I decided in 1936 to go back east and see if I could get located in some kind of forestry work.

HKS: So, the East was still where you wanted to be.

MBD: Yes. Primarily because of my family being here.

HKS: I look around your study here, and I see a lot of western stuff. That's later in life?

MBD: Yes, from my years in Montana.

HKS: Would a Ph.D. have taken another three years?

MBD: At that time they said two to three years, because I needed some things that I hadn't picked up at Connecticut. As I recall it Dean Mulford said two more years would be sort of a minimum. That just looked out of the question to me, so I gave up and came back east. I took a temporary job with the Northeastern Forest Experiment Station, which was then located in New Haven, and became well acquainted with Ed Behre, the station director. He was very helpful in getting me placed. I worked for awhile on an experimental forest in Massachusetts. Then I received an offer of a civil service appointment with USFS, Region 9. As an aside, while I was at the Northeastern Station in Massachusetts, my partner was Norm Borlaug, then an undergraduate in forestry at Minnesota.

HKS: Wow. He went on to big things, didn't he?

MBD: Yes. He received the Nobel Prize in agriculture. Norm and I did field work together and we became lifetime friends. He certainly had no thoughts of going on to become so famous.

HKS: Were you surprised that he went on to earn the Nobel Prize?

MBD: I was surprised, although I knew after that summer in Massachusetts that he had tremendous capability—a keen observer and great determination. Norm had a year or more to go at the University of Minnesota, so he went back there, where he did graduate work in plant pathology. Forestry lost a great scientific leader when Norm switched to agriculture. A few years later I ended up out in Minnesota. At that time he was really on the go. It's been a friendship that's continued over the decades. He comes to town two times a year and we get together. We've had some close family ties as a result of those earlier years. It's a friendship I have valued tremendously.

After the work on the experimental forest in Massachusetts, I went on to USFS, Region 9 on the Manistee National Forest and became a technical foreman in the CCC camp. It was a camp that was all blacks and many were prison escapees. It was an experience that I will never forget.

HKS: The stereotype of CCC is white, middle-class, but it wasn't all that way obviously.

MBD: No. There were several all black camps on the Manistee at that time.

HKS: I never heard about prisoners. I know they had black regiments and they had Indians.

MBD: Some were escapees of the state penitentiary I believe at Battle Creek, Michigan. When the Forest Service took over the camp, the state police were there to keep order. Things kind of settled down after awhile—except when the Joe Louis-Max Schmeling fight came along; Joe Louis was defeated, so the boys decided to tear the camp apart, but we intervened. After about six months as a technical foreman, the Northeastern Station offered me a transfer back to that station at Cooperstown, New York.

HKS: So that began your long association with Cooperstown.

MBD: Yes, and it continues to date.

Northeastern Forest Experiment Station (1937–41)

Forest Cooperatives

MBD: Work at the Northeastern Station with a forest cooperative was tempting. This was a new venture in the field of economics, trying to help the farmers manufacture and market their timber and at the same time practice good forestry on their woodlots. The station's major interest, of course, was the management of the woodlands.

HKS: Is that something that State and Private Forestry would do now?

MBD: No. There is still much research needed on how to organize a forest cooperative to service farmers. What system of forestry is best adapted to farm forests? The organization and operation of a forest cooperative includes many unknowns. Cooperatives, of course, are not new. There were several agricultural cooperatives at that time in the Lake States, marketing farm products: milk, cheese, grain, etc. The thought was that the cooperatives that had developed to serve agriculture might well serve forestry. That's why the station became involved. They helped organize the cooperative, and with federal funds they built a modern wood processing plant including a sawmill, dry kilns, and other facilities.

HKS: Was there any opposition to this?

MBD: Some. Ideologically there was opposition, but the fact remained that there just wasn't a market for timber in that area. There were only one or two small sawmills.

HKS: You were too far from the paper industry there?

MBD: Probably sixty miles to the nearest pulp mill. It was kind of a no man's land from the stand point of timber markets. With so little timber cutting, the second growth stands had a good start. There was more timber volume and better quality timber there than many areas. Another factor was the high tax delinquency. Farmlands were becoming delinquent and there just wasn't a land market. The state had a small acquisition program, but basically the market was limited. Often opposition would surface but nobody got real anxious over it. Several industrial people would come around and look at things, throw questions at us, but they never got really concerned.

HKS: Farmers are a pretty independent lot.

MBD: That's right, and this organization had some of the best farmers in the area helping out. That's the way the cooperative developed quite a large membership. I was there for about four or five years with the cooperative and was primarily concerned with the manufacturing and marketing of hardwood timber. We had several staff members, others became involved in the forest management activities.

HKS: So, it was a step beyond farm forestry.

MBD: Yes, at least a broadening of earlier concepts.

HKS: Was the project successful?

MBD: It depends on who you talk to. The financial end was very, very difficult because of loan restrictions. These were substantial and made it difficult to use the funds that normally would be available to a private corporation. I left there in 1941 and went on to study other cooperatives in the Lake States area. One of their biggest problems, other than the limitations of capital, was how to recruit and keep management. Somebody that knew the manufacturing end didn't know the forestry end. Trying to put the two together made for continuing difficulty. What really got them off and going good were the markets developed during World War II.

HKS: Oh, sure.

MBD: Then of course, you could sell most anything. After that, I don't know just what happened, but sometime after World War II, a fire burned the sawmill, and that brought everything to a close. Now, fifty years later, there are still people around Cooperstown that talk fondly about the forest cooperative.

HKS: Was that tied into your graduate program, marketing?

MBD: Sort of, but redwoods and second growth hardwoods do not have much in common. What I enjoyed most was the opportunity to work with the farmers and local people. They had lots of meetings to get people to join the co-op. I got well acquainted all through the county with farmers, and most of them were very, very sincere about getting markets for their timber and managing their wood lots. There wasn't a problem of salesmanship with the farmer. They were with us all right. The problem was to make the organization work.

HKS: Somebody had to pay the dues.

MBD: That's right.

HKS: Pay the rent or whatever it was. Somebody had to be in charge of it. Was it just you by yourself or was it a large project?

MBD: No, we had a staff of four or five people. A forester by the name of Charlie Lockard was head of the project. I was his number two man. We had three other fellows that did woodlot cruising, timber marking, and mill studies. The station was also responsible for the administration of the loan. That became quite a chore. In fact, for one year I left the Forest Service and took on the job of managing the plant. One year's experience taught me that it was not the thing for me to do. It was shortly after that that I transferred out to the Lake States Experiment Station in 1941.

LAKE STATES FOREST EXPERIMENT STATION (1941–44)

HKS: I visited a forestry co-op in Japan. The women knitted sweaters and they had a store. The whole community, the whole family, was involved.

MBD: This is the kind of thing that I became involved in out at the Lake States. Raphael Zon was the director there at the time and was interested in co-ops as a social venture. In fact, he had several staff members working with local groups.

We had several co-ops in the Lake States. There was one on the Chippewa National Forest that had been started by a station member named Paul Zanegraf. Paul was from Denmark and had a background in co-ops there. He got several small pulpwood producers organized in a co-op, and I took over working with them. There was another small co-op in East Tawas, Michigan. Gordon Fox, who was a district ranger on the Huron National Forest, had been interested in starting that one. In the Lake States, particularly the northern part, the Finnish people were the real co-op people. The whole community got involved. The families, the communities, all the organizations there—everybody talked co-ops no matter what they did whether it was selling milk or buying clothing or marketing timber. It was all done the co-op way. Co-ops were just a way of living.

That was about what it amounted to and has much to say for it. Those fellows in Minnesota did very well in marketing pulpwood cooperatively. I worked with them for about a year and a half. Then the war began to accelerate. Scarcities and shortages here and there of equipment, critical products for the war effort became an interest of the station. The War Production Board financed studies and eventually set up a group at the station. Some station people went on into the military services. Others went on to writing and analyzing, and completing reports.

WORLD WAR II

HKS: So the Forest Service basic budget was reduced, but there were war funds that filled in.

MBD: Yes, the station regular funds were probably cut 75 percent, as I recall. Something like that. Just the hard core of project leaders were left. Then the War Production Board came in. Of course, what they were interested in was trying to increase production of various kinds of timber, pulpwood, other products. Within the station we set up a unit to provide the War Production Board with basic information from all over the region. This was done countrywide. Lake States started earlier, because some of the problems surfaced there sooner. We had a staff of maybe eight or ten people reporting monthly to the War Production Board on production trends and problems, I had the job of summarizing the reports and making a monthly report on significant changes in the production of forest products in the Lake States.

HKS: Was it interesting or was it tedious?

MBD: Some things were tedious. I had questions about some of the answers I was getting repeatedly. But it was interesting, too. I had a chance to do several field studies of production bottlenecks.

One that comes back to my mind, I was up in northern Minnesota where I stopped into a lumber mill. I noticed they were setting up a shingle machine to manufacture wood shingles and I said to the operator, what are you going to use for wood. This was up in the pine country. Oh, he says, jack pine. We're going to make shingles out of that. At that time shingles were almost non-existent. You couldn't buy them any place. I kind of looked at him, hesitatingly, and he said, I know what you're thinking, but jack pine shingles are better than no shingles at all. Those were the kinds of things you ran into.

HKS: I suppose the so-called substitutes like asphalt shingles were just as scarce during the war.

MBD: Oh, yes. The big shortage was in transportation. You couldn't move cedar shingles from the West Coast or from any place. You had to rely upon native woods for all kinds of construction.

HKS: Seems like it would have worked pretty poorly...

MBD: I never went back to look. We had several experiences like that. Of course most of the problems frequently were just the red tape, plus some hoarding that one would encounter. At least we felt that we were doing a fairly good job. The war production work kept increasing our funds, they wanted more information. So I spent about a year and a half on that project at the Lake States.

There was always a matter of working and living arrangements. In the big cities, you couldn't get places to live; you could get jobs. Up in the north country you could get a place to live and a have reasonable income, a lot of the people were happier that way.

HKS: Were there major war industries in the Lake States?

MBD: There was a huge munitions plant in Twin Cities. There were lots of other industries too, plus, of course, several large pulp mills up in northern Minnesota and even more in Wisconsin. At that time the Lake States was in the forefront as a pulpwood producer.

RAPHAEL ZON AND CARLOS BATES

MBD: I probably learned more from Raphael Zon than any other person I worked for in my thirty-nine years in the Forest Service.

HKS: Did he speak with a accent?

MBD: Always. Always with an accent. He took advantage of it. But I'll say this, he was sure hard to work for, a tough taskmaster. I'll never forget the first article I wrote for publication. It had something to do with trends in lumber and pulpwood production in Michigan. I worked real hard over it for a week or two, then took it down to him. I'd never had such a depressing reply in all my life. He was just ruthless. He took it to pieces. You know, this isn't right. This isn't right. You should have said this. This is the way he would react. Well, I learned a lot out of it, but I guess I'd have to say that my respect changed a little too.

HKS: Just yesterday, I was scanning through the McArdle interview that we did twenty years ago. Thornton Munger at the PNW Station treated him the same way. Maybe it was that generation.

MBD: I think that's right. One thing that characterized Zon more than anything else was his love to have an argument. I remember one time when I was first at the station, Bill Greeley, a former Forest Service chief, came to visit with Zon. Zon would always ask somebody to drive to meet visitors so I took him down to St. Paul to have dinner with Bill Greeley. I was just a young forester, of course, sitting in awe of two great people. All they did that whole evening was argue. [laughs] They got pretty close to calling each other names disrespectfully, but they argued. When it was all over, everybody was happy. That was Zon's way of having an entertaining evening.

He was a fluent writer, loved to write. He could sit down and write about any-thing, and then he'd say, you check the figures. One thing I remember particularly was an article he wrote for *Encyclopedia Britannica*, and while I was at the station it came up for a five-year revision. He said, Dick, you have to check this over real carefully. Here's what I've written, and now I want you to check all the facts, fig-ures on area growth and volume. So, I spent many an hour and a weekend—my wife helped me—looking up all data to be sure everything was right. I took it in to him. Fine. Fine. No question. About three months later, he called me down to the office and said, I just got a check from *Britannica*. It was two or three hundred dollars. He says, you did most of the work on it. You ought to have most of it. I said, no, Mr. Zon, it was just part of my job.

HKS: You called him "Mr. Zon"?

MBD: Yes.

HKS: Everyone did?

MBD: Oh, yes. Most everybody called him mister around the station. Some of the real old-timers would call him Raphael, but most everybody called him mister.

Anyway, he said, you know, while coming to work this morning, I got to think-ing about what to do with this check. Just previously the word came over the radio that Stalingrad had been taken by the Germans. I think maybe we ought to give the check to the Russian relief. [laughs]

HKS: He was Russian, right? By birth.

MBD: So that's where the check went. Yes, he was Russian. In fact, his wife came to this country before he did. Incidentally, she was a very brilliant and able person Zon was very much of a liberal, so much that it caused some difficulties.

HKS: He must have had a lot of words with H. H. Chapman and Emanuel Fritz and those kinds of guys.

MBD: I guess so. He was always writing letters to carry on arguments and try out new ideas. Some people differed with him strongly.

HKS: Not in terms of his scientific work.

MBD: No. It was in his political views. To talk about his writings, he frequently made liberal use of material written by others. I don't think he misused it, but he prob-ably used materials from others more than most of us would.

HKS: I can see that could bother in a scientific sense.

MBD: Yes. He was an interesting person, a stimulating person to be with. He had all kinds of ideas, all kinds of interpretations of events that most of us would not have. He had a tremendous influence on the younger people in the station. You just have to look around to see that some of the people that came out of there and how well they did in forestry to see that his influence was part of their career development.

HKS: I've seen I guess four or five different photographs of him, and in every one he has a cigar.

MBD: Oh, yes. That was part of the image making. The cigar was lighted once in awhile.

HKS: He and Carlos Bates with his pipe must have gotten along fine.

MBD: Carlos was there when I was at the station. We had offices side by side. Carlos used to work nights, and I worked days. [laughs]

HKS: Why was that?

MBD: It was just his preference to work that way. He would get all wrapped up in interpreting shelterbelt data, which was his specialty. And he wouldn't quit. He'd work all night. Maybe go home at two o'clock in the morning or later. Just a very intense person, he would concentrate on something and was not going to give up. In later years, he was working on the influence of shelterbelts on crop yields. He developed a theory that shelterbelts so high had such an effect upon soil moisture and crop yield. He put up several large baffles out on the great plains to work out a wind velocity-moisture relationship. He was still working on the data after retirement. What he needed most was the computer capability we have now to handle his complex data.

HKS: So, he didn't assume that maybe his hypothesis was false. He thought he was screwing up the data.

MBD: Yes. He thought he just didn't know how to work it out. Of course, nowadays we could do what he wanted to do on a computer in a short time.

HKS: Sure.

MBD: He had stacks and stacks of plot data. He just couldn't accept the fact that the relationship didn't work out. Since then, I've been kind of interested in the subject and find the Russians did a lot of similar research about the same time. I assume that he must have had some contact with them through Zon. Anyway, some of the Russian studies indicated a close relationship between crop yields and shelterbelts—the height, width, and densities. In this country, I don't think any-body came up with anything as conclusive as the Russians did.

This was an interesting period in my life, one that I look back on as having probably more influence on me than any other assignment, the association that I had with Zon and Bates. When I first went out there, he was working on Pinchot's book.

HKS: *Breaking New Ground.*

MBD: *Breaking New Ground.* He was editing it.

HKS: The book was published in '46 or '47.

MBD: Yes. Zon was working on it in '42. I did some searching of reference material for Zon. He'd use younger staff members as trainees. He used to say, I will call you in and dictate a chapter or letters. You had to scribble it down the best you could. I remember one day we were sitting there, and he said, you know, I suppose you think that this is awful to sit here and take dictation from me. But I can't stand dictating to a secretary, 'cause she won't argue with me. Well, to be very frank, none of us argued with him very much either.

HKS: Do you have any sense that he was editorializing *Breaking New Ground* or just checking it over? A lot of people claimed Pinchot didn't really write the book.

MBD: Well, I'll say this, my impression was that he did considerable writing. When I was there, he was rewriting, and I'm not sure who had done most of the initial writing.

He had a tremendous command of the English language and of several foreign languages. I never could figure out why, this must have been just native ability, because his spoken English wasn't that good, but he could sure put thoughts and ideas into the English language that most of the rest of us would struggle with long and hard.

Certainly Zon was one of the more outstanding and controversial people in the early history of forest research. I don't know whether you've seen it or not, but there's a small slip of paper that Zon wrote to Pinchot saying in effect that in order to facilitate research he recommended the Forest Service establish a series of forest experiment stations throughout the country. That I found in the files of the Lake States station.

HKS: Maybe you can't answer this directly, but how come Clapp, the patron saint of research, is in the Washington, D.C. track, and Zon's up there in St. Paul with his career? Is that by choice or the wrong temperament, too liberal?

MBD: It may have been the latter.

HKS: Too liberal.

MBD: The Lake States was the place to be for liberals then. Senator La Follette was active as a liberal then. My impression from working with Zon and others was that his liberalism was better handled out there than it was in Washington.

HKS: Forestry has more conservatives than liberals.

MBD: Yes.

War Production Board (1943-44)

HKS: In part you were becoming more senior through passage of time, but in part World War II took its toll on the other supervisors. What was happening?

MBD: I left the Lake States in late '43 and came into Washington on War Production Board work. My assignment was to work as a consultant on the problems relating to pricing of forest products. At that time, there were ceilings on most forest products, and the question was what was the effect of a price ceiling on production.

HKS: In the history of the Lake States Station, you are quoted saying, price ceilings didn't work.

MBD: That's right. In my opinion they weren't effective except in a very general way.

HKS: Is that because supply and demand can't be taken independently?

MBD: Probably, it's very hard to tinker with prices effectively. Also everything had to be done in such a hurry that you didn't have time to get the price picture on a comprehensive basis. With the computers we have now, you probably could do a much better job. I was on the price work for a year and a half. It was kind of a discouraging job, because just about the time you'd get something worked out, the demand would be shifting.

HKS: Were you working on all forest products?

MBD: Only on a few forest products, mainly lumber and pulpwood.

HKS: Were there any rationed forest products?

MBD: Not that I'm aware of, but if it was it was on specialty products. One of the jobs I had was to study a price dispute on bobbins for knitting mills in Great Britain. Much of the clothing for service men was made in Britain to save shipping space. So, the bobbins and pickersticks became scarce items.

HKS: What are bobbins made of?

MBD: Always from dense hardwoods such as dogwood.

HKS: I don't know what the curing time is, but you could really turn out bobbins in a hurry if you put your mind to it.

MBD: You have to have a quality product. Bobbins have to be made of a very hard wood, as they get a tremendous amount of vibration in weaving.

HKS: You look at what eastern Europe and Russia's going through now, centralized planning doesn't work too well. That's what we were doing.

MBD: We were lucky that the war was over so soon.

HKS: I talked to some lumbermen who were involved in WW II, dollar-a-year types. Was it one big network or was it so bureaucratized that you didn't see the rest of it.

MBD: You didn't see much of it. It's hard to get the full picture. About the time that you'd get a price ceiling worked out for, say spruce pulpwood, some other species would be in short supply. Then another staff member would be brought in to develop another price regulation.

HKS: How many people were there like you in Washington, D.C. working on forest products?

MBD: Oh, I don't have any idea.

HKS: Was it like fifty or was it like ten?

MBD: Oh, probably ten or so.

HKS: Were they all well-trained?

MBD: Various backgrounds. You had a great mixture of people. One of the things that we worked on was stumpage price control. Maybe there were one or two regulations that were put out, but it's almost impossible to put a ceiling on stumpage because of the relationship to harvesting cost, transportation, all of these things that go into a stumpage price. We were just lucky the war was over when it was, because to me the whole price structure was going to cave in.

HKS: You didn't feel that you needed to go over to the Washington office of the Forest Service at all. You didn't feel a part of that. You were with the War Production Board.

MBD: Well, no. We worked out of offices in the Forest Service and made much use of their historical material on production and prices.

HKS: Oh, you did.

MBD: But every day we were over at the office of the War Production Board or some other agency. The WPB and the Office of Price Administration just didn't have enough space for us. Forest Service offices which were formerly occupied by research became available to us. Ed Crafts had the job prior to my coming into the Washington office.

HKS: I didn't know Ed very well. Most people said he was abrasive, but very smart.

MBD: Yes, an exceptionally sharp mind. There were those who would say he wasn't so smart, because he couldn't get along with people. He was abrasive all right. I always got along fine with Ed. I had a few run-ins with him, but you had to respect his ability.

After I'd been in Washington about a year and a half, George Trayer, who was heading Forest Products, came in and said, I'm looking for somebody to go to Italy. We have to have a forester for the Allied Control Commission. Is anybody interested in going? I said, I'd sure like to go to get out of this place.

ALLIED CONTROL COMMISSION (1945–46)

HKS: Italy had been retaken.

MBD: The Allied armies had occupied much of Italy by that time. The fighting was still going on up in the Po Valley and on up into Austria. But the Allied Control Commission, which was the bridge between military government and civilian agencies, was set up with their own organization in Rome. I was sent over to work in the agriculture/forestry section and spent a little better part of a year and a half there. My staff assignment was to get the Italian Forestry Corps operating again.

HKS: Did we worry officially which side they were on—whether they were fascists or not?

MBD: Yes, much so. We could not include any of the former fascists in the new forestry organization. So we had to recruit outside. One of the interesting aspects was that some of the foresters in northeastern Italy were of Austrian-background and when the war broke out they went over into Austria. They were not tagged as fascists, so some of those fellows were the ones that we put back in top jobs.

HKS: Were there actual documents, I mean, was the bureaucracy still in a physical condition that you could get names and lists of people?

MBD: Yes and no. They were not complete.

HKS: All the bombing and so forth.

MBD: You did it pretty much by word of mouth. This fellow would say, there's a good man. Get a hold of him. That's about the way you went. My counterpart was a major in the military government, Major Bump. We worked together. Some Italians came back. You see, the many foresters were in an elite group under Mussolini, and they had special privileges. Those people were out completely. We couldn't use any of them. But even in the year and a half that I was there regulations were eased, because there just wasn't any other talent available. The Austrian group across the northern part of the country helped a lot, but down in the south, we just didn't have people to draw upon. Actually, the Italians did very well on their own. I can't claim very much credit.

HKS: Just the thought that there's a major war. The country's all shot up.

MBD: Yes.

HKS: To get the local economy going again; because they have to have salaries, they have to buy groceries. The transition to peace must be quite a trauma.

MBD: It sure is. I would have to say that I had great admiration for how well the Italians got things going again in the short time I was there. I met a lot of local foresters. I particularly got acquainted with the people at the University of Florence where the forestry school in Italy is located. Dr. Pavari, head of the forestry department, was a fine person, high principled, and helped much with ideas on how to get on with the job.

One thing we were involved in was bringing fuelwood down from Austria into Trieste. Trieste was a port of entry to the old Austro-Hungarian Empire. It was later a part of Italy. Many ships were sunk in the Trieste harbor. After the war they couldn't get a fuel line in, so we had to bring wood down from Austria.

HKS: Austria was occupied by the Soviets at that time, right?

MBD: The eastern part of Austria was. Austria was split at that time as I recall. But anyway, that was just one of the more specific assignments we had.

HKS: I keep referring that Lake States history. In it you said forestry was a good way of helping control the restless—that's your term—the restless Italian population. I suppose unemployment was a critical problem.

MBD: Oh, it certainly was. Food supplies and jobs were short, so we were sent up to Trieste to get a forestry project going to provide employment for the people. Learning the techniques of reforestation in Trieste was a shock to me. On the steep slopes there was practically no soil. You had to take an ax, a pick, and other tools you could get to make a hole in the ground to put a tree in and then go find some dirt to put around the roots. That was the way we reforested some of the barren slopes around Trieste. I went back there on one of my IUFRO trips, and surprisingly some of those plantations looked pretty good.

HKS: So Yugoslavia was off limits.

MBD: It sure was. There was a lot of local fighting there. There was a zone A and a zone B. Trieste was A, and most of Yugoslavia was B. Anyway, the fighting went on after the war stopped. It wasn't unusual in the morning when we were going out onto the slopes to go along where there had been several hangings.

HKS: It's going on today.

MBD: There is a long, long history of antagonism, bitterness. After Italy I went on a short assignment to Greece. Again we were trying to get some forestry projects going to provide employment.

HKS: Had the war been hard on Greece? Had there been a lot of combat losses?

MBD: Yes, many.

After a few months I returned to the U.S. and to the Forest Service in forest economics. Earlier I had a telephone call from Ed Crafts. He said, either come home or get out of the Forest Service.

HKS: You were in Athens then.

MBD: I was in Athens.

HKS: Did you cross paths at all with Arthur Ringland? He was in a lot of UNRRA type things out there.

MBD: No.

HKS: It's a big world. I understand that.

MBD: I expect he was up several notches higher than I was. Anyway I decided it was about time to get back home. My wife had received clearance to work to do Girl Scout work in the displaced persons camps in Estonia, Latvia, Lithuania. When she arrived in London, she couldn't get clearance to go further, so she received clearance to come on down and stay with me until arrangements were settled. So she went to Greece with me and helped organize programs similar to the Girl Scouts just on a volunteer basis in some of the Greek villages.

HKS: I'm still trying to visualize what it was like for you. You got off the plane in Athens. What currency do you use? You didn't have the infrastructure then that you have now.

MBD: Having currency was the least of your concerns, because currency wasn't good for anything. The day that I went into Athens, the exchange was, I think ten drachma to a dollar. Within two weeks it was ten-thousand drachma to a dollar. So, having the drachma didn't mean anything. It was what you could take in in terms of physical goods. If you could take in sugar or flour or canned goods, you could buy anything on the open market. That was still the situation when I left Greece in '46. The same thing happened in Italy with the lira.

HKS: I'm still wondering how you got your job done, because you had to eat and sleep and you had to get to work.

MBD: UNRRA, United Nations Relief and Rehabilitation Administration, had set up bases in these countries, so when you came in you always went to the UNRRA office. This was where you received marching orders so to speak. The group I went into Greece with included four or five Americans, and we each helped one another.

When I was ready to come home from Italy and was sitting in a hotel one night, Fred Renner, who was a former Forest Service researcher from California, came in with two fellows. Right away, we got together and Fred said, why don't you go to Greece with me? The one fellow that didn't come in our group was a forester. And I thought, I'll go, Fred, if you can get clearance for me. He said, I'll get on the phone tomorrow. Next morning, he called into the chief's office in Washington and received the necessary clearance for me. So, that's the way I picked up the job at Greece. I was there and knew how things operated, and he was anxious to have a forester with him.

HKS: A lot of improvising.

Northern Rocky Mountain Experiment Station (1947–51)

MBD: That's right. When I came back, I went to the Northern Rocky Mountain Station at Missoula where I was put in charge of the economics and forest survey programs.

HKS: Why Missoula?

MBD: Ed Crafts said, that's where you're going—Missoula. That's about all I know. I think it was because there was a vacancy. Anyway, that station was one of the smaller ones. They had a few staff people working in timber management research, some in water, fire, and economics. The director was Charles Tebbe. Charlie had only been there a few months when I came. He had come out of the chief's office, I believe. Anyway, he didn't have a background in research, but was an extremely able fellow, very conscientious. Shortly after he got there, he was called off on a detail in some other part of the country. I don't know where. Anyway, he asked me to take over the acting director's job, because I hadn't really gotten involved in project work by then.

HKS: When Tebbe came to the station, was everything in chaos because of the war?

MBD: Yes. You see, after the war, funds became available in various ways. The funds didn't come back the way they were cut off. They were reallotted. I don't know how it all operated, but the effort at the station was largely one of reconstituting a program and putting things together again. As I've said many times in the Forest Service, I was becoming a jack-of-all-trades. I moved into most anything that came along and enjoyed it. I have always enjoyed working with people, and that made it a lot easier.

HKS: Did you ever want, at that stage in your life, to go off into National Forest Administration, or was Research really where you saw your future?

MBD: I enjoyed research administration. It wasn't as complicated and as rigid as National Forest System is, and I got into some very interesting things in a lifetime in the research organization with emphasis on research administration.

I stayed at the Northern Rocky Mountain Station for about four years. Of all the thirty-nine years I was in the Forest Service, I enjoyed the Missoula years the most. I was at the age that I could enjoy outdoor life—skiing, fishing, camping—and enjoy the association with many local people. My wife was then teaching at the University of Montana, so we had lots of ties with the university staff. Many of things that you see around in the room had their origin in Montana.

HKS: That antelope, is it one that you shot?

MBD: No. I'm not a hunter. A friend of mine in Montana gave that to me. I became interested in Charlie Russell paintings and met some people who had worked with Charlie Russell in Montana. This was just one more thing that added to the charm of living in the area.

ECONOMIC STUDIES

HKS: Tell me a little bit more about the specifics of some of those economic studies. What were the issues?

MBD: One question in Montana was industrialization. Should they have a pulpmill or not? I would say that probably 75 percent of the people would say, well we're enjoying life here. Why do we need a pulpmill? I remember we had a study going at the station by Blair Hutchison that grew out of the Forest Survey, knowing timber supplies, growth, and mortality. So one question was, is there enough timber east of the Continental Divide to support a pulpmill, and if so where should the mill be located. After we drafted the study, we made copies and sent them around to various communities. There were several meetings, one in Bozeman. That one I remember particularly because we presented a draft proposal for a pulpmill, what wood it would require, what employment would come about, and that sort of thing. We sure got roasted. We don't want a pulpmill here. We don't want the air polluted. We don't want the water polluted.

HKS: So, the chamber of commerce people weren't in favor of this.

MBD: They kind of egged us on to start with, but they pulled back fast.

HKS: So, what do you look for? A wood supply obviously, but transportation, infrastructure.

MBD: Transportation, infrastructure, labor supply, community interest, acceptance and support, that sort of thing.

HKS: Was the mill built while you were there?

MBD: No. A larger mill was built subsequently in Missoula. Such studies were a natural sequence to the forest survey resource analysis.

HKS: Do these kind of studies always lead to publications, station papers, and so forth?

MBD: No. There were several studies in one form or another, but such analyses can show limitations as well as opportunities.

HKS: I see.

MBD: Economic studies often provide ideas for various industrial developments in the wood industry. I gather the feeling in Missoula now is that there is too much timber cutting.

HKS: There is. The town's almost all motels.

DIRECTOR OF LAKE STATES FOREST EXPERIMENT STATION (1951–65)

MBD: Tourism pays off quite handsomely. Unfortunately my Missoula days came to an end, and in 1951 I went back to Lake States as director of the station at Saint Paul, Minnesota. I remember when I was appointed director, I had to come in and talk with Lyle Watts, who was then chief. Lyle said to me, "Dick, I'm going to give you one instruction: go out to Lake States Station and shake it up." He didn't say it just that way, but it was clear that was what he wanted. By that time Zon had retired. E. L. Demmon followed Zon, and I followed Demmon.

HKS: What was Kotok's role in all of this? He was assistant chief for research.

MBD: He was assistant chief for research.

HKS: I would have thought he would be the one to give you this pep talk.

MBD: No. I remember going and talking with Ed, but he didn't give me any particulars. He was leaving, to be followed by Les Harper.

HKS: Who do you think picked you to be the director?

MBD: I think Ed Crafts primarily. He was on the chief's staff at the time. But Ed, like other people, wanted his men placed in key spots in the organization. When Ed was in charge of Economics and Forest Survey in the Washington office, I was at the Northern Rocky Mountain Station. We worked well together in developing programs.

HKS: Answer this any way that is appropriate to you. Was Kotok a good leader for research—for the time, for the '50s?

MBD: I really didn't know Kotok well enough to respond to your question.

HKS: When did Harper come in?

MBD: Harper came in '51.

HKS: Somebody said in one of the interviews—I can't remember who it was—that Kotok could never forget that he no longer lived in the West.

MBD: I would agree. His whole background had been California. He didn't associate the problems of the Northeast or the southern states with the kind of things he had associated with in the West. Consequently, he just didn't get through to the research organization. At least that was my observation. I was in and out of Washington on the forest survey work, and with many of the top people you just do not get acquainted.

Well, with that charge from Lyle, I wasn't sure what I was supposed to do.

New Programs

HKS: Does Lyle say, "I want you to do this, Dick, and here's a bunch of money?"

MBD: No, he sure didn't. "Do it with what you have. Get more when you can." In '51 we had about twenty-five staff members, a budget of $250,000. The Lake States Station had not developed like others. It was next to the smallest in the country. I just started getting acquainted, working with groups around Lake States, the timber industry people, Izaak Walton League, and all kinds of local organizations. We had some excellent research talent, but few on the staff had been interested in public relations. One of our early moves was in genetics work where we felt there were opportunities to improve timber productivity.

HKS: What else was going on in genetics other than at Placerville?

MBD: There were a number of individuals in stations. There were no other institutes like Placerville. Paul Rudolph, for example in the Lake States, had been doing much on selection of superior trees, and he had plantations that were very useful in providing seed sources for further study. Others involved were Leo Isaac in the Pacific Northwest, Phil Wakely in the South, and Ernie Schreiner in the Northeast. Anyway, what we finally did was to set up an institute at Rhinelander, Wisconsin, where the Lake States had some cooperative work with the state of Wisconsin. We set up this institute with limited funds and grew as we went along.

HKS: Are there significant battles over turf? You have Placerville, the preeminent genetics place that Congress gives money to each year. Now comes along some competition for the genetics dollar. Is that an issue?

MBD: To some extent, although I think generally additional institutes were supportive of one another. The strength of an organization is always the regional and local ties.

HKS: I can see you at the appropriations committee testifying. Why do you need all these labs?

MBD: No, station directors do not testify. The chief and deputy chief for research handle congressional testimony.

HKS: Would Congress ask these kinds of questions?

MBD: Sometimes. Once in awhile we would be asked questions like facility needs, more as justification than as criticism. The origin of the Placerville Institute was different than the southern and northern institutes.

HKS: Mr. Eddy.

MBD: Mr. Eddy. I don't know what his part in Placerville was.

HKS: It was privately owned, and he gave it to the government.

MBD: That was a little different than the other two institutes.
    Then one of the other things that we started was recreation research in the Boundary Waters Canoe Area. Much controversy was going on up there over recreation use. Fortunately we had a very able research scientist trained in the social science studies. Through his own initiative, he started interviewing people to get their perspective of recreation problems.

HKS: Bob Lucas?

MBD: Bob Lucas, he had the capacity to understand that there were many social implications in recreation and forestry.

HKS: Was this novel in research at the time, to have someone who wasn't trained as a forester to do research?

MBD: Yes. This approach was questioned by some people. I became well-acquainted with Bob before we brought him into the project. I had confidence that he could do it. Recruiting staff for recreation research is a challenge, because you don't know whether you want more of a forester or more of a sociologist or what you want. It kind of has to develop around what the parameters are of the particular project. Bob's studies really paid off. The people who were involved in administration and research programs came to respect Bob very much. I remember one of the studies early on, when the staff was trying to define wilderness: where does the wilderness begin? Often they listed Duluth, although the BWCA was many miles north.

HKS: You've talked about genetics and recreation research. You've given me a list here on the outline. You want to talk about one of those other projects?

MBD: Let me mention watershed research. That was a project we started at Grand Rapids, Minnesota. We were interested in the movement of the water in large swamp areas of northern Minnesota. That is where the water comes from that supplies the Mississippi and other rivers flowing out of northern Minnesota. Not much is known about the quantity and quality of this water.

HKS: How do you trace how water moves? Do you put dye in it or something?

MBD: Yes. We had locations where we would put dye and then go half a mile or so and get a reading on water movement. There had been no previous watershed studies in the Lake States of this type. Zon had been interested and he had received some material from Russia on similar conditions there. Engineering research was another new venture. Along with a few other stations we initiated engineering research to lessen harvesting costs and to use more of the low quality second growth hardwoods.

HKS: Is this timber harvest type engineering or road building engineering?

MBD: Mostly timber harvesting. Some manufacturing equipment, too. But very little relating to roads. There were other projects studying harvesting under conditions much different than in the Lake States. Each project had a distinctive mission. The one that we drew on at Houghton—Michigan Technological University—concentrated primarily upon harvesting small timber and uses of chipwood. We started a small engineering unit at Houghton, two engineers. One is now director of the Forest Products Laboratory, John Erikson, a highly competent engineer.

Wildlife habitat was another new activity we started, studying the vegetative cover requirements for different wildlife. This was part of meeting Lyle Watts' challenge to shake things up. What we were doing was to reduce the emphasis on timber growing silviculture and shift to some of these other research areas. This was quite timely from the standpoint of the Lakes States, because their potential for growing timber in terms of national supply is limited. But their potential from the standpoint of general, multiple use is high. So I was quite pleased and quite energetic in pursuing some of these new areas. We had much interest from groups around the Lakes States. Like recreation, a lot of the forest industry people, particularly, were interested in this.

HKS: How about the universities. Did you have a lot of co-op work with them? Was the diversity better for them?

MBD: Universities had rather limited research activities. We put some funding into research cooperative projects both directly and indirectly—Universities of Michigan, Wisconsin, Minnesota, Michigan State, Michigan Tech. Most of these schools developed cooperative projects with us. We benefited from the standpoint that it helped us train and develop people for advanced degrees. This was one of the activities that we got into with the universities, to get a well-trained staff through the funding of student research. It worked out to the advantage of both.

One of the things I'm proudest of is the number of highly competent and successful people that we were able to recruit in this expansion activity during the '50s. I mentioned John Erikson, director of Forest Products Lab. Three station directors came out of our group—Bob Buckman, John Ohman, and Roger Bay.

HKS: You were director of that station longer than most directors are at stations, right?

MBD: Yes. I was director at the Lake States Station for thirteen years. Zon was there a little longer.

HKS: I haven't made a study of this, but just looking at a Charlie Connaughton type. They put him as director of a station as part of his seasoning for higher leadership. So there's a lot of rotation in directorships.

MBD: Yes, to some extent.

HKS: But maybe that's just an impression I had.

MBD: I have to say though I resisted moving.

HKS: It might give you a real advantage if you got to know your congressional delegation.

MBD: Well, that's right, it helps. In fact the last time when I moved, Les said, "You had it too good too long."

HKS: I asked this question before, but in a different way, about turf. Your program was growing faster than others. Is there any jealousy? Is there ever a problem?

MBD: Oh, sure.

HKS: How come he gets this and I don't get that.

MBD: I remember I used to be jealous of the Southern Station, because that was a time when the congressional delegation was strong in the South. We were able to get things turned around at least to some extent. We had several congressional members that helped us tremendously. Senator Humphrey was one of our best and closest friends, a good one to have on your side.

NEW FACILITIES

HKS: As long as you're talking about this, let's go right into that. Congressional delegation.

MBD: All right. Senator Humphrey was there at that time, and he was interested in our research work at Grand Rapids.

HKS: Humphrey—I want to dwell on him a little bit, because he's so important in the broader Forest Service. Was he a creative guy or did he absorb other people's ideas and make them work? What was his genius in getting things done?

MBD: I would have to say he had lots of imagination no matter what he got into. He was creative too. But even more than that, if you gave him just a hint of an idea, he would pick it up. I recall one time in northern Minnesota, we'd had a timber management project near Cass Lake for many years. Our facilities were just a small office on the second floor of a local building. Anyway I saw Senator Humphrey in Grand Rapids one day, and he said, Dick, how are you doing up here now? I said, well, pretty good, senator, but we sure could use laboratory facilities. I'll remember that, he said. And it wasn't long before things started to happen.

HKS: Now, theoretically you're not supposed to do that, right? You're going around the budget process. Congress doesn't like you to do that.

MBD: You should keep others informed of what you are doing and what you see as problems ahead as well as long-range needs. I don't think this is going around the budget process.

HKS: The location of these labs in Rhinelander and Bottineau, did they result primarily because that was the logical place to have them or because you had congressional support there? What makes things work?

MBD: We had a master plan for development of research facilities. We had on going research at both Rhinelander and Grand Rapids back in the '30s and continuing on into the '50s. This plan identified locations and research projects. The implementation of these was guided by long-range program needs.

HKS: They had to have the favor, but they're looking for projects.

MBD: They're looking for high priority needs. At least the ones I mentioned like Rhinelander.

HKS: I realize that a lot of your research was physically in cooperation with the university. That's where the physical plant was. The story we're going to talk about right now is where you're beginning to build your own laboratories. What are all the advantages of having your own structure? Other than adding space. That's an obvious one. Is being off campus good?

MBD: Many kinds of the research you need have to be located near the forest. If you go back some years in research, everything was centralized at station headquarters. In the summer the staff would move, and they'd be gone for three or four months and then come back to headquarters. This was costly and disruptive. I think there were many advantages of having facilities located near where the field work was located.

HKS: So location is a major consideration.

MBD: Location is certainly a factor. You know, from the standpoint of people, many of the forestry families are happier in a community like Rhinelander than they are being in a large metropolitan area like Twin Cities. This varies family by family, but it's a factor in recruiting staff. I've often heard Bob Buckman say that the happiest years of his life were at Grand Rapids. That's where he and his family settled and grew up, and that is where he did his most productive research.

HKS: Your station was the first to have its own headquarters building.

MBD: That's right.

HKS: How significant is that? Other than the physical convenience. Were your people scattered around the campus?

MBD: Yes, the forestry building at the University of Minnesota had a third floor that included space for headquarters offices for the Lake States Station. The station kept growing, and we located staff in other campus buildings. We had groups off campus. We also had groups at some of the field locations. So there were several factors involved in getting an organization together at one location. At Minnesota, the university was always very cordial and receptive to the station, we had fine relations over the years. When we talked about having our own facilities there, I remember the dean saying to me, I couldn't be more supportive. We were proposing a building right along side their forestry building. They had the land. He said, that's the way that the universities and federal government ought to work together—having a major unit at a central location. They were very supportive.

HKS: The universities are growing too, during this period. They probably want space.

MBD: They urgently needed the space we were occupying.

HKS: They want to get you out of there.

MBD: They did. In fact, they rebuilt that whole building after we moved out. And they added another building too.

HKS: Frank Kaufert was dean.

MBD: That's when Frank was dean. Frank was a wonderful fellow to work with. He came there a little before I did, but we were there together over the thirteen years I was at the station, and I don't think we ever had a major disagreement. I had a very high regard for Frank and the good relationships that we had.

STAFF RECRUITMENT

HKS: Staff recruitment. If you're going to shake things up, then you're going to recruit some people.

MBD: Sure enough.

HKS: Did you have a good selection? I mean, the universities are starting to turn out Ph.D.s in large quantities.

MBD: Part of the reason they were turning out more Ph.D.s was because of the support that the Forest Service was giving to them. At one time at the Lake States Station we had the highest number of Ph.D.s of any station in the country. About half of our professional staff was involved in graduate work.

HKS: It seems to me that the other stations would be competing for the same Ph.D.s for the same reasons you were.

MBD: They were, but we had early contact with undergraduate students. We'd hire them for the summer. Some of them would continue working for us part time. This is one of the advantages of a station being located on a university campus, you have frequent contact with the students. Roger Bay became director of two stations. I got acquainted with him when he was an undergraduate and worked at the station. When we set up our watershed project, I happened to be in Green Hall and ran into Roger. I said, what are you doing now, Roger? He said, looking for a job. I said, you want to start in watershed research? Yeah, I'd love to. We recruited him through this contact. Just that kind of informal contact made the station attractive. One of the professors said, where did you get all these good men? Well, we just keep looking around.

HKS: The hiring of Ph.D.s, this is really significant in terms of, for lack of a better term, maturity of research as a rigorous operation. As opposed to being descriptive, it's now more theoretical.

MBD: That's why you have to have fellows who have had good scientific training, people who have the training and background to carry on independently in research.

HKS: You recruited Bob Buckman.

MBD: Yes, I recruited Bob. He was up at our Grand Rapids laboratory for some years and did some very good research.

HKS: What was there about Bob that made you appoint him director of the Northern Conifer Lab, whatever it's called?

MBD: The Northern Conifer Lab then.

HKS: You have all these guys. They all have Ph.D.s and so forth.

MBD: He did an excellent study on management of red pine. But beyond that, Bob was a great fellow for reaching out to the community and making friends. That's all part of the job of a research forester, making friendships, contacts, working with people in the community, not just on forests—especially if you want others to know and use your research.

HKS: Was that unusual?

MBD: Bob was stronger at it than some. That's essentially why he did such a good job later as a station director. He could reach out to people, and he had excellent scientific training. This is the kind of scientist that you try to get to head up programs.

HKS: The pioneer units. Did you have any of those?

MBD: We had one at Rhinelander. Phil Larson, a plant physiologist, a highly competent physiologist.

HKS: Did those things evolve or did you have a plan, you really wanted to have a plant physiologist?

MBD: Yes. There were different ways such units came about. Basically when you identify outstanding competence in a particular subject area that fits into your program, you consider ways to give him responsibility. It's the individual that you're interested in, you set him up with much freedom to select his studies. He has to be an outstanding scientist. There were not very many of these units nationwide. I don't know whether they still have them or not. But at one time there were several around the country. I think Phil Larson, at the Lakes States, was one of the most productive. I understand that now he's retired, he continues to do research.

HKS: Buckman made a comment about pioneer units. He grew to think that they weren't a good idea, because what do you do when the guy retires and you have this whole infrastructure built up?

MBD: That's only one way to look at it. If you have a competent man and you want to give him an opportunity, let him do what he can in the years that he has to do it. And if the study folds up afterward, so what. Much has been gained already.

HKS: Universities go through this all the time.

MBD: That's right. That doesn't bother me particularly. Sure, you just don't hire another scientist to take his place.

HKS: Talk a little bit more about recruitment. Obviously there's salary and there's location. People have families, schools, and all the rest. Was it hard to get talent? I mean, were you really competing with universities or could you go out and pick good people without any difficulty?

MBD: No, there was a scarcity of good, well-trained scientists. By and large, in research, you have to recruit on a personal basis. You go to the schools and get acquainted with the graduate students. You get leads from the faculties. You take students on for summer periods, temporary employment. This is the way you get acquainted with the capabilities of an individual.

HKS: At the same time, were you losing good people to universities?

MBD: We lost some to universities. We lost some to industry. We lost some to other stations.

HKS: I don't know how you can generalize this, but what kind of questions did they ask you about? Academic freedom? I don't know what they might ask you other than the cost of living sorts of things.

MBD: The cost of living was always a question. Beyond that each individual had his concerns—location, education for children, community life, etc.

HKS: Did they want to be able to design their own projects?

MBD: I don't think this was of much concern. They wanted most to get into a particular subject matter area. That's what they were after. If they wanted to, say, get into silviculture versus genetics, they would make it pretty clear, that's the direction they wanted to go, and if that isn't what you were looking for, you look elsewhere.

HKS: So they wanted to make a career of research.

MBD: Yes, most were thinking of a career. And another thing, the university professors generally have a tremendous influence on their students and particularly on graduate students. Often what the professors have encouraged them to go into is what they're looking for.

IMPROVING STATION PUBLICATIONS

HKS: One of the things that caught my eye while reading the station history is that you upgraded the appearance of the station publications. My question is, what audience were you trying to reach? What was the rationale? Now, I don't know how ugly they were to start with. They are pretty jazzy things these days, the color and all that. You're obviously going beyond the scientific community.

MBD: Probably, one never knows.

HKS: Review by the station editor. I can see that, that it should be intelligible. But the aesthetics of publication, that certainly is...

MBD: If you look at some publications back in the thirties and even the middle forties, they didn't have much appeal. What good is a publication if it's not used?

HKS: If you were fully successful in transmitting your knowledge, who would be your audience? The field forester? Who?

MBD: Generally, it's a group. It's the supervisory personnel, and it's those on the ground, field people. A publication on management of northern hardwoods. Who's going to use that kind of a publication? The supervisors, state foresters, industry foresters. They are the ones who should be interested in the publication. You know, nowadays the pile of paper that comes across your desk in terms of publications is substantial, and if you've got one with an interesting title, a good layout, and easy to read, that's probably going to catch your eye.

HKS: Was this, changing the appearance, agency-wide?

MBD: It was going on elsewhere, there was much variation.

HKS: It costs a lot of money to make attractive publications.

MBD: It sure does. But there's a lot of money wasted if they are not used.

UNIVERSITY RELATIONSHIPS

HKS: Relationships at the universities. Start with some general statements.

MBD: When the McIntire-Stennis program got under way, it was obvious that there was going to be a larger research effort at the universities and desirably so. We sensed the need to develop more cooperation and coordination. Unfortunately the McIntire-Stennis program did not develop anywhere near as rapidly as was hoped. Consequently the universities were often in difficulty when entering into a cooperative project. Eventually, as McIntire-Stennis started to take off, that problem was overcome. Certainly back in the 1950s as the Forest Service was expanding rapidly, it was difficult to have a meaningful and productive relationship between the universities and the stations. Not that the desire wasn't there, but if you don't have the wherewithal to do it, then it's pretty hard to progress.

HKS: Were you involved in the passage of the bill itself?

MBD: No, I wasn't. To the extent that we could be supportive we certainly were willing but most of the initiative had to come through those who handled contacts with congressional committees.

HKS: I'm not sure the best way to ask this question. You worked with forestry faculty and also the graduate students. Were you ever concerned in any significant way that the professor might be more interested in getting money for the grad students than turning out quality research? I mean, students can flunk out of school. Universities have a little different mission than the Forest Service. Was that ever an issue as far as you were concerned?

MBD: Not as far as I know.

HKS: Forestry school research was good quality.

MBD: Yes. The biggest difficulty with the universities was that the research had to be pretty much within the confines of the period of time that the particular graduate student would be there. When he leaves everything goes, so your choice of projects is circumscribed a little more. In contrast, our station research might continue for ten or fifteen years. This made it hard to put the two together.

HKS: So an extensive degree of reliance on graduate students was a factor for project assignment.

MBD: Yes.

HKS: I hadn't thought of that. I asked that question, because when I was a very junior scientist in Portland, there were some complaints that some of the money the Forest Service was spending on campus was being used to fund graduate students first, getting research second. So, I always wanted to ask the question, and you answered it thoroughly.

MBD: It varies somewhat from school to school, but I think that's a fair generalization.

OTHER FEDERAL RESEARCH PROGRAMS

HKS: Are there other examples of a federal agency being on a campus the way Forest Service is?

MBD: The Forest Service and the Agricultural Research Service both had people on campuses, but generally in ARS their staff is attached to individual departments, in contrast to the Forest Service. My observation was that the Agricultural Research Service found it difficult over the years with that kind of a tie. An individual loses his affiliation with the Agricultural Research Service. I can't speak for today, but that was the case several years ago.

HKS: I would suppose when it comes to recruiting talent, for some the chance of teaching a course once a year or sitting on a graduate committee would have a great appeal.

MBD: I'm sure it did with certain individuals. The Agricultural Research Service also had several field laboratories just like the Forest Service. They have got some large installations too.

HKS: The only Forest Service facility that I've seen on a campus is the one at Corvallis. Is that typical? Is that on a more elaborate plane?

MBD: It's more elaborate than most of them. An other one that's on a campus is at Fort Collins. The Rocky Mountain Station has a very nice facility. It is also a station headquarters building like the one we have at Saint Paul in contrast to the Corvallis one. I have always felt that the PNW station ought to be up at Corvallis, not down at Portland, but that's a personal opinion.

HKS: Many of the scientists that were there in the sixties when I was there are now senior scientists, but they're now in Corvallis, so it's almost happened.

MBD: That's right. That's almost happened. Right. A lot these things just take time. They shift around.

HKS: What is the significance of the creation of ARS, the Agricultural Research Service, and I don't know what CSRS is?

MBD: Cooperative State Research Service.

HKS: How were these functions performed before these agencies were created? Was it sort of ad hoc?

MBD: Generally the ARS is a consolidation of many research groups in USDA.

HKS: Some of that came over to the Forest Service. Entomology came over.

MBD: Only forest entomology and pathology. This was back in the early fifties, around '53 I guess. There was a general shake up at that time in Agriculture, trying to recombine and get units of like subject matter into the same organization. There was a group in the Bureau of Entomology and Plant Quarantine centered on forest entomological studies. That's what came over. There was still the Entomology Bureau that went into ARS at that time, but there was a part of it that came to the Forest Service. Same way with pathology. The bigger part of the pathology organization went to ARS, but some forest pathology work came to the Forest Service.

CSRS is a different organization entirely. This receives and administers appropriations for research at state agricultural experiment stations. The money is passed on to the state agricultural experiment stations. CSRS sets the standards and criteria under which funds are allocated to the individual agriculture experiment stations. CSRS as such did no research.

HKS: Sort of like McIntire-Stennis, but for agriculture.

MBD: Yes, McIntire-Stennis is administered through CSRS. It's under a separate law, but it's administered within the CSRS organization.

HKS: I don't want to trivialize this, but in a sense, it's housekeeping. They create a little better bureaucracy, so that there's better control and consistency and all that.

MBD: True.

HKS: Do you think the forest research model may have somehow influenced the need for that or did the time just finally come after the war and all?

MBD: I can't talk with much authority on that subject. In one form or another, the functions of CSRS have been carried on in USDA for many years going back before the Forest Service research legislation.

HKS: Is the Forest Service the only agency that has a major research program of its own?

MBD: There are a few small research units in other agencies.

HKS: I remember Ed Crafts argued against putting the Forest Service and the BLM together. He said, look, the Forest Service is not like other agencies. There is research, and it makes it unique. You don't just move it around like you do other agencies. I didn't know if that was a valid argument or not that he made.

MBD: It's valid all right. I think that it's probably the reason for the strength over the years of Forest Service research. It is a strong, well-organized, central organization. When you look at some of the problems that some of the other agencies have had, you were kind of glad you were in Forest Service.

RESEARCH QUALITY

HKS: Shortly after I got to the station in the early 1960s, there was a major study by the National Academy of Sciences, or some prestigious group of that stature, that had evaluated all federal research and had characterized Forest Service research as second-rate and pedestrian. Do you remember that study?

MBD: I remember that very much.

HKS: We went over to some auditorium in Portland, and we had a pep talk about it. It was not true, but we had an image problem, and I don't remember the details. You were a station director when that thing came down. You must have paid attention to that.

MBD: To some extent we did have research of that type. It was largely a result or an outgrowth of studies that were started in the thirties.

HKS: Lots of descriptive studies.

MBD: That's right. Lots of descriptive work. I remember a case in fire research at the Lake States. The staff member had all kinds of data on fire behavior. He was trying to make something out of it. The data had accumulated over the years in a certain pattern, and he couldn't analyze it. This was prior to the computer age. Later we learned how to design studies to get something out of all such material. This was much the same type of a problem as I mentioned with Carlos Bates.

HKS: This was an observation I made being fresh out of grad school. We had a course in experimental design at Washington.

MBD: Yes.

HKS: I saw that several of the old-time studies had no hypothesis. You just collect data and write it up. But this was changing, because of all the Ph.D.s that were coming in who were trained in research.

MBD: That's true. What was happening at that time by and large was shifting from a lot of empirical studies based on field plots to doing research in laboratories. Having to go through some of these situations helped to get the emphasis away from just collecting data and get it into the laboratory. Design what you're going to study and carry it out. There may be some necessary field work too, but the whole nature of research has changed over several decades. More time is now spent on analyzing what the problem is. What are you going to study? Why are you going to study it?

HKS: The primary reason I was hired, we were closing down a twenty-year descriptive study on the effect of slash fires on vegetation. This was the early 1960s, and the project started in the late 1940s. I looked at the project file, the names didn't mean anything to me at the time, but there was a critique of the project when it was proposed, we'll say 1947, about that time, from the Washington office. This study should not be undertaken because it might show the Forest Service policy for slash burning is incorrect. Was that a generational thing? Did you ever run into that?

MBD: I've heard of a few instances like that. But certainly they grew less as research was strengthened and accepted.

HKS: It really wasn't an issue?

MBD: No. There were some people who felt that way. But you know, in any organization you are always going to have some discontents.

HKS: So, the regional forester didn't get on the phone and say, Dick, God damn it, you have been publishing this stuff and it embarrasses the hell out of me?

MBD: No. More often than that, the regional forester called me and thanked me for having done something. I remember the head of timber management in Region 9, he was having problems in getting rangers to adopt certain procedures in silviculture. The station had been doing research on this, and I said, Herb, well, why don't we just take the scientist, and for a month or two just have him go to each national forest in the problem area and let him put on some training sessions. Herb said, I'm willing to try anything. I've tried to write dissertations, orders, and everything, and I've never gotten anyplace. We did that, and about a year later, when Herb was retiring, he came over and I've never seen a person as thankful for what we did in putting the scientist on the ground to work with his people. Now, you can't do this with every problem. Obviously, you'd get no research done. But there are times when this should be done.

HKS: So, the field people basically were receptive. They didn't think researchers were weird somehow.

MBD: No, they were receptive, very receptive.

HKS: I imagine you could extrapolate this to complaints about the *Journal of Forestry*: too damn technical, it means nothing to me.

MBD: Yes.

HKS: But you're saying Forest Service research was well received by the field.

MBD: Yes. I think so. Very much certainly in the Lake States. And I think it was generally in each region.

HKS: That's part of your rationale for making it more attractive and more readable.

MBD: Yes, that's right.

Contentious Issues

HKS: We haven't talked about contentious issues. How independent is research really from what's going on in the real world? Talk about Ashley Schiff's *Fire and Water*.

MBD: I met Ashley Schiff once on a field trip.

HKS: Prescribed burning was part of your research. So how did you feel about that?

MBD: We argued, traveled together, and disagreed. He was poorly informed and had a very limited background in biological science.

HKS: He's a political scientist?

MBD: He might have been a political scientist. Where he had gotten his information from I don't recall, but his basic problem was that he just wasn't well informed. Some of the things were kind of, as you say, half-truths. But you've got to separate them out.

HKS: You knew he was writing a book, or the book was already out?

MBD: I'm not sure if it was out, obviously he'd done considerable work on it. He was quoting things, and I was aware that some of his references were to fire research that Harry Gisborne did at Missoula. Harry was one of the original fire researchers in the United States. He probably had the best basic knowledge in his years as anybody might have in terms of fire behavior and control.

HKS: Schiff takes Zon to task.

MBD: Well, Schiff was an interesting chap. He just hadn't taken the time to sort things out and to understand them. I would like to think, if he had more days like I had with him, he would have done a much better job in his book. There's certain areas that he could be critical of, but the way he stated them was always inflammatory.

HKS: I'm not trying to be falsely naive here, but Harvard University Press certainly gives peer review to manuscripts. How does he get this book through peer review? Maybe they didn't send anyone who was involved in the subject.

MBD: I don't know. Every once in a while, a book like that comes along and you just wonder how.

HKS: So, in your experience in research, Ashley Schiff's and similar books and their criticism generally of resource management and forestry, you didn't see that as a scandal or something?

MBD: No.

HKS: You didn't feel you had to defend against these books?

MBD: No, those kind of books put you on your toes, make you look a little closer at what you're doing.

HKS: Congress reads it or staff does. Do you ever get questions?

MBD: I never had questions about Schiff's material.

HKS: Another contentious issue is wilderness. Bob Lucas was working on this. Wilderness certainly was a controversial issue nationwide.

MBD: That's right.

HKS: What was going on in wilderness research? The Wilderness Bill was passed in '64. You were still a station director. Supposedly, that resolved at least some of the controversy by having that bill go through. Was Lucas on the speaker's circuit? I mean, what happens in a situation like that.

MBD: No, we continued to study emerging problems.

HKS: You need a working balance.

MBD: You can't spend all your time on one study. Just having a free exchange between people puts you on your toes a little more. I think it's good for researchers to get out and have to talk about their studies. On the other hand, as I've always said, you can't talk about something unless you've done something to be well informed.

HKS: Did Lucas have five sociologists working for him? What was going on by this time?

MBD: At this time, Bob was pretty much a one-man show with a few temporary helpers. He had helpers doing the interviewing, but he was the scientist in charge and designed his studies. He was an exceptional individual, he had very good judgment about how far do you pursue problem; is this worthwhile or am I wasting my time? These are decisions research people have to make if they are going to come up with some answers to problems. He had good judgment as to what areas to explore in order to solve the questions he was looking at. This judgment varies from individual to individual. Some work far better when they've got two or three others right with them that are taking issue.

Forest recreation research, this was a new field, wide open. You could go in any direction. You had to pick certain spots that you thought would give you answers to the critical problems that were facing the administrators of public lands. That essentially was the direction we were working in the Boundary Water Canoe Area. You're dealing with a limited resource. You've got so many acres of water, trails, and campgrounds. How are you going to use these in order to maintain the quality of the areas and meet the needs of the people? That's in essence the crux of recreation research.

HKS: Was Lucas collaborating with Canadian scientists?

MBD: Yes. He had many contacts with them. I don't know how often. The Canadians had a lot of controversy too. Of course, part of it was a different historical background, different laws, and a different form of government.

HKS: The provinces have a lot more to say about it.

MBD: Yes, sir. A lot more. The provinces are stronger. I am under the impression that the Boundary Waters management programs have pretty well settled down. Now, maybe they're boiling again. That's the history of the Boundary Waters. It goes along about ten years, and then things blow up again.

HKS: I suppose you had experiences with newspapers.

MBD: We had some of that. But you just take it as it comes.

HKS: How about newspapers? That's really a marvelous entrée for you to the public, and yet everyone complains about journalists.

MBD: Sure enough.

HKS: Do you complain about them, too?

MBD: Oh, sure.

HKS: But isn't research a little safer, less controversial?

MBD: Less controversial just by nature of the fact that research tends to be more factual and less governed by daily emotions.

HKS: Would the editor tend to call the regional forester or you about some controversy?

MBD: I guess I'd have to say, if the editor was in Milwaukee he'd call the regional forester, if he was in Minneapolis he'd probably call the station. We had many contacts with the press in the Twin Cities area and generally had very good treatment. They would ask some peculiar questions, but once we would get a staff writer out on a trip to see what was going on, we'd get a much better story.

HKS: Did they jump at the chance, or did they worry you were going to co-opt them? Were they cynical or what? Or a little of both?

MBD: Oh, I don't think you could generalize one way or the other.

HKS: We joined the Society for Environmental Journalism. And the biggest problem they have—this is in their own bulletin—is we don't understand the environment; we're journalists and we worry about it now. How can we trust the sources? All the special interest groups are feeding us data. And there's conference after conference these guys are going to to try to learn what's going on about global warming and ozone layers and all this stuff. I imagine it's much more sophisticated now than it was a generation ago.

MBD: You can do so much better when you have a person in a staff writer who has been exposed to a lot of these things. We used to try every year to get somebody from the press out on a field trip for a week or so. Over a period of years after that, you would get questions back from them by phone.

HKS: Whose job was it at the station to issue releases? Does the station editor do this, or did you do this?

MBD: We had an editor in the later days—but earlier anybody in the station might draft a release and then talk it over with me. We tried to have writers develop their stories based on discussions.

HKS: Jim Sowder was in charge of administration. Would it be one of his responsibilities?

MBD: Yes, Jim just naturally gravitated to public relations. He was good at it. People had confidence in Jim, whereas others would have difficulty. Well, who was that fellow we talked to last week? Pretty soon you'd say, Jim Sowder. Yeah, I want to talk to him again. That's the way it would go.

THE WASHINGTON OFFICE (1965–67)

HKS: We're about to move you to Washington, D.C., to stay. What was the mechanism by which you wound up going to Washington?

PROGRAM DEVELOPMENT AND EVALUATION

MBD: Several things were involved. One is I was getting to the stage in my career where moving was probably a good thing for me and for the organization. An opportunity came along to participate in a USDA research planning team which would include representatives from Agricultural Research Service and Economic Research Service, the Cooperative State Research Service, and the Forest Service.

We were set up here in USDA to project a long-range program for research in the department. There was the continuous questioning in congressional committees about the direction the department was going and the magnitude of the program that we could foresee as being desirable. With that interest, particularly in the Congress, the department felt that it was desirable to set up a group once removed from affiliation with agencies. That's the way that the job originated. This appealed to me, because we had been involved in developing programs and recognized the need for continuing coordination.

HKS: Do you hear of an opportunity like that, you get on the phone, and you call someone? How do you actually get notified, selected, and accepted?

MBD: Selection was done by each agency. Harper called me and asked if I would be interested and I said, yes. So after further discussion, he went ahead and moved me in here. Within a month after I was here, other representatives from the department were selected. We had a group leader who came from ARS.

I might add this comment as an aside. The Forest Service had been involved in research long-range planning from way back in the 1930s, ever since the Copeland Report came out. That early report, mostly by Earle Clapp, gave direction to future research. Since that time, there have been a series of studies and research projections of program needs. This type of undertaking continued on over the decades. During the 1960s there were several revisions.

The other agencies in the department had not been involved so continuously and as completely as the Forest Service had. ARS had had several studies for special subjects such as livestock, crops, etc., but these had not been tied together as a comprehensive plan. With continuous reorganization going on in the department, it was difficult to get continuity in the planning effort. Consequently when we started I probably had the best background information available of any of the groups.

The main question was, how do we evaluate our needs in relation to the others so that there is a comprehensive plan that's balanced between all of the particular subjects in USDA. This was the main subject we had to confront early on. Largely you rely upon judgment. There isn't much more that you can draw upon. Take the ARS for example. They had programs for corn, for wheat, for livestock, and so on. In forestry, how do you structure a parallel program? Is it just only trees or is it timber, water, range, recreation, engineering, disease, insects, etc.? Or do you go by forest types? There are many categories that you can use. This was a continuous source of difficulty in projecting a program in forestry in contrast to agricultural research. We either did an aggregation at a higher level or a lower level. It depended upon what classification was used. We just used the judgment of everybody concerned. Each agency presented their comprehensive views and projections for a ten-year period. We worked by projecting what we called SMYs, Scientist Man Years. Now, it would be Scientists Years.

HKS: You betcha.

MBD: Then we converted the scientist years into dollars and came up with cost estimates. I recall we worked on five and ten-year periods for programs and consulted with various specialists as to needs. We eventually came out with a plan which was recommended by the work group. There's an awful lot of education that goes into one of these studies. Fellows who were in there from the other agencies had no background in forestry, and I had little or no background in the agriculture. We also had discussions with the congressional staff people to be sure that we were directing the program development in a way which would be useful to them. They made several helpful suggestions, but generally they were willing to be guided by our judgment.

HKS: Intellectually this is a little bit like RPA.

MBD: Sure enough. It's the same kind of a thing.

HKS: Congress is involved in this, so supposedly there's some compatibility at budget time.

MBD: That was what everybody was shooting for—both the agencies and the congressional committees. Budget-wise the plan would be a guide for developing programs and for priorities in building laboratories. Then, you would inevitably ask, well, what's next? We ended up with a massive amount of material and put it into several smaller reports hoping all this would be useful in guiding the budget process.

HKS: This was published?

MBD: Yes. The subject became somewhat controversial toward the end from the standpoint of individual agencies. The department's budget requests generally were lower than the study projections. The result was that the plan was never implemented fully. It was used only as a guide.

HKS: How about people in the field, the scientists. Do they think this is good or do they think you're just fiddling around?

MBD: I don't think they were too concerned, too interested one way or another. It was mainly an exercise from the standpoint of the agencies involved, although we did bring in some of the field people for consultation.

HKS: Are the stations as autonomous or not as the regions are to the Washington office?

MBD: I don't know how to answer that. I would say that the stations are fairly autonomous. You have to stay within certain guidelines, but individual studies are selected at the field level. Major new projects are cleared with the Washington office. I don't know what the arrangement is at this time, but I know that approval of studies then were the responsibility of the station directors.

HKS: The stories I hear about Charlie Connaughton, he's sort of legendary.

MBD: Oh, yes.

HKS: Larger than life I'm sure by now, telling Ed Cliff, you do your job and I'll do mine.

MBD: Well, I'm not quite sure of this. I don't remember ever telling Harper that. [laughs]

HKS: Let me go back. Harper invited you, but he was getting close to retirement.

MBD: He retired about a year after the long-range study was finished.

HKS: When you came back here in 1965, did you buy this house then? Did you see yourself staying in the District one way or another?

MBD: I didn't particularly. I bought this house.

HKS: I guess you have to. The IRS makes you buy a house within two years or pay capital gains on it.

MBD: That's right. I bought it when it was about half built and have lived here twenty-seven years. And it turned out to be a pretty good investment.

HKS: Having been involved in this study must have been very valuable to you in the subsequent years.

MBD: Yes.

HKS: You have a sense of what's going on and what ought to be done. You did a lot of your thinking in advance. What kind of a workday did you have? People at your level don't work forty-hour weeks. Yet with car pools, how do you work those extra hours?

MBD: That was one reason why I located where I am, because I was going to have irregular hours. At that time we had a good bus service, so you could go back and forth with little trouble. As for work schedules, we adhered pretty much to the eight-hour day except when you get into the final push on some of these office jobs, and then it gets a little hectic about what time you work and don't.

HKS: My mental image is that at four o'clock your phone would ring and it would be a congressional staffer saying, stop by for a drink. We want to talk about... Did this really happen?

MBD: The congressional people seldom came to us. We had to go to them when we discussed things.

HKS: They wanted to talk things over with you?

MBD: No, no. At least not with staffers at my level. They undoubtedly may have with Jemison and Harper. That I don't know. I'm sure that they kept informed of what was going on.

HKS: They had to car pool, too, I guess.

MBD: Oh. That's right. Of course, car pools often control your activities. That's part of the reason why I located out here. Having lived in Washington once before, I realized the importance of transportation facilities.

HKS: Okay. To summarize, you don't really know how to measure the effect of this study.

MBD: No, I don't really have any idea. One of the principal benefits of it was that it developed an organized plan that the budget people in the department and the budget people on the Hill could refer to and use and consider in evaluating the overall progress of the various agencies.

HKS: Also it was a marvelous way for you to be introduced to the process.

MBD: Yes.

HKS: What was your title then?

MBD: Frankly, I don't remember, probably staff assistant or something like that.

HKS: You saw what was going on and you liked it here.

MBD: Yes. You know, one of the really good things that came out of that effort was the contacts that I developed in the other research agencies in the department.

HKS: So, those contacts really hadn't been routine in the past? They were hit and miss?

MBD: They were hit and miss. Every day for the best part of two years we were meeting and working together. And some of the people that were most helpful in subsequent years are those that I met from other agencies, particularly ARS.

HKS: So this mechanism—I don't want to call it the social mechanism—but it did remain in place.

MBD: Very much so. Particularly with the people in ARS, which we had more in common with than the other agencies because of the type of research that we were involved in.

Associate Deputy Chief for Research (1968–72)

HKS: Is there more than one associate deputy?

MBD: Today there are two associates. Back in the mid-1960s there was one. I'm not just sure when we shifted to two, but it was probably in the late sixties.

HKS: Is the relationship to the deputy in any way similar to the associate chief to the chief? What was your function when you became an associate deputy?

MBD: Certain areas were considered the area for the associate, such as the details in developing budgets. Allocation of funds to the stations were pretty much handled by the staff under the direction of the deputy. That was one area, the associate had to take a pretty strong lead. The other included the details of operating with the stations. This often channeled into the associate rather than the deputy. There wasn't any clear cut line as you suggest between chief and his associate.

For example, I was the associate for Jemison. He was heavily involved as president in the IUFRO organization and was traveling much. Then the Food and Agriculture Committee on Forestry, and several other broad areas required that the deputy chief be in travel status. So, the associate just moves in and carries on. And when the stations have to have a decision, you make it instead of the deputy. That only happens once in a while, but you're the man behind the plate when the deputy's out of town. That's the way it operated.

HKS: Jemison had been Harper's associate.

MBD: Yes. The pattern was well-developed when I came along.

HKS: So Jemison must have had a job description in his head, what the associate did. He probably wanted to change some things.

MBD: Oh, yes. One thing that I would emphasize is that Jemison made a tremendous contribution while he was associate to Harper. In all the years that Harper was developing activities, George was carrying on. In fact, as a station director, my calls as a station director were probably more frequently to Jemison than they were to Harper. Harper was off with this and that, and George knew the details of the problem. He was a fine person to work for, always helpful.

The Harper-Jemison Legacy

HKS: Let's talk about Harper. It may have been as much the time as the man, but research really grew and matured during his administration. More, in a sense, than it did under Earle Clapp, who we all think of as one of the great leaders.

MBD: There's a difference there. Clapp was cutting a path through the forest that nobody had gone through. When Harper came along there were a lot of paths out there. He had to decide which ones to improve. Policies that Clapp developed have carried on and are basic to the organization today. Harper was involved policy-wise, too, in terms of trying to implement a broader program, trying to strengthen the organization so that the scientist had more opportunity to pursue his interest. Harper gave the scientist a lot of encouragement with his general philosophy that the scientist at the project level should be the key person in research. He wanted to keep administrative levels at a minimum, so that there was not a lot of dictating on how and where research should be done. He followed that principle very well. He also had good acceptance and a recognized stature in the department working with the other research agencies. I think that helped a lot in getting ahead with the total program, getting funding, and in attracting scientists.

HKS: Characterize him as a person if you could.

MBD: He was a person with a great capacity for work and to look ahead. These I would say were his outstanding characteristics. He had a certain amount of reserve, and if you didn't know him well, you might consider that as being sort of self-centered. But I never found Harper that way. He was always a personable individual with those that he worked as far as I was concerned. He had some very strong convictions, things that he wanted to do, and he worked at them. I think that it's good to have some objectives in mind, specifically what you want to accomplish.

HKS: Do you think you had a easier job as associate than Jemison had being an associate?

MBD: I had a different kind of a job, but yes, I think it was probably easier just because the pattern had been cut out to some extent. But also you have to recognize that times change, and during the Harper period there were many kinds of expansion going on in research. Research was the fair-haired boy in Harper's time. Not that monies and people came easily or quickly, but compared to other periods when research was just kind of brushed aside. Harper would take advantage of this period of growth and influence growth along certain channels that are important.

By the time I came to work with Jemison, the growth of programs had pretty much crystallized into a certain pattern. From there on it was a question of trying to give more effective direction to what we had. One of the big items when I was working with Jemison was the laboratory construction program. We had many, many requests for laboratories in various parts of the country. This took a lot of time, consultation with the people on the Hill, identifying where the most urgent needs were. Some of this started back late in Harper's term, but it really snowballed during the Jemison period.

HKS: Congress was aware that you had this list of say, fifteen. Did they say, get real. You're only going to get two?

MBD: They would say, you know, you're not going to get all these. We'd say, no, we're not. But these reflected current and future needs.

HKS: The Corvallis lab was built in a couple of stages. Apparently there was a master plan.

MBD: One thing that happened that influenced funding was a ceiling on costs for laboratory construction. If we had put in funding for the whole Corvallis laboratory, there wouldn't be funds for other locations. So we developed several of the laboratories on a staged basis. That was good. Maybe it cost a little more, but it certainly made people realize you weren't going to get a million dollar laboratory just by having a number of scientists at one location. You had to have a long-range view as to what the research requirements were going to be for facilities. The second phase you wouldn't need for several years hence, so it gave pause to think about future needs.

HKS: You mentioned earlier that Lyle Watts asked you to be station director, shake the place up, but didn't give you any money. What's the role of the chief in research? Did you ever really work with the chief much? You worked under Lyle Watts, McArdle, Cliff, and McGuire as chief.

MBD: You attend the chief's staff meetings every day, every morning. If you've got some problems that you want some advice on you bring them up in the staff meeting, and it gives the other members of the staff an opportunity to offer suggestions. Once in awhile the chief would give you guidance on priorities.

HKS: McArdle came out of research, didn't he? Did that make him more sensitive or less sensitive?

MBD: I think McArdle was easier to work with because he knew research, and he knew the ways of research people. You didn't have to spend time backgrounding him on the particular problem. He sensed right off what it was. Mac was a most unusual person to work with. He made you feel that you were good at your job and gave you lots of encouragement and advice. He was very helpful in that respect. Mac did a lot for Forest Service research, not just in building research, but in giving encouragement to scientists. He often spent time in stations when he was on a trip. He would talk with the scientists and talk their language. Also, I think it's worthwhile to note that because of his knowledge and background in research he probably didn't spend as much time on research as some of the other chiefs would. Mac was always most supportive.

HKS: What was the process by which you became associate deputy chief? I mean, were there candidates for the job, or did you just move into it? George said, I'm going to be deputy, and I'd like you to be my associate?

MBD: I really don't know.

HKS: It just happened.

MBD: All I know is that I was selected. Nowadays, there's a much more formal process of having to put names in and make recommendations. The only time I talked with Harper was about coming in here. From then on up I was never a party to decisions and discussions about me. Probably just as well. [laughs]

George pretty much continued the policies of the Harper-Jemison team, that's the way I like to refer to it. And he was confronted with a slowing of program growth. About that same time there were several retirements on the Hill, changes of key senators. Senator Humphrey became vice president, so he moved out of the Senate area. Then several of the senators in the South retired. I can't remember all the individuals, Senator Russell was one.

HKS: Is that why there is a fire lab in Macon?

MBD: I suppose so. That decision was made before I was in the Washington office.

HKS: But Russell would've been very supportive of it.

MBD: Yes, I expect Russell would've been supportive. As I mentioned already, Jemison was heavily involved in IUFRO which took much of his time.

HKS: How did that happen? Do you know? Harper was, too. He was vice president, I think.

MBD: Harper got to the point where he was confronted with the retirement policy of the Forest Service. Age was a factor then. When Harper retired, it was natural that Jemison, who had done much of the staff work for Harper and was well acquainted with the IUFRO organization, would take over from Harper. The plans for the IUFRO congress in Munich were well advanced.

MINORITY HIRING

MBD: One of the more difficult things he encountered (and the others of us subsequently) was this push to employ more minorities in forestry. The Lyndon Johnson era came in, and assistant secretaries were given the job of rapidly boosting minority employment. The message was: you don't take anybody else on in the higher grades until you get some minorities in there. Well, I can appreciate the need for a strong approach, but when you start out to hire in upper grades with no background in forestry, you have second thoughts about how far you should go. We did locate a few minorities, particularly where you had a need say, in a fire laboratory, for a physicist. Then you didn't need a forester.

HKS: I'm not minimizing the problem, but it seems to me, research had an easier job than the rest of the agency.

MBD: Yes, but even then it wasn't easy.

HKS: Towns like Corvallis are larger, and you didn't have to hire foresters. You could hire a physicist or hire a pathologist or you could hire an entomologist. The kind of skills that women and minorities might tend to go into as opposed to forestry. But you are saying, still it was difficult.

MBD: It still was difficult because minorities just were not coming out of the forestry schools. Even though a meteorologist, a physicist, or a social scientist, you could hire in research, still the hardcore of research needs required forestry-trained people. We started making grants to Tuskegee and other black schools for people interested in coming into research. Sure, we made progress, but to do it as fast as was wanted was another thing. On the whole, Jemison was exceedingly conscientious in trying to do the best he could. Department people, I think, recognized that, but they just kept pushing.

HKS: Did you have to file a report every year with someone in the White House on how you were doing?

MBD: I don't remember how that was done.

HKS: I suppose it would go through the secretary's office.

MBD: Probably it would go to the secretary. And agriculture was pretty low on the list of departments in terms of percentage of minorities employed.

HKS: What I read today, this is manyfold more difficult now.

MBD: Oh, yes.

HKS: The thrust is to make the Forest Service the employer of choice. That's the fundamental problem. People don't want to work for the Forest Service. It's not that the Forest Service won't offer a job; you can't force them to come to work for you.

MBD: Nowadays I understand there's the hangup on forestry. Some people think forestry is destructive and not the kind of an organization that they want to be affiliated with. This is a hangup that the profession is contending with, and I hope we're making some progress. And also, what was started back in the 1970s and 1980s working with the various schools and encouraging minorities to get into forestry is paying off now. It did in the late 1980s, and you see more signs of it now.

HKS: Maybe this will emphasize the difficulty you had in the 1960s. When I was a graduate student at the University of Washington, in the Puget Sound section of SAF, we had a speaker after dinner from Congress. His wife wasn't allowed to sit in the room at the head table. She had to wait outside. SAF did not allow women. This is in the 1960s.

MBD: Today I'm impressed with the progress that's been made when I go to a SAF convention, the number of tremendously able women that are there and presenting material, doing even better than some of the men are.

HKS: I guess there isn't much more to say about minority hiring other than you worked on it.

MBD: You just worked hard recruiting.

HKS: Is that something you did as opposed to George?

MBD: No, George pretty much took the lead on this. Sure, we handled the details, but that was a subject that George personally felt needed his attention.

HKS: I can't put the date on it, but it'd be in the mid-1970s. I had a good friend in the experiment station in Portland, and I dropped in to chat on my way through town one day. He showed me a letter that he had written to someone at the station at Bend about minority hiring. If you don't hire them, you're going to lose your job. It was a pretty tough letter. I was amazed.

MBD: You know the top people in the department were just plain tough. They realized that they had to make progress. When you look at the history of the 1960s and social problems that were rising country-wide, you realize why they were that way. It was probably good for all of us to have this pressure. We certainly wouldn't have made the progress that we have today, if we hadn't had the pressure we had then.

HKS: You build up a momentum.

MBD: That's right.

TRAINING PROGRAM

HKS: We haven't touched upon the extensive training program that started under Harper and accelerated under Jemison.

MBD: This was both retraining of older members and of younger graduates—sending them back to school under various training arrangements and assisting graduates so that we got a much better trained group of scientists than previously. By the time we got into the late 1960s, early 1970s, we had more Ph.D.-trained people than any of the other agencies in the department. We made a lot of progress rapidly. Both Harper and Jemison recognized that the level of scientific capability in Forest Service research had to be raised. Training was about the only answer. We had a tremendous program going on. One time in the Lake States, I figured that over half of our professional people were involved directly in graduate work. That was made possible to a considerable extent because we were located on the campus at a major university. But we were also bringing people in, moving them around from field locations, so that they could get on in their training. This was a subject that George spent considerable time on while he was deputy, trying to facilitate training activities.

HKS: When I was preparing to interview John McGuire, he gave me his personnel file. In it was a rather strongly worded letter from Keith Arnold, "McGuire, if you want to amount to something in the agency, you're going to have to finish your Ph.D." [both laugh]

MBD: Mac put the pressure on all of us for more training. It was part of building a research organization, a scientific group with much capability.

HKS: In 1964 or 5, I was authorized—I'm not sure if that's the right word—to leave the station and go to Yale for a Ph.D. It was very tempting. They even found a job for my wife in New Haven. Instead I resigned, went back to school, and studied history. A three-year free ride as it were—tuition, salary, fringe benefits, if I would get my Ph.D. I understood that for the three-year Ph.D. program, I would be obliged at least morally to work for the agency nine years afterward. Is that—?

MBD: I'd never heard that, it is new to me. When you select people like that for special training, you talk with them, and they're cognizant that they are expected to come back, but there's never anything binding.

Contentious Issues

HKS: We talked about this earlier when discussing your St. Paul years, contentious issues and whether or not they effect research priorities. I jotted some thoughts down and you didn't cross them out. So let's talk about the clearcutting controversy that was really getting to be rather heated about in the late 1960s.

MBD: The first one I recall was the Bitterroot controversy in Montana. That became very much of an issue, because the timber harvesting there was followed by furrowing of the steep slopes and planting. It looked kind of barren after it was all done. A few years later I went back out there, and I was impressed by the extent of the forest cover that had developed in a few years. The people who were upset by the clearcutting on the Bitterroot had shifted their concern to other areas by that time. But the research group was not often directly involved in such controversies. There was one in Wyoming also.

HKS: So, you or George didn't call up the station director in Missoula and say, hey, have one of your guys go take a look at that?

MBD: No, I didn't, a station director in the area was more likely to take an interest.

HKS: What would the regional forester say if you'd done that? Would that have been a turf problem?

MBD: No. Usually in that kind of a situation the request for help would come from the forest supervisor. Our researchers often worked with the rangers and supervisors. One thing I think you have to keep in mind is that there were other criteria that were developing in the minds of the American public. For example, landscape viewing became a value, and back a few years before that, it wasn't a concern.

HKS: We had the so-called Church guidelines on clearcutting. Here you have a heavy hitter, a senator, who's high-profile on clearcutting. It must trickle down into the agency somehow that this is a priority. The U. S. Senate is on our tail on this thing.

MBD: Oh, surely. You know there were a series of hearings held by Senator Byrd on Monongahela clearcutting. I recall that Ed Cliff spent one whole day, testifying. One person he took along with him was Carl Ostrom, who was then in charge of timber management—silviculture research. I was always proud of Carl being selected for that job. He was one of the best silviculturists in the U.S. He went up to the hearing and gave a scientific perspective to a controversial subject.

The reason I bring up this changing of views and attitudes on clearcutting involves a question: is your primary interest growing timber, maximum sustainable volume, and to what extent and how do you take watershed yield and landscape viewing into consideration? These are values that were not so prominent fifty years ago. So, the question of clearcutting is not just one of how do you grow the most timber through clearcutting or shelterwood or selection systems. But other values that must be considered.

HKS: It's interesting how times change. When I worked for the Forest Service on a ranger district in the late 1950s in Region 6, we laid out clearcuts. I did, day in and day out, and we left strips along public highways to screen them. And in our minds, in our heads, this was done to enable us to carry on what we thought was appropriate silviculture, but also being sensitive to what the public wished. But when I read about this now, it's always put in the most cynical light, that we're trying to hide something.

MBD: Oh, yes.

HKS: There's been a shift in public perception. NEPA, the National Environmental Policy Act, with its requirement for impact statements. Scientists must have been involved in that, because of all the impact statement requirements. The field people didn't know enough for certain things, right?

MBD: I don't know. We were not directly involved in preparing NEPA statements. This was the responsibility of another group in the Service, but research people participated in discussions by providing reference and background material. After all, we wanted to have as good a job done by the national forest staff as was done by research. It was never a contentious subject as far as research was concerned.

HKS: There's a lot going on about this time, around 1970, with Earth Day and creation of EPA and Council on Environmental Quality. Overall there was a tangible increase in awareness on the government level.

MBD: There certainly was and increasingly so.

HKS: We had to change the way we did things.

MBD: We just had to change and recognize new values.

HKS: But in effect, doesn't this help you in Congress. You need more money and Congress is more sensitive to it, or did Congress not react?

MBD: The standard answer was to stop what you're doing and do the new job instead.

HKS: I see.

MBD: You just change your priorities. I'm not sure how well that can be done, you don't just change an assignment of an individual. You've got to be concerned about what the individual's capabilities are.

HKS: We obviously need a lot more science in order to, not repudiate, but at least deal with the accusations that we didn't know what we were doing. But to some extent they were true, right?

MBD: Sure, but one also must give adequate recognition to what has been accomplished.

HKS: We just didn't know.

MBD: And there are a lot of things we still don't know today, too.

HKS: Jemison retired sooner than expected? Is that true?

MBD: Well, sooner than many of us expected.

HKS: I mean, you look at someone his age, you figure they're going to retire at age sixty-two or something.

MBD: Of course, George was getting up there. I think that he felt he had put in about as much of a career that he should, and at that point, he decided to move on.

KEITH ARNOLD, DEPUTY CHIEF

HKS: So, Keith Arnold becomes deputy.

MBD: Keith Arnold becomes deputy for research.

HKS: I understand that you asked not to be considered for the deputy position because of your wife's illness.

MBD: Yes.

HKS: Keith was dean at Michigan. You must have heard some of this through the grapevine or maybe directly from Keith. Ed Cliff was calling, and he had a real problem. MacNamara was playing volleyball or swimming or something with—who was the secretary of agriculture?

MBD: Oh, Orville Freeman.

HKS: Orville Freeman. I guess they played handball every Thursday night or some such thing. And Freeman was saying, you know, I need a new deputy chief of research, and I don't have one on-line. The heir apparent has deferred. MacNamara says, I've got fifty guys, any one of them which could be your deputy chief of research. Ed Cliff heard about that and he got on the phone and said, Keith, you're the only one that's qualified at this moment, at that level you have all the civil service requirements met. Is that a true story?

MBD: I don't know.

HKS: Keith wasn't opposed to the idea, and he was happy at Michigan.

MBD: He had some difficulty moving in, because he had left the organization earlier.

HKS: He had been station director at Berkeley. Had he been in the Washington office?

MBD: Yes, he was in the Washington office. At that time we had division directors. He was the head of forest protection research which included insect, fire, and disease research. That was a new combination which Harper and Jemison put together. The only part I had in Keith's coming back was that he called me and said in effect, Dick, if you want that job, I don't want it. He understood my reason, and we left it that way.

HKS: There is a lot to that job. The fact that he was out of the agency for a half a dozen years was significant.

MBD: This was part of his problem in getting back into the organization. You know, many of the station directors had known him earlier and figured he'd had his shot at the job and had moved on.

HKS: Was there any sense of lack of appropriate loyalty to the agency, because he had left?

MBD: I don't think so.

HKS: I mean, being dean at Michigan is not chopped liver. It's a pretty high ranking job.

MBD: It sure is. After Keith got here and settled in, he soon had the support of all concerned. He moved along quickly.

HKS: He what, operated at a more theoretical level than Jemison? I don't know how to characterize the difference.

MBD: There's a difference in the individuals in the first place. Keith is a person who makes decisions and moves on with them quickly and effectively. He's not one to spend a lot of time rationalizing pros and cons.

He had to continue minority recruitment which was still pressing us in the late 1960s. He was also involved in international activities with IUFRO, FAO, MAB. Those things took considerable time. The big subject that he was involved in was the question of the organization of research. We had been through a period of tremendous growth. We had a structure with strong scientific leadership here in the Washington office. At the station level, under the director, we had station divisions. At the Lake States, as I recall, we had six station division directors because we had six functional programs, some large and some small.

Harper had developed a concept that the project leader was the one who should be the key scientist in the organization. He should have the freedom and the liberty to carry on his research activities. Then, there was a great imbalance. You know, the organization kind of grew like topsy for ten or fifteen years. Big expansion in fire research for example, that's when the fire labs came in. Then there was forest entomology and pathology which had come over to the Forest Service from ARS. Some of the stations had only two scientists in a project, others had a dozen or so.

There was obviously a need for reconsideration of the organization structure. This was the job that Keith took on. The staff in the personnel office made several rather complete analyses of the workload in the station division organization. Some had two or three projects; some, twenty projects. How to get this back to a more even workload was a problem. Keith spent the best part of a couple of years on this particular subject, while the rest of us were carrying on the day-to-day activities.

HKS: He has a great deal of confidence in his ability to be a quick study.

MBD: Yes.

HKS: I could see that that would be a little nerve wracking, working for someone like that. I don't know how you judge someone's grasp of the situation. Is that a fair characterization that I made?

MBD: Yes, I think so. This is what I was touching on earlier, he made decisions rather quickly. I couldn't fault the decisions. It's just that was his make-up. He wasn't going to spend a lot of time mulling things over. He had it well fixed in his mind where he wanted to go, and that's what he did.

HKS: You've been involved in the history of research off and on for many years, writing letters to people. Someone characterized Keith, who left the agency in '73 to go to Texas, as not—what was the term—not willing to put up with the hurly burly of federal research.

MBD: I expect that the Texas job had its attractions compared to the daily demands of the Forest Service organization.

HKS: It sounded like it's a pretty hard-nosed operation, and you get beat up a lot.

MBD: Working in the federal service is a lot different than university life, that's for sure.

HKS: Who beats you up besides Congress?

MBD: Well, the work load first, and second the large organization that you're working with. You have to sort out, shall we say, the small issues and the big issues. Those decisions are not made easily or quickly at times. You have to decide when a problem here needs attention. Some things inevitably fall by the wayside.

HKS: So Harper and Jemison, in terms of the growth of research, were pragmatic. They took advantage of opportunities.

MBD: Yes.

HKS: The growth wasn't uniform. So their legacy was a greatly expanded research program. And was Keith's role to try to start stabilizing and balancing it out?

MBD: Yes. I would say that his main role in the three or four years he was there was trying to get organization responsibilities at the stations clarified and placing some scientists in other assignments. Either one of those is a big task in itself. Particularly the latter one, and of course, that question of getting people functioning in a new assignment is one that you don't resolve over night. It takes some time, and some people never do quite make it.

HKS: I've been so impressed interviewing two chiefs and now three deputies, how much time you guys must spend on personnel matters.

MBD: Sure enough.

HKS: Of making sure there's people in line, as station directors retired and moved on.

MBD: Several alternates must be available and trained. The importance of this has developed more and more as time has gone on. In the 1930s, 1940s, and 1950s, we were not getting the large number of retirements. The big recruitment period in the Forest Service was back in the 1930s during emergency programs. Those people began to retire in the 1960s and 1970s. So the importance of getting people in line to fill vacancies accelerated during the 1960s and 1970s.

Deputy Chief for Research (1973–75)

HKS: Keith resigns fairly abruptly.

MBD: He resigned and left for Texas in a few weeks.

HKS: I guess Steve Spurr put on the pressure to get him down there.

MBD: Yes, I understand that is what precipitated it.

HKS: So, were you made deputy immediately or were you acting for a while?

MBD: The usual procedure is to be designated as acting until department approval is received. I don't know where such appointments go for approval.

HKS: Were you interviewed by anyone at the secretary's level?

MBD: No, I wasn't.

HKS: So, McGuire decided to appoint you deputy chief although officially the secretary did it.

MBD: I guess that's right. I believe John was the one that made the decision with the approval of the secretary.

HKS: You wrote a letter to Bob Buckman recounting this era. John asked you to "let things cool off" following Keith Arnold.

MBD: Yes, reorganization was well along and the staff needed to settle down.

HKS: Anything specific?

MBD: No. John was talking about the job. I told him I wasn't going to be there for too many more years, I was getting near to the retirement age. He said, "As I review the research organization, the thing that we need most is to get settled down and get on with our job." We're still getting the impact of reorganization which Keith implemented in terms of getting people into the spots where we wanted them. We had, as I mentioned, six or seven division directors at some stations. The combinations we made in the reorganization just didn't have so many division jobs.

The other thing was to get adjusted to another concept of how this new organization was to function—that the project leader was to be the scientist and leader. We were eliminating one organizational level. In place of the division level we had two assistant directors for continuing research. We had two assistant directors in a station in contrast to six or seven before. These were largely administrative and not scientific positions. When we put in an assistant director for research planning, this was a new concept. Also we set up assistant directors for administrative services. There was not only physical placement of people, but getting mental adjustment to a new kind of functioning in an organization in which we wanted to have the scientists be in the forefront.

I think that's what John had in mind mainly in this settling down. We had had tremendous growth. We had had reorganization. We had new programs. All three things needed to get settled down. In some situations this settling down came quite easily, others I don't think ever did. Having been in the Washington office there for some years, I knew the key people at all the stations which was a big help in making staff adjustments.

HKS: John came out of research. He was certainly sympathetic with research. We've talked about this with Cliff and McArdle—did that help you any to have a chief that had been in research?

MBD: I'm sure that made it easier. Yes. I guess one of the things that impressed me was that Ed Cliff had always been in the National Forest System, and we never thought he had much interest in research. When Ed got into the chief's job, he promptly became interested in research. It wasn't just a passing interest. He dug right in to what some of the problems were that we were confronted with and this helped out a lot. McGuire didn't have to go through that phase, because he was already there, but it sure helps when you a get a chief who is interested in research activities.

HKS: You'd been associate for quite a while.

MBD: Yes.

HKS: So now you're deputy. Were there any surprises? You changed offices and secretaries.

MBD: No, I just moved to another office and had worked with the same secretaries. No, there weren't any big surprises.

HKS: But there was essential responsibility that you didn't have before. The buck stopped with you.

MBD: You just kind of go along with it. That's it.

HKS: Who was your associate?

MBD: There were several changes. Most of the time, Bob Youngs, who later became director of the Forest Products Laboratory, was the associate. By that time we had two associates. Carl Ostrom was in there for a while, and Herb Storey, plus several others acting.

HKS: Where's Buckman?

MBD: In research, we had two or three staff assistants. They were the ones that took a hold of a specific job. Bob Buckman had one of these staff assistant jobs. He had been in a staff position in Washington for some time. Later he moved out to PNW station as director.

HKS: Was he ever your associate?

MBD: For a short period.

HKS: But he became the deputy when you retired.

MBD: He became the deputy when I retired.

HKS: Did you have a hand at that selection?

MBD: Oh, sure.

HKS: I mean, when you were getting close to retirement, you sat down with McGuire...

MBD: I sat down with McGuire and his staff. At that time they had quite a complete list of prospects for all deputy positions. Bob was one of the candidates, and as I recall, the staff was much in favor of his selection as deputy of research.

    I started to get on with this settling down business. I was kind of surprised when John mentioned this as the first thing he wanted. He's a very keen observer. Apparently he had felt this need and realized that Keith had been pushing fast and hard to get reorganization in place before he left. John sensed that there was a need to slow up. Some of the stations never did quite take on the standard reorganization pattern. Some of them went for one deputy director instead of a series of assistant directors, and the total number of staff was the same, but the setup within a station had variations within. John said, fine, let's do that, and we will work toward the standard pattern as we go along rather than trying to jump from this stone to that stone at once. Which was an astute way of keeping things going but not stirring them up.

A Typical Day

HKS: Describe a—I know there isn't one—a typical day. You're fifteen minutes from work. You get in your car, you drive in, you park, and you go in. What do you do?

MBD: The first thing each morning is the staff meeting in the chief's office.

HKS: Every day?

MBD: Every day. If the chief's out of town and the associate is tied up, once in a while you don't have one. But generally every day you had one. It's kind of a show and tell activity. What are the bugs that are bothering you today and what should we do about them?

HKS: You go around the room.

MBD: You go right around the room. There are always two or three contentious issues. I'd say that about 80 percent of them involve national forest activity. Research doesn't have, or didn't have at that time, much to bring up at the staff meeting.

HKS: Did your associate attend also?

MBD: Not unless you were detained otherwise. I know one thing, you learn never to miss a staff meeting. [laughs] Strange things happen at times, and you want to be there or you sure want to be represented.

HKS: This was what, a half hour every morning?

MBD: Oh, sometimes fifteen minutes and sometimes two hours. You never knew. You keep every morning open for a staff meeting, and generally you didn't make any appointments until after eleven o'clock. That system worked very well. We tried for a time to have a research staff meeting right after the chiefs' staff, but the timing of it was difficult, because you never could know when you were going to get out of the chiefs' staff. Your staff had to make their appointments for the day, so we finally gave up on it.

HKS: I find that amazing. If you count up the salary hours. I realize that it's important and all that, but when you think about that.

MBD: I don't know how else you'd do it.

HKS: Were the minutes taken and distributed?

MBD: No, there were never any minutes taken that I knew about. On some specific issue, somebody might take them, but there were never any minutes that were written up for circulation.

HKS: So John would lead off and say, this is what I am doing today?

MBD: Generally.

HKS: Is there a hierarchy? Would people sit around the table in a certain sequence, always in the same chair?

MBD: [laughs] I guess it depends on who you talk to. I never felt that there was a hierarchy, but I noticed certain individuals always went to the same chairs.

HKS: The deputy chief for national forests strikes me as a pretty high-profile job.

MBD: It kind of depends. Individuals are individuals. Tom Nelson was in charge of the National Forest System as deputy while I was a deputy for research. Tom had come up through the research organization and he was always supportive. You're kind of a minority in the outfit because you've got a deputy there for state and private and national forest, and other deputies who had come up through the

National Forest System. Tom was a big help to us in research, and his attitude passed on out through the whole National Forest System. You could see that where we were getting help in the regions it was an outcome of the interest and support that Tom Nelson was giving us. We tried to do likewise to the National Forest System.

HKS: One of the most valuable archival records that I have encountered is called the "Minutes of the Service Committee." In these minutes you can see, in the early 1920s, wilderness coming across the stage, and all the debates in chief and staff. Wilderness, was it good or bad? That's why I was asking if there were minutes of these meetings.

MBD: I think maybe the general content of the staff meetings changed over the years, because the subject matters are always the things that were under the gun and had to be taken care of that day or week. It wasn't always a long-range perspective. It's who's going to do what today.

HKS: Okay, so the chief and staff meeting's over. Then, what, do you have mail to catch up on?

MBD: Mail to catch up on and probably a couple of dozen telephone calls waiting for you to return. You pick the ones that are most urgent and get staff assistants to handle the others.

HKS: Did you usually have a working lunch where you...?

MBD: No, no, we never had a working lunch. Generally with one or two of the staff members, I'd go down to the department cafeteria. I just don't think it's healthy for people to be involved eight, ten, twelve hours a day in a particular subject. You've got to back off from it a little while.

HKS: You read about the so-called power lunches in Washington, D.C.

MBD: That's right.

HKS: Your associate over in EPA that you meet at lunch or something. But that really wasn't what happened?

MBD: No. No, I wouldn't say that was what happened at all in the Forest Service. Once in a while it would, but it wasn't a common occurrence.

DEALING WITH CONGRESS

HKS: How much of your time was spent dealing with Congress and preparing for testimony and that sort of thing?

MBD: That varies a lot. There was a period in February and March, then it shifted to May and June, in which budgeting and appropriations took much of your time, not just with Congress, but on the other end with your field organization, because you need their input also.

HKS: Did you testify in Congress on matters other than budget? A bill comes through and you're brought in because of your expertise?

MBD: No. Once in a while you may do so, but I don't recall ever going up on the Hill to handle other than research activities. That was usually just once a year when you'd go up for the budget hearings, and then you were subjected to questions. Julia Butler Hansen was chairman of the committee while I was in the deputy spot. She always had some sharp questions for research.

HKS: You mentioned earlier that you didn't do liaison with the staffers ahead of your testimony routinely.

MBD: No, I didn't do that routinely. Quite often, if the committee staffers came up with a question, I'd send fellows like Buckman up there. Obviously if Julia Butler Hansen called, I would go up. But some of the staffers on the committee and our staff people would get together. Some of our staff people were up there frequently and on the phone answering questions about specific programs.

HKS: The staffer has to prepare questions for the person to ask.

MBD: They'd get an idea on the question, then they'd call me and ask, does this kind of a question bring out this subject? Some of our staff people would work it out with them.

HKS: Were you ever in an antagonistic situation?

MBD: You could tell there were some antagonistic questions.

HKS: But that might be more political, Republican versus Democrat.

MBD: No, I don't think that anything was very indicative of "they're out to get you" at all. Generally, you know, the people on the Hill are competent individuals and have a lot of tact and background.

HKS: How long would you testify roughly? Is it an hour? Is it a day?

MBD: On the budget hearings generally I'd say you might have anywhere from a half hour to an hour. It depends on what the particular subject is at that time. A lot of things are resolved with the staff people before you get to a hearing. Sometimes the hearings are very perfunctory. Maybe you have five minutes to answer questions.

HKS: But the research budget is presented as a part of the Forest Service budget.

MBD: That's right.

HKS: And the chief is there. McGuire is there officially dealing with the testimony, but he would turn to you sometimes?

MBD: It kind of depends. I've been there at times when I never testified. McGuire would handle the questions, and that would take care of it. If you testified, it was to a specific question asked by the chairman of the committee. And you don't talk unless you're asked. That's the way it operates.

HKS: How about people in the secretary's office?

MBD: There's usually one member of the secretary's office at a hearing.

HKS: Do they testify very often?

MBD: Usually right at the opening there's a brief statement that they give. Beyond that, I don't recall anyone from the department becoming involved at the hearings in the research items. In the national forest, at times, there were some comments back and forth.

HKS: The research budget; some casual observations. Roughly 5 or 10 percent of the total Forest Service budget, in that neighborhood. Does Congress spend all of their time on National Forest Administration and then take a look at Research?

MBD: The activities on the national forests are of the moment, and they often get the attention. The controversies that are going on, that's where all the questions come. Generally you don't have that kind of thing in research. Research is not a controversial subject that engenders questions, it's more talking to the justification for the budget proposal. Very seldom do you get much back and forth questioning. I'm sure that sometimes there were substantial problems to discuss. But at least during the period I was there, I spent a lot of time getting ready for the hearings, and I was disappointed because I didn't get much time. [laughs]

HKS: This so-called eight-hour day you put in. You must have done a lot of homework at night.

MBD: Oh, sure.

HKS: To prepare for this. Memorized the numbers.

MBD: Memorized the numbers and carted them back and forth home to study them. But after you've been around there a few years, you know they're not going to pin you down to the last dollar on a budget item.

HKS: But you want to appear knowledgeable.

MBD: Oh sure, you've got to be knowledgeable about it.

HKS: Keith told a story. He came back from the Soviet Union Saturday night and testified Monday morning, and he wasn't over jet lag. I guess you went out to his house that Sunday and worked on the budget. He was in pretty bad shape, then. [both laugh]

MBD: That's right. Well, that happens. But, you know, you develop a staff that has a feel for these things and they know what to have ready. Changing dates for hearings is common, so you never know where you're at. If something comes up in Congress that committee members have to be there on the floor to vote, the hearing is cut off until later. Then it's shifted around, morning and afternoon.

HKS: I've asked others the same question. Where do you park? How do you go up on the Hill. Do you walk a mile?

MBD: If it's a nice day like this, some of us would walk and some might take a cab.

HKS: But you don't want to be late for your testimony.

MBD: No, most of us would go up in the cab, and then take a relaxing walk coming back. But it varied. No, you've got to be there, and you want to be there ahead of time.

HKS: Would you ever call on the station director, say there was major line item for a new lab, to have them sitting there just in case?

MBD: No. The hearing room in the House was so small that you couldn't even get all the deputies and a few assistants in the room. So anybody in the field would not be there. In the Senate they had a larger room. The people who were there for testimony were all in a row up front, any others would sit in the back of the room and listen. But you're not asked questions if you're back there. Just because of the physical arrangement, I couldn't turn around and say to Buckman, "What is the report on this issue?" You just can't. You are on your own. That's the best way to express it.

HKS: I never realized that the chief was there throughout the whole thing, because I thought the deputies would go over for each major program.

MBD: The chief is the one, and the only time that the deputy talks is when the chief turns to him. That's the procedure.

HKS: It's still a mysterious process. [both laugh] You think of the immensity of government, and from the congressional staff point of view, to try to keep a handle on it. It's daunting. Your job, it seems to me, was easier than the staffers' side.

MBD: Probably.

HKS: Because they're handling so many agencies in agriculture.

MBD: Julia Butler Hansen was a pretty sharp person. I always enjoyed meeting her when she was chairman. She had been with it for years, so that she knew what was involved.

3-Bug Program

HKS: Let's get more specific here. The 3-Bug Program. Why three bugs as opposed to three diseases or three something elses?

MBD: Let's put it this way. In the late 1960s and early 1970s, nationwide we began to have a series of insect epidemics. The most common one, I guess, was the tussock moth. Then we had the spruce budworm and the mountain pine beetle, gypsy moth, southern pine beetle. There were a whole series of infestations. Some of the entomologists related the problems to climate. I think generally foresters were more inclined to relate it to the general age class of the forest. You were getting the forests that were at the age in which they were susceptible to southern pine beetle, and same way with he tussock moth, and other insects.

Sure there were disease problems, but those were more likely to be endemic than epidemic. And we had, well, heart rots, for example, various rusts. There were cases made for needed research for each of these. We had some research going on several insects, but the magnitude, the number of insects that began to surface as epidemics were so great that it was beyond the capacity of the Forest Service research organization and beyond the capacity of the forestry schools and state forestry departments. So, as I said, it kind of grew over a period of, well, starting in the '60s and kept becoming more common. We had discussions with the Cooperative State Research Service and through them to the state agricultural experiment stations, trying to see what others could do on the particular problems that we were hearing so much about including questions from congressmen.

It was outgrowth of the discussions of this type that we decided we had reached the point where we had to make some kind of proposal for additional funds. We couldn't manage with the small group of scientists we had handling the insect problems. The assistant secretary, Bob Long, covered the Forest Service but not CSRS. He got quite excited over the complaints that he was forgetting about timber losses from insects and suggested that we try to rationalize some kind of a program to go to Congress.

So internally we did this and talked with the stations and the stations with the agricultural experiment stations around the country and decided that three most important insect problems we ought to take on first were the tussock moth in the Northwest, the gypsy moth in the Northeast, and the southern pine beetle in the

South. Also, Long was favorably inclined to approach the need budget-wise through the Forest Service rather than going through ARS or another agency. Part of his decision, I'm sure, was because the Forest Service had a straight line organization so he could identify just where the research was going to be done and budget-wise just how to handle it. I had several discussions with Long on this, and finally we decided that, well, we've got to get one person in here full time on this. At that point in the Forest Service, Keith Shea was then in charge of our forest protection research—insect, disease, and fire.

HKS: He's a pathologist, right?

MBD: He's a pathologist. That's correct.

HKS: Did it make sense to have a pathologist in charge of bugs?

MBD: I don't think at that level it is particularly significant. You're looking more for an individual who can provide leadership, who has an understanding and capacity to work with various diverse groups, and Keith had that. He worked easily with any organization. He didn't let the little things bother him.

I had known Keith when he was an undergraduate forestry student at the University of Minnesota. I had confidence in his capacity. So, we detailed Keith, put him over under the assistant secretary's office, out of the Forest Service, because you had other agencies to deal with that are not under the assistant secretary, then you've got the problem of how do you coordinate.

HKS: I know Ken Wright. He was for the tussock moth.

MBD: Ken was a good leader, he worked well with other organizations. And he knew the Northwest.

HKS: We used to play volleyball, so that's how I'd see Ken in those days. He said, you know, I'm over in the secretary's office, now. And I thought, gee, is that ominous? I didn't know if that was good or bad. [both laugh] So, he was the tussock moth guy in the 3-Bug Program assigned to the secretary's office.

MBD: Yes. In each one of these bug programs we had a program leader. We had a comparable leader in the southern pine beetle and in the gypsy moth. Keith spent a lot of time trying to pull together a program that would meet the needs of the problems that they were confronted with and make maximum use of all of the agencies that had a capacity to work on the problem. It was a big challenge to try to pull this together country-wide, and he did a tremendous job.

HKS: Was it successful? I mean, apparently there's still a bug problem.

MBD: Yes, there are still insect problems, but I think there has never been a forestry fact-finding effort as well organized and executed as it was on these three insect programs. You don't get a simple answer. It's always complicated by the ecology of the forest and ecology of the insect. You have to piece things together and try to work out an interpretation that gives you a control. Certainly with the gypsy moth and the southern pine beetle, we have had a continuing effort for many years.

What this particular 3-Bug Program was to do was to bring everything together, summarize it, put it together in one place where all of the facts could bear upon the subject, and try to interpret in terms of a program that would give you answers of what to do. Certainly you just don't say, okay, we've been working on these three areas. These are the facts. This is the answer. That isn't the way that insect problems are resolved. You get good leads, good direction, and test them.

HKS: The red-cockaded woodpecker had been listed by this time. So you have habitat combat between the bug people and the bird people.

MBD: Many things are always involved. Of course, when the bug program was taken over by the secretary's office, I kind of was on the sidelines from then on. The Forest Service had been important in getting some action going, but from then on it was a joint USDA-state effort. There were several committees or councils and so forth around the country that reviewed their activities and decisions. But at least from where I sat, I had a feeling that this was one program that was really moving and had the support of everybody. You know, we had the Universities of Idaho, Oregon, and Washington all involved in the Pacific Northwest on the tussock moth. Never before had this particular kind of an effort been made. It was interesting to see how stimulated people got to make things move. The final publication, I expect you've seen it, it's a book about like the Sears-Roebuck catalog or thicker. It has a tremendous amount of material in it on the tussock moth.

HKS: Did it give you the notion that this is a good strategy, let's try to find other areas?

MBD: Let me say this, this had been tried a couple times before and was not so successful.

HKS: Oh, okay.

MBD: During Keith Arnold's time—I think it was in watershed, maybe in fire—anyway, there were attempts to put together package programs of this magnitude, probably not as comprehensive as the bug program. But we never seemed to get through. We would get them together and they had a lot of promise, but we never had the push to get the necessary money. You see, in the 3-Bug Program we received a special appropriation of six million dollars.

HKS: That's a big percent of your insect budget.

MBD: It was a big, big step. Of course, if you put it in terms of the *total* research budget, for all the organizations involved it is closer to 2 percent. One of the things we observed in the earlier efforts to package programs was was the need to get all parties knowledgeable about a proposal so that as the program progresses you have their interest and support. We were fortunate in the 3-Bug Program to have had so much publicity about timber losses. This kind of a program could well be the basis for some future proposals.

HKS: How do you name things? Over in fire they had Operation Skyfire, Operation Firestop. The 3-Bug Program, that's not very sexy.

MBD: I don't know how those developed. I expect probably Keith Shea had more to do with it than anybody else. You know, sometimes a simple title is very appealing.

RESOURCES PLANNING ACT

HKS: It actually explains what it is. The next on my list is RPA. My assumption is if the Resources Planning Act is going to work, research projections have to be a part of the assessment, because you have to have knowledge in order to carry it out. Is that correct?

MBD: I'd say you're correct in terms of the basis of the law itself. How it operated organizationally is that the RPA work group did their own projections using some of the earlier research material. They drew up their own guidelines for the projections and set up their own systems. Actually the RPA was just getting started when I was there, and I had very little to do with it. They were setting up their own organization. They took some people from research and put them into the RPA organization.

HKS: But when they made the assessment and they're identifying problems somewhere, research has to study the problem.

MBD: People at various levels participated in the discussions on the specific programs. But the responsibility was with the RPA group. We were helpers in getting their job done. I would point out that the USDA long-range study of research needs provided much of the material for research projections.

HKS: By your observation, did research benefit from this?

MBD: I wasn't there long enough to know. As time goes on research should benefit.

FOREST PRODUCTS LABORATORY

HKS: Okay. The Forest Products Lab. We haven't discussed that really at all except who the director is. Let's start with a philosophical question. How do we as a society justify having the government do research that benefits industry? I mean, why don't we say to the industry, do it yourself? It seems to me that Ronald Reagan would have said that.

MBD: Probably.

HKS: And Eisenhower.

MBD: A lot of people did say it, too.

HKS: Is that just talk? Has the propriety of the Lab been questioned?

MBD: I've heard it discussed many places, many times. But you have to look back. The Forest Products Laboratory started in 1910. That was not when forest industries were doing very much research. In fact there was an awful lot that was not known about many of the various tree species and the qualities of their wood. So they started at a time when there was a wide open field to get information on the utilization qualities of wood and the uses that were a possibility. They grew to be a tremendously competent organization and worked very closely with several industry groups. Our scientists worked with their scientists.

Generally we did not have serious questions raised about the propriety of being involved in the particular research areas. In fact, the Forest Products Laboratory tried to be one step ahead of the research that was going on in terms of say, better use of hardwood species in wood pulp. Most research back in the 1920s was centered upon the use of conifers in making wood pulp. And hardwoods were just beginning to be considered. There was some work on hardwoods in the 1930s. The competent people gave more credence to the Laboratory's effort than any other particular attribute. Of course, in addition they had fine facilities at the Laboratory at Madison to carry on their research.

HKS: When I read through chiefs' reports and literature over the years, there's the research program with its experiment stations and it's always "and the Forest Products Lab." It's never considered part. It's always considered unique or separate in some way.

MBD: Just the name is different.

HKS: Yes.

MBD: Now we have eight stations and the FPL. Also we have laboratories at Corvallis and several others around the country. They never seem to take on the identity that the FPL did. It doesn't separate them from the organization particularly. It's just that the origin of the name FPL that goes back sixty years or so.

HKS: At the station at Portland we had a forest products program. I had no idea what they did. How is that linked to what the Lab does?

MBD: Let's go back a little. During World War II, the Forest Service did service work for the War Production Board. An outgrowth of this WPB work was the establishment of what were called FUS units at individual stations—Forest Utilization

Service. Every station had two people involved in studies of regional utilization problems, and often these studies were taken over by the FPL. Gradually at some stations the staff moved more into specific projects, where others were truly service units. What you were seeing at the Portland Station was what we called an FUS unit. As I understand it, these units are all closed out now.

HKS: Yes. I've noticed that a couple of the guys I knew at Portland in utilization are now at the Forest Products Lab. I don't know if that was normal moving up through the ranks or because they were closed down.

MBD: I think that as funding became tighter on into the 1980s, there was a movement to bring these individuals—some were retiring—but younger ones were brought into the Forest Products Lab. Some of them continued doing the same sort of thing at the Lab that they did at the stations, servicing various industry groups.

HKS: Buckman tells a story. When he took the assistant secretary, John Crowell, out to the Lab he said, this is all very fine, but it's too far in the future. We can't justify it.

MBD: Yes. [laughs]

HKS: Was that a typical reaction on the part of the secretary's office? You must have taken people on tour, too.

MBD: Sure.

HKS: Why are we doing this?

MBD: That's a common question, why are we doing this? to put us on the defensive, justify what you're doing. It would come up now and then. Particularly, I think, with the kind of program at the Forest Products Lab. But just what his comment was, that it's way out in the future, is the reason for the importance of having the Forest Products Laboratory. There's got to be somebody out there beyond the individual company doing research on unexplored subjects.

HKS: You didn't have a sense that the folks in OMB were waiting for a chance to whack that thing out of there. It won't benefit the current president, so why have it?

MBD: No. They were looking for anything that they could to cut out of the budget. If it happened to be the Lab and if we didn't have a good justification for it, why, they'd probably get rid of it. There were times when that type of a question came up.

HKS: In an administrative sense, was the Lab a maverick? I mean, did it run its own show? Because it was unique. It didn't have any competitors as it were within the agency. Did you have any problems in terms of administering the Lab?

MBD: No. I would say that we recognized their differences, we recognized their competence, and there were strong individuals on the staff. I worked easily with the Laboratory and greatly admired what they were doing.

RESEARCH LEGISLATION

HKS: The need for research legislation.

MBD: The McSweeney-McNary Act, I think it was 1928, provided the legislative authorization for the research organization in the Forest Service. The act listed a number of experiment stations, maybe ten, eleven. It listed authorizations of funds by subjects. These were generalizations. They weren't restrictions, but when you got around possibly to closing or moving a station, some of the local people would say, well, here, in your legislation it lists a station at Missoula, Montana, or Tucson, Arizona, or elsewhere. Also listed were several types of research: timber management, watershed, and range management, etc. There were several questions about the interpretation of the language.

One omission that always seemed to bothered us was authorization for work in international areas. As I recall the act, there was only one mention of work in international forestry, and that was Forest Products Laboratory shall do work on tropical timbers, which it did. I got to thinking about this and talked with legal services, and they pointed out that we were free to use other legislative authorities that the department had in international areas. Development of IUFRO was not a subject of question from the standpoint of authorization. It was there through other legislation that was available to the Forest Service. During this time there was reconsideration of various acts that Forest Service had authorizing programs and the decision was made to try to bring these all together into one act. At that time, I mentioned some of the questions we had about the McSweeney-McNary Act.

One thing specifically, as we got more and more involved in international organizations like IUFRO, FAO, and so forth, we just were not sure just how far we could go. This act was mentioned as a subject for the legislative group to consider. One of the specifics that came up was forestry attachés. There are agricultural attachés at many of the U.S. embassies. With all the activities that forestry was getting into, we thought that maybe having attachés, say, one in the Pacific area and one in Europe, maybe one in Africa, or something like that, would be helpful in servicing forestry activities. I got a rather negative reaction from the State Department on this. They said, just depend upon the agricultural attachés. They can help you. You're all in agriculture. Why do you need somebody else? Well, of course the obvious answer was that very few of the agricultural attachés had forestry background.

HKS: Yes.

MBD: Then there were other questions of the AID forestry programs and several others around in the government that were involved in forestry activities. As consolidated forestry legislation was considered further, the Research Act came back into consideration. That was about the time I retired. Eventually research was put into one act with the National Forest System and State and Private Forestry. The authorization for forestry research was very general. It deleted the names of stations, and some of the other items that we had considered. But it was still at least a couple of years after I had retired before the act went through. I viewed this with a mixed feeling, because in some ways the title of the McSweeney-McNary Act was a good point of identification for forestry research. And just saying in a more general act that you should do research doesn't give anything specific to guide you in developing programs.

HKS: My understanding is that the Farm Bill carried the authorization for a new deputy chief for International Forestry.

MBD: I believe so.

HKS: It was a way of getting it through.

MBD: This is the very kind of a thing that the lawyers in agriculture referred us to. These special authorizations occur in various kinds of legislation as amendments and riders. The authorization may be for the Department of Agriculture, so why
· not use it in forestry as well as agriculture? I guess my concern relates to my early association, identifying McSweeney-McNary as the forest research authorization for a federal program. Then all of a sudden it's replaced by more general language. This kind of bothered me. [laughs]

INTERNATIONAL FORESTRY

HKS: Obviously International Forestry is important to you, because this is what you have been working on so much as a retiree. What was happening in International Forestry when you were deputy chief? What were the issues other than could you do it?

MBD: Quite specifically, we had had a staff group in research that was financed through the AID Program, much as they have today in the forestry support program. This did several specific tasks that the AID program wanted. During the late 1960s, early 1970s, that program kept contracting, and we got down to a small staff, two or three, maybe four people, so that our capacity to do work in the international area was much more limited. The interest, though, was growing, and our problem was to find a way to service it, and eventually that developed.

HKS: Why doesn't International Forestry in terms of its place in the organizational chart fit more logically in State and Private Forestry, which is an extension concept? My understanding is that International Forestry is largely an extension concept, only it goes outside of the United States.

MBD: I guess there's some rationality there that's sound, but when you look at it from the standpoint of the agency, international activity in research was a subject which had been active for many, many years. IUFRO itself is an international organization and participation is largely by research scientists.

One thing leads to another. FAO has what they call a COFAL group. This is a committee on forestry at FAO that reviews the FAO forestry department program periodically. With the research group's other activity, it was natural to pick up one more thing. Not that there wasn't some discussion about it, but the research people were already involved.

HKS: It strikes me that IUFRO must have been significant in terms of the mentality of the players.

MBD: That's right.

HKS: And the World Forestry Congress, which is broader than research, doesn't have all the activity going on that IUFRO does.

MBD: The aggressiveness and effectiveness of IUFRO had much to do early on with the international activity being in the research group in the Forest Service. I am sure Les Harper could give you as much better background on this than I have. I came in as more of a Johnny-come-lately. We made a special effort to bring people in from State and Private and National Forest System into the various activities where they had the interest and capabilities, so it wasn't just a research fraternity. It was service-wide. And it wasn't of the magnitude that you could, with any good sense of judgment, come up with a deputy position or organization to carry on, not at that stage. Since then, the international activities have taken on much greater significance.

HKS: With the recent collapse of the Soviet Bloc, International Forestry has taken on a whole new dimension. But in your mind, at the time, did you draw a line between International Forestry and tropical forestry and Third World forestry?

MBD: No. Many of the assignments that I went on were short term. In Africa, in Europe, in Asia, all in the same trip and for various organizational reasons. The subjects that I gave attention to just didn't fit all into one slot. You have to operate on a broad basis to represent them. Part of it was the fact that foreign travel is fairly expensive, so you want somebody to go to a meeting in Dehra Dun, India, for forestry and somebody that can go on to Pakistan for a different type of a session. You have these geographic problems, getting people around.

While I was deputy, Bob Long was very, very strict on foreign travel. I remember one time planning to go to a IUFRO meeting, and I sent three names up for approval. He looked at them and said, Dickerman, one person from the Forest Service is going to go to that meeting. We had two IUFRO division leaders—Bob Callaham and Herb Fleischer—and I was on the executive board. I said, okay, Mr. Secretary, do you have any feeling as to which particular position should be represented? He said, you'd better go. Well, I had some very disappointed individuals in the organization, because they were members of a board that was governing IUFRO, and they couldn't go to the meeting. I mention this because that was how limited foreign travel funding was.

HKS: Explain something to me in terms of the bureaucracy. Why are you asking an assistant secretary this? Why didn't you ask your chief this question?

MBD: Because at that time all travel and foreign travel had to be approved by the assistant secretary.

HKS: So, there was a major crackdown government-wide.

MBD: That's right. It was government-wide. It wasn't just the forestry...

HKS: So trip by trip. They didn't say, you have fifty thousand dollars this year for foreign travel. Each trip—.

MBD: No. Each trip went up for special authorization.

HKS: How long did that go on?

MBD: I don't know. [laughs] Probably for a year or so.

HKS: It seems rather incredible to me in terms of management. Obviously there was some abuse there.

MBD: There was some abuse and there was some easing off from it as they went along, but that was the way the game was played at that time.

HKS: The rules that you couldn't take annual leave when you were on international travel.

MBD: We ran into some abuse of leave on travel.

HKS: This in a sense is before your time, but you may have some feeling for it. There are some retirees in the Durham area that meet for lunch. They worked in Africa on various assignments. They said, the big mistake we made as a nation was thinking that the Marshall Plan could apply to Africa. The Marshall Plan was a First World plan. Do you have any thoughts on this?

MBD: I really don't.

HKS: Africa at that time had independence. There was a whole layer of issues.

MBD: I expect that's true. One thing I found in international activities, different parts of the world just work differently. It isn't a standard pattern even in IUFRO. Try to get leaders for certain subject matter materials. You just have to approach it differently.

HKS: In International Forestry it turned out to be very significant, it's the Institute of Tropical Forestry. I interviewed Wadsworth last month. Somebody sent him to places like Borneo, but he doesn't know the forests in Borneo. Does that make sense in retrospect?

MBD: Well, probably no one other than Frank had been involved in tropical forestry for many years.

HKS: Or what else would you do?

MBD: What else do you do? You look at your alternatives. You don't have anybody who knows the species or the ecology of Borneo in your organization. Apparently some group needed or wanted somebody from the States. Frank has been involved in international activities on a scientific basis for years. He has a better appreciation than many of what is involved. To come up with the best candidate, you've got to recognize that even the best has limitations as well as good capability.

HKS: He reported directly to the deputy chief for a long time, but now he's under the Southern Station. Did that happen during your time?

MBD: No. That arrangement changed several times. I believe, when I was first in Washington, in the late 1960s, Frank was reporting to the director of the Southeastern Station, because Florida was under the Southeastern Station and was closest to Puerto Rico. And in addition, he became involved in the experimental forest there that is part of the National Forest System. So he was also administering national forest lands. I think probably that had more to do with it than anything else in shifting of responsibility into Washington and away from Southeastern Station. I don't know when, I think it was while Buckman was deputy, that they moved the administrative responsibility down to the Southern Station which makes more sense to me than maybe any other arrangement. It's always been a problem, because of the national forest and the research responsibilities, and each with a very small staff. You know, maybe three or four at the institute and a couple of them paid in part out of national forest monies. At one time, we also had a scientist there funded in part from State and Private Forestry.

HKS: You had a program in Hawaii about the same time, right?

MBD: The program in Hawaii came later.

HKS: Okay.

MBD: The one in Puerto Rico goes way back into the 1930s.

HKS: When I lived in California, Bob Callaham was speaking, maybe to SAF, and talking about his areas of responsibilities, which included Guam.

MBD: That was later. That came even after I left. Buckman must have worked that out. They tried to use the same pattern in Hawaii that we developed at Puerto Rico in the sense of having the Pacific Island areas included.

HKS: At coffee this morning, we talked about Frank, the person, and the fact that he's had his current position for thirty-five years and how unique that is. This bothered Ed Cliff. You weren't getting enough out of Frank, he had more potential, and the Forest Service was not taking advantage of his experience.

MBD: I pointed this out to Ed Cliff.

HKS: Do you have anything you want to put in the interview about the Forest Service not knowing quite what to do with the institute for a long time?

MBD: I think that's kind of obvious, the way Frank kept shifting around organizationally, that he didn't feel very secure in any particular arrangement. My own judgment is that probably the most useful arrangement is what they have at the present time, operating under the Southern Station. I can't see operating out of the Washington office. There's just not the magnitude of program.

HKS: Conceivably it could move under International Forestry now with the new deputy.

MBD: Yes. That is a possibility. I don't know enough about what they're leading into with their new organization.

HKS: Frank gave me, in preparation for the interview, a thirty-seven page vita. I don't know how many people do this, but he had listed every meeting he had ever gone to. For the promotion he never got, I guess.

MBD: Yes. [laughs]

HKS: It was incredible how many hours that man spent in the air and in airports.

MBD: And it's increased over the years. You know, the last few years there's been more and more travel for him. He had the same problems I mentioned earlier, travel funds, Frank couldn't get out of Puerto Rico unless we transferred some money to him. He just didn't have enough money in his budget to take on many tasks. We just couldn't get additional funding. Probably the restraint was in the department more than in the Congress. We just couldn't get new budget items out of the department that included Puerto Rico.

HKS: I would assume that the Department of Agriculture, in terms of farm technology, must routinely do tropical agriculture.

MBD: Oh, yes. There's quite a program of agricultural research in Puerto Rico. I don't know what the scope of it is or subject matter, but there is a lot of research going on there.

HKS: It will be interesting to watch and see what happens.

MBD: Yes.

ENDANGERED SPECIES

HKS: One item left on my list is endangered species. Research must be involved in studying the habitats, ecology, and physiology of endangered species. Is that correct? Like the woodpecker and the owl?

MBD: I don't remember the year that the Endangered Species Act was enacted. What was it, in the 1960s?

HKS: Yes.

MBD: We did not get involved in this activity until later.

HKS: Someone asked me, why doesn't the Forest History Society do a study of the history of the red-cockaded woodpecker, the issue? So, I dabbled a bit. One thing that amazed me was that the bird was listed in 1969 or thereabouts, and in 1970 there was a conference co-sponsored by a lot of groups including the Forest Service on the state of the art. There was an incredible amount of information available on the life history of that woodpecker. People had been studying it for a long time.

MBD: We didn't get involved with the Kirtland warbler habitat in Michigan. We didn't have a specific project, but we had some habitat studies, observations going on with the ranger in the area, and had developed a fairly good understanding of the habitat requirements. But this was a subject that was just beginning to come into focus in the 1960s and 1970s. I don't recall having a budget request for endangered species.

HKS: Every time a new species is declared, there's a sense of urgency.

MBD: Sure enough.

HKS: And people are going to say, well, we don't know how we are going to have timber harvest, for example. Certainly research must get bumped, hey, why don't you study this right now.

MBD: Well, you get many suggestions.

HKS: But it also creates a need for different specialists.

MBD: Oh, that's right.

HKS: Ornithologists in research.

MBD: Yes, why not?

HKS: Or would you ordinarily work cooperatively with organizations that already had ornithologists?

MBD: It works both ways. At the present time, some of the stations have ornithologists and projects on the endangered species. There are other organizations in some of the state groups that are working on the endangered species habitat relationships. We set up a wildlife habitat project, back in Harper's time. That habitat work went into subjects other than ornithology. Just a couple months ago, I was out at St. Paul and heard a very good presentation on some of the research work they were doing on habitats. So there's some research going on now. But you take the spotted owl. Jack Thomas, who is the long-time research scientist, made a lot of contributions and got involved in many controversies too. [laughs] But it was an outgrowth of his broad interests that he got into the spotted owl controversy.

HKS: Someone suggested to me that there's a whole generation of very senior, influential scientists like Jack who are about to retire.

MBD: That's true in several forest science areas.

HKS: Because of the recruitment spurts in the agency's history.

MBD: Sure, an even flow of recruitment or retirement is seldom achieved. You go along like this and then you go along like that. [gestures]

HKS: The idea was that the agency ought to have us interview some of these people, because there's a transition going on in the agency of another kind.

MBD: Maybe, sometimes you get some good things that way, but I'm kind of philosophical about it, that you get out of an individual the capabilities and productivity that he has and we he goes, they go with him. And that's it. Interviewing might be worthwhile, at least you'd get some history out of it.

HKS: The implication was that the Forest Service might not recruit Jack Thomas today. They're looking for different kinds of scientists.

MBD: Oh, yes, always.

HKS: It's more and more specialization.

MBD: More and more specialization. That certainly is what's going on and will continue to do so.

Decision to Retire

HKS: One thing we haven't talked about is the decision to retire. Had 1975 always been your target?

MBD: In the later years, yes, '75 was the limit that I had in mind. Had it been particularly relevant to organizational needs, I would've been receptive to retiring a year or two earlier. But '75 was my outer limit.

HKS: You never seemed to be a part of the decisions made to promote you. [both laugh] It happened. You got a phone call or something.

MBD: That's right.

HKS: Can you shed any light on how Buckman was selected to be your successor?

MBD: About the only comment I can make is that his name was one that was involved in the long-range placement planning. Buckman was obviously one of the candidates for a top position, so that he was well up in the list when the need to select a replacement came along.

HKS: For a situation like that, do you call the candidate to come in for an interview, or do you know enough about him already?

MBD: In this case, I knew enough about him. I'd known Bob since he started in the research group in the Forest Service. I didn't need to follow up any further. I suspect that some of the others in the staff who were involved in the decision probably talked with him some, but I didn't feel the need to do so.

HKS: All right. So, you retired. Did you take any time off and play around?

Society of American Foresters (1976–80)

MBD: I didn't have the SAF particularly in mind at the time that I retired.

HKS: Had you been active in SAF?

SCIENCE ADVISER

MBD: Off and on. I'd been very supportive wherever I was located. Out at the Lake States I was involved in several ways. I always had a very deep regard for the profession. So when I retired, I took several months traveling around and doing odds and ends. I wasn't particularly anxious to get relocated right away. In fact my pattern of thought had run that I was going to take a couple of years to just kind of knock around before I settled into something. But at the SAF Carl Ostrom was working as a science adviser. He left suddenly and I became interested.

HKS: Is that a real job? I mean, he has to go to work?

MBD: It's what you make out of it really. Basically the primary responsibility is assisting in the development of the working group programs for the national convention each year. That was the one solid activity when I went out there. Carl left in late '76. Then Hardy asked me if I would help out. In fact, I only worked there part-time for, the best part of two years, just one or two days a week.

HKS: That's a compensated position? You received a salary?

MBD: Yes, there was a salary. It's like all those jobs. The longer you're around, the more demanding it gets. [laughs]

HKS: What's the difference between what the science adviser does and the outfit that runs *Forest Science*, the journal? To my mind, you have through *Forest Science* some pretty high powered people involved with the society.

MBD: During this time, there were a number of international forestry organizations developing. IUFRO was well along. IUSF, which is the organization representing the societies of foresters in the various countries, was getting organized.

The job of the science adviser was to provide leadership in SAF for science activities, especially in the international activity. Also it involved representing SAF at organization meetings, in addition to the regular SAF chores. I remember one conference I went to at the Forest Products Lab that was concerned with the utilization of tropical timbers. That was a real high powered conference. It took a lot of time on the part of SAF. The science adviser at that point was serving as a right hand man to the executive vice president in all of the international activities. That was about as broad as it was. You never knew what was next.

HKS: The title's a little misleading. At least it is to me.

MBD: Well, no, the title is apropos to the extent that you realize that the development of the national convention programs by the working groups was quite demanding in getting good programs developed, in selecting outstanding scientists for programs not only from the SAF but outside. So that part of the job is quite relevant, but as happens in all societies, there are lots of odd jobs around which somebody has to pick up. I suppose you could come up with other titles. I don't know whether they would be more appropriate or not.

HKS: You made a general statement about the role. Are there some specific topics you can recall that as science adviser, specific assignments you had?

PEOPLES REPUBLIC OF CHINA

MBD: One was development of the science and technology exchange with China. That was one that became a major operation before we went very far. The SAF had had some contacts with the Chinese Forestry Society, and it was obvious that they were most anxious to re-establish their contacts with other societies. There were several individuals in the Chinese society who had gone to school here in the States. Dr. Wu, who I worked with closely, was a graduate of Yale and Duke.

HKS: I believe I met him at a Duke reception somewhere.

MBD: He's been through here several times. He particularly was trying to reestablish contacts. Apparently he had very good contacts within the Chinese government, so that he didn't have the difficulties some of the other individuals and groups had. But the development of the exchange and all the arrangements were the responsibility of the science adviser.

HKS: At the IUFRO congress in Kyoto, there were Chinese there.

MBD: Oh, yes.

HKS: And the thing that struck me is that the younger scientists, we'll say under thirty five or something, wore western clothing. The elders wore Mao jackets. That ideology must have been difficult to deal with.

MBD: That's another thing that you had to bridge. Just getting correspondence through in the late 1970s before the U. S.-Chinese relationships had opened up was a chore.

HKS: Nixon had been to China.

MBD: But the U.S. hadn't yet recognized China as such. There were several contacts in forestry going on informally.

HKS: Are you working with the State Department?

MBD: Yes. We worked through the State Department. We also worked through OICD, Office of International Cooperation and Development, in USDA. This office had been set up to coordinate the activities within USDA. The Chinese offered to pay all of our expenses once we got off the airplane in Beijing and until we left. They would take care of the housing, transportation, meals, quarters, everything like that. But we still had to face up to the question of getting the airfare back and forth.

HKS: No Super Savers to Beijing at that time.

MBD: No. It was an expensive deal. OICD came through on that one and helped us very much. I might just mention the name of the individuals that went on the trip. Since SAF was the key organization in developing this exchange, we decided right off that the team should represent a cross section of all of the various forestry organizations in the U. S. And yet we had to keep the number down to six. There were some criteria that had been agreed upon between the two countries in terms of who would go, how many and where. The individuals who were selected were Carl Reidel from the American Forestry Association; Bryan Clark from research in the Forest Service; and Jim Yoho and Carl Gallegos from International Paper Co. They split the time there, because they couldn't take the full time. They represented the industry. Then Donald Duncan from the University of Missouri and John Gordon from Yale from the forestry schools; and Doug Leisz from the National Forest System. I wound up as the team leader from SAF. This selection was made by the SAF and concurred in by the OICD office.

HKS: Roughly, what was airfare?

MBD: Oh, I would guess around $1500 round trip to Beijing.

HKS: You can get to Tokyo easily. It's that last—.

MBD: Yes, it was that last part, Tokyo to Beijing, that was difficult to schedule. We landed at Beijing, and the Chinese almost had a band out there to welcome us. [laughs] Then we traveled extensively to the far north and the far south, seeing forestry activity and problems.

HKS: Was this the first scientific group?

MBD: This was the first official forestry team to the Peoples Republic of China.

HKS: Fred Knight went with a bunch of entomologists.

MBD: What happened was that after this initial group, we identified several subject areas of mutual concern, then subsequently there were a number of teams that went to China. The entomology group was one, genetics another, and a few others.

HKS: I know a Canadian group went over.

MBD: Yes, there were Canadians. There were quite a few. In fact, the Canadian foresters came very shortly after we were there.

HKS: Of course, Canada recognized China for a long time.

MBD: Yes, and there were other countries later.

HKS: So you arrive at the airport, and you are warmly welcomed.

MBD: Warmly welcomed and dined extensively. We spent a few days in Beijing, and then took off by plane to Heilongjiang Province, which is the northeast province of China, up against the Russian border. We flew up to Harbin, the capital of the province.

HKS: This is west of Inner Mongolia? I'm trying to visualize it.

MBD: No, it's northeast of Beijing and east of Mongolia and northwest of North Korea. And the northern boundary of Heilongjiang Province is mainly along the Amur River. Then from Harbin, we took an overnight train north, stopping in several villages where there were timber utilization plants, timber harvesting, timber cutting. We saw quite an area of coniferous plantations.

HKS: Were the forests extensive?

MBD: The further north you get, the more forests there are. A lot of the Korean pine which is similar to our eastern white pine. And big trees, not as big as our West Coast ones, but good sized. They had a large timber harvesting program going on. One of the things that kind of caught my eye, they apparently had done a lot of scouting to locate logging equipment, arches, and various kinds of tractors. They had some equipment there that was just far too big for the kind of timber in the area. I'm sure they changed to smaller equipment before long.

HKS: So they were thoroughly mechanized—.

MBD: Oh, yes, they were mechanized. The Finns had been there with a group and had helped them in equipment use and maintenance.

Coming back to Harbin, we stopped in several places where we saw plantations and various and sundry operations of tree planting. One of the interesting places where we visited was the Northern Forestry School at Harbin. They had a huge building, bigger than, I think, any forestry school in this country. The Chinese are much like the Russians in the sense that everything they built was big. The president of the school was a very affable, friendly, courteous, and helpful individual. We spent the best part of a day there at the school. They had activities going on much as we do in this country.

HKS: What was your sense of the quality of education?

MBD: At that time, they were struggling so to get going again. They had lost all their books and equipment during the Cultural Revolution. The Red Guard had taken all the books, and most educated people and students had to go out and work in the fields and not spend any time learning in the classroom.

From Harbin, we flew down to Shanghai and spent a day around there.

HKS: Did you fly in propeller planes or jets? [pause] That's a very long haul.

MBD: These were jet planes. Most all of their planes were Russian made.

And from Shanghai, we took the train over to Nanjing, where the large forestry school in southern China is located. We spent more than a day around there meeting the faculty and finding out what they were doing, what they were interested in, what they saw as forestry problems. Each place we went, we tried to emphasize, what is your basic problem that you're most concerned about in the management, harvesting, and utilization of your timber?

HKS: Could most of the faculty read English, so they could get access to the scientific literature?

MBD: No, I would say maybe only 20 percent could understand us. Every school you went to, you'd find somebody who had been in the U.S., so we had no trouble talking about their interests and activities. Nanjing had a very impressive layout in facilities and faculty.

Then we headed over to Changsha in Hunan Province, which is Mao country.

HKS: Way inland.

MBD: Well, it's inland somewhat. We didn't go over to Mao's home, which was still further west. Changsha is a big city. We visited several forest experiment stations in that vicinity. They had some interesting plantings of species selection and selection of the superior trees.

HKS: My vision of China is it's pretty arid.

MBD: They had done a tremendous amount of reforestation there and had large plans for continuing.

HKS: The pine.

MBD: Pine. Their reforestation is most intensive. One of the things that bothered me, every time I looked at a plantation, every tree was there. You don't see such complete plantations in this country. [both laugh]

HKS: So if one died, they replaced it.

MBD: Yes. I said to Dr. Wu, how do you do this? He said, if a tree dies, somebody in that village goes up there and puts another one in, see, so we've got it fully stocked all the time. They had the labor to do that.

HKS: No economists screwing up the work that was going on.

MBD: No, sir. You didn't have to worry about the economists. [both laugh] We must have visited four or five areas around Changsha, and then took the night train on down to Guangzhou (Canton) and spent some time around there. We went out to an experiment station where they were testing eucalyptus.

HKS: Mei was with you on this trip?

MBD: Oh, yes. She was on the trip all the way. She was our interpreter and helped much in developing an understanding of our interests. The Chinese had an interpreter along, but as usual, with a technical subject you need a couple of interpreters to decide upon the proper translation. She had grown up in the Fujian Province which is right across from Taiwan, so it's south central China. There's a mixture of dialects in the area, and she got along fairly well with the various dialects.

HKS: But the written language is uniform.

MBD: Yes.

HKS: So the literature is readable by everyone.

MBD: Yes. We spent the Fourth of July in Guangzhou and had an exciting experience. Some of Mei's relatives came in and visited with us at the hotel and gave us a little local color. That evening some of our crew went out to celebrate. [laughs] Some way or another somebody had some firecrackers. They tossed them away and lo and behold, one started a fire.

HKS: Uh-oh.

MBD: [laughs] Things had to move pretty fast to get the fire out. It was kind of a tense minute. Anyway, we got through the Fourth of July in Guangzhou and had a good time about it. Then from there, we flew on a small Chinese plane down to Leizhou which is on the Gulf of Tonkin. We were there two days. That is the area where the Chinese had tried at one time to establish extensive rubber plantations. But apparently the climate and soil were not right, although there are remnants of the plantations there now.

HKS: Don't they grow bamboo?

MBD: Yes, of course bamboo is grown many places. In fact that was one of the interesting things we saw up near Nanjing. One of the profs there has spent a lifetime studying bamboo, both the selection of the superior kinds of bamboo and the planting and management of bamboo. He had a very interesting story. It was fascinating to me to see what knowledge he had brought together under difficult circumstances.

We saw extensive plantings with various spacings and several kinds of pine. And we also visited the dune stabilization plantings along the coast of the South

China Sea. They put on a very nice luncheon for us there. I guess there was at least a watermelon for each one of us. [both laugh] It added much to the day, because of course when you're in southern China, it's hot and drinking water is scarce. As I said, the day before was the Fourth of July, so we were right into the heat of the summer. Out in some of the plantings, they put on a beautiful luncheon. All the way through they seemed to enjoy entertaining us. They always did the best they could with what they had and then even a little more.

HKS: Did you have a sense of any security restrictions?

MBD: No, not at all. We could see anything we wanted.

HKS: Were you free to wander where you want and all of that stuff?

MBD: You could wander as much as you want. It was up to you. It was a question of whether you thought you were going to get lost or not. [laughs]

HKS: Sure.

MBD: I had been on a similar assignment in Russia back in '74. There we had all kinds of security. In fact, we had a security officer traveling with us all the time, and frequently he would go into the restaurants before we would and scout out the surroundings.

HKS: This was while you were still with the Forest Service?

MBD: Yes, when I was with the Forest Service, but it was the same kind of a mission. The objective of these science and technology exchange trips was to identify and develop programs of mutual concern and interest to the respective countries. So you develop a pattern of working. In Russia we had two groups, one group in fire and one in entomology. We traveled with the Russians to the respective areas of interest. The reason I started to mention the Russian exchange was there we had several security problems. In fact, the State Department could tell before you got there what room you were going to be in, as it was supposedly bugged.

HKS: Did we reciprocate with equal pettiness when the Soviets toured the U. S.?

MBD: No, not in forestry. One of the things that we decided early on, we're going to be wide open. If they wanted to go to town and buy a lot of clothing or something, go ahead. They couldn't quite believe that that was the case.
    On the other hand, on the China exchange, everything was open all the way along as far as I know. We didn't have any security people traveling with us. Maybe some of the individuals may have had instructions, I don't know. But anyway, we didn't feel that we were constrained at all and were free to roam around and see what you wanted to and ask any questions of anyone.

HKS: When you were developing the itinerary, were you involved in all the places you could go?

MBD: To some extent. Not as much as to places, but rather as to subject matter.

HKS: Okay.

MBD: We would indicate we were interested in tree improvement work they were doing, for example, in chestnut blight in China—subjects of that type. Then they would pick out the locations. And I'm sure that part of their selection was to be sure that we got around the country as much as possible, because going from the northern border of China and Russia down to the Gulf of Tonkin is a far distance. Some of the locations were probably not the most essential to see, but it did give us an excellent perspective of the forest areas.

HKS: Do they have a central bureaucracy like the Forest Service, or are the provinces relatively autonomous?

MBD: Well, I guess I would say somewhere in between.

HKS: Is it more like the U. S. or more like Canada, I guess the question is.

MBD: My hesitancy is that we were there just as China was opening up. And many of the provinces were just constrained. They did not have transportation and other facilities they needed. I'm sure by now the provinces are stronger than they were at that time. I find it hard to characterize the situation in 1980. Many places we went, the kids and older people would come out "An American! Where did you come from?" They just hadn't seen an American for many years, it was almost a shock to them.

   One day I was hiking along the top of one of the walls around Nanjing and came upon an old Chinese stone mason. He saw us, he couldn't speak any English. He motioned in some way so that I could understand that he thought that we were Americans. He clapped and bowed just like God had come to his country. It would give you a great feeling of pride. Part of that friendliness is inherent in the Chinese people, they tend to be friendly and very polite to anybody.

HKS: Nanjing. That's where Chiang Kai-shek holed up in World War II, right? So that would be sort of a stronghold for the older generation.

MBD: The older generation, yes. That's an interesting city, old walls and many historical places. The trip through China was more relaxing than it was to cross Russia into Siberia and back to Moscow. Both were stimulating as well as tiring.

HKS: This may sound like I'm a victim of the nightly news. In those days, when you're planning a trip like this, does the CIA approach you, saying, go over there and look for this? Maybe you couldn't even answer the question if supposedly this had happened.

MBD: I can say frankly that nobody—CIA or otherwise—approached me. The only thing that we had was a little cut-out from the OICD office on how to handle problems.

HKS: You mean, if you get in trouble what you should do?

MBD: Yes. What you should do. I don't think anyone in our group was contacted by the CIA.

Soviet Union

HKS: Do you want to elaborate a little bit more on your earlier trip to the Soviet Union? Was it the same kind of fact finding trip? Was that the purpose of it?

MBD: It was different to this extent, John McGuire and Keith Arnold had been to Russia as a forerunner to our trip. They met in Moscow and went down to Kiev. I don't know where else, but they didn't travel as extensively as we did. Their main mission was to come up with an agreement to guide for exchanges, not to identify the subject matter except in a very general way. Our trip followed theirs, we had two research groups—one in entomology, the other in fire. I think, all told, thirteen or fourteen people went to Russia on our exchange.

　　We had a couple of days in Moscow. After that, the group separated. I traveled north and east with the fire group, the insect group went south. We were looking at specific studies in fire research which would be of interest to both countries. One of my most memorable days was when we were way north of Leningrad near Lake Omega. They took us out to give a demonstration of fire control technique. When we got out to a spot over comes a great big army helicopter. The Russian foresters use the army helicopters frequently, the military makes them available to them. Anyway, out jumped a crew of half a dozen or so, they came down, all lined up, marched over to shake hands with me and the rest of the group. Then one reached in his backpack and pulled something out and handed it to me. I kind of looked at it. It was a stick about twelve inches long, maybe an inch or so in diameter. It was a stick of dynamite. [both laugh] These fellows had been jumping out of the helicopter with packs of dynamite on their back.

HKS: OSHA wouldn't approve of that at all. [both laugh]

MBD: Well, I was really quite shocked when they did that. But anyway, then they took this dynamite and went out through the woods and set a fire line putting their sticks along the way, then detonating them. And lo and behold, they blew a good fire line right down mineral soil all the way along where they had blown the ditch with the dynamite. Of course, they had knocked out some trees in the process. [both laugh] It was a startling exercise to see. But also very effective for remote areas.

　　Then after that, we headed east to Siberia, flying to Lake Baikal area near Irkutsk. That trip gave me a great appreciation of the tremendous size of Russia and of the vast forest resources.

HKS: I know, twelve time zones.

MBD: You look at the map and think how large it is, but when you get on the plane at midnight and you fly all night long, and the next noon you get off over in eastern Siberia. You've been in the air close to twelve hours and not yet to the Pacific Coast. All you saw was flat, what looked like timber country, much of it I understand was marshland. It's quite impressive when you see something like that. A friend of mine reminded me of this when I came back. He said, the biggest problem Russia's has is holding that country together. It's so big. [laughs]

HKS: Transportation costs would be enormous to bring the natural resources in the east to where the people are in the west.

MBD: That's right. Many of the areas in Siberia have tremendous forests remaining and there just hasn't been any way to get the timber out of there over to the Pacific Rim countries where there's a tremendous market. The Trans-Siberian Railroad is way down on the Chinese border and the timber is further north. We had an opportunity to go out on Lake Baikal.

HKS: Did you see the famous pulpmill that gets so much bad publicity?

MBD: We saw some of the air pollution coming up from it. We didn't go to the mill. I guess the other thing that impressed me, from Irkutsk we flew north to a city, one of their new cities called Bratsk. Anyway, on the Angara River at Bratsk there's a huge hydro plant. It was built, as I understand, at least in part to supply electricity for a new electrified railroad that was to parallel Trans-Siberian Railroad across the northern part of Siberia. One of the Russians told me, our biggest problem is to find a way to use all the electricity that's generated. We have only half of the generators operating now. With the pulpmills and the sawmills that we have here, we don't use much of the electricity. With only one half of the generators installed, we produce more electricity than all of Grand Coulee. So, it gives you some idea of the generating capacity on that one dam.

    Another thing I recall vividly is traveling around there by car. As I said, this was a so-called "new city." I would imagine within, say, a couple of hours of travel, we must have passed fifteen or twenty weddings. [laughs] I commented on that to one of the interpreters, and he said, well, what do you expect? All of these people here came from other parts of Russia. They came here as youths, and the weddings are going on all the time. [laughs] I was fortunate on this Russian trip to have as an interpreter a forester who had worked for me in St. Paul. A Russian Pole, Alex Vasilevsky. Alex had worked on the forest survey in the Lake States there, and he had lived in what is now Russia, but was Poland then. He spoke the Russian language fluently and got along well with the Russians. If it hadn't been for Alex, I'm afraid it would have been a dull trip. He kept things moving and interesting.

HKS: The Soviets had been able to impose the Russian language uniformly across the country? It's not like China where you have dialects.

MBD: I guess I can not answer that. We went across the north. As far as I could make out there was a common language. But I suspect when you get further south, you get into what are now separate countries. There you undoubtedly have more dialects.

HKS: What was the state of technical knowledge?

MBD: Highly variable, I'd say. They had done very little in tree improvement work, because of the background.

HKS: Lycenko.

MBD: Lycenko, yes. Just outside of Moscow we visited their central forest laboratory. They had some tremendously advanced work, especially the effect of air pollution upon vegetation.

HKS: But the basics. If they planted a forest, the trees survived?

MBD: Oh, yes. They had good reforestation techniques all right. The Russian foresters are well trained, highly competent. They were having the same problems that the Chinese were of not having equipment and transportation. They picked up ideas real fast. You see, after we were there, a Russian group came over to this country, and we took them to the Pacific Northwest. They would pick things up and leave some ideas, too. So many times we tend to think of others as being less developed, and actually in many of these countries they already have the knowledge; it's the problem of getting it into use.

HKS: We've had so much publicity in recent years about the illegal sales of computers to Russia.

MBD: Yes.

HKS: Presumably, they don't have the technical capacity to actually manufacture them. They want to buy them from the West. I suppose that same limitation is throughout.

MBD: I believe so.

HKS: Although forestry doesn't use a lot of high tech.

MBD: They don't use a lot of high tech yet, but they pick things up pretty fast.

HKS: I'm working with an Austrian friend right now through IUFRO to put out some proceedings of the recent Berlin congress.

MBD: Oh, yes.

HKS: And all the western papers come on floppy disks, but the eastern papers are typewritten.

MBD: Those are some of the common things you run into in international contacts. Of course some of this you overcome by these exchanges. Getting together and seeing what the techniques and equipment are. Before long they find a way to get it into their country. Both the trip to Russia and to China were highlights of my participation in international forestry work.

HKS: What happened when you got back? Did you bundle up cartons of technical literature and mail it to China or what?

MBD: Oh, yes.

HKS: Or did they have all that stuff already?

MBD: Yes and no. Some things got through and you'd wonder how it ever got there. You'd also be amazed what they didn't have. So it's very spotty. I'm sure that by now, things have opened up so much that a lot of the literature is getting through without any difficulty.

ACTING EXECUTIVE VICE PRESIDENT

HKS: Okay. The next thing on the list is Acting Executive Vice President of SAF.

MBD: I had been out at SAF for two or three years part-time. The Renewable Natural Resources Foundation was just beginning to operate. Things were moving along quite rapidly.

HKS: That foundation was inherent in the Tom Gill bequest?

MBD: That's right.

HKS: It was always part of the master plan.

MBD: Hardy Glascock, as the executive vice president of SAF, was a very vigorous individual and did a tremendous job of getting the RNRF moving. At that time the SAF had its largest membership ever—about twenty-two thousand, I think. Many activities going on and a large staff, I suppose, thirty people or more there at Wildacres. At some point Hardy decided that he was going to leave the SAF and go over and become the general manager of RNRF.

HKS: Was he of retirement age?

MBD: He'd been talking about retiring for some time, and I don't know just what prompted his decision—probably his broader interest in the RNRF than in just the SAF.

HKS: Was there some problem between Hardy and the council? Was he eased out?

MBD: Hardy was an aggressive individual. To some he was kind of abrasive, but he had the drive needed to make an outfit go. Yes, he had problems with the council. I attended some council meetings, but I don't recall any particular incident. The SAF had had a search committee for his successor and had identified two or three people. For some reason or another, they were not interested. So, by the time that Hardy was ready to leave, a successor had not been selected. I was there, and a possibility, not that I wanted it at all. Bernie Orell was president at the time and talked to me. I told Bernie that I wasn't interested in the position, but if I could help him out for a few months, I would do so. The council proceeded to ask me to take over. I agreed to stay for a short time. I had some other plans for the summer that I wanted to follow through.

Also, before we go further I'd like to make it clear that when I took over the acting responsibilities at SAF, I had an understanding with Bernie Orell, then the president, that he would handle RNRF-SAF relations. I was not to become involved as I was to be acting for not more than three months. I'd had no previous exposure to all the arrangements, and Bernie thought this would be the best arrangement.

HKS: Cooperstown, no doubt, was a part of your plans.

MBD: That was part of it. Mei and I also had plans to take a trip to Asia.

HKS: So Hardy now was over at the foundation.

MBD: He was over at the foundation.

HKS: There was a physical plant there. They had a building.

MBD: There was a building, newly finished. There were plans and plans and plans for subsequent developments, and he was working on those. It was obvious that there were some rough spots along the way, problems between individuals in the various members of RNRF and the Society. After I left, Bill Bromley took over for the rest of the year.

HKS: Also as acting.

MBD: As acting, yes. Bill got quite exercised over some of the arrangements that he found or arrangements that supposedly existed that he couldn't find. I don't know what he did personally. I know Bernie Orell was quite concerned over some of the problems that existed. Bernie twisted my arm hard to stay for the rest of the year, but I'd been under the gun for enough years. I decided it was time to get away from things.

World Forestry Committee

HKS: One final topic here under SAF, and that's the World Forestry Committee.

MBD: This committee is one of the oldest committees in the SAF as I recall. It was set up back in 1928 or thereabouts. Sam Dana was one of the early chairmen of this committee.

HKS: Is that right?

MBD: I went back through the records of the committee to see what they did, and there was a great diversity of activities. One of their main things in the early days that continued was hosting foreign visitors, planning their itinerary, and seeing that entertainment was provided for them. Now, there are many foreign visitors coming through Washington and visiting out at SAF.

HKS: Of course, the Forest Service picked up a lot of that slack.

MBD: Yes. The Forest Service took on the main workload of guiding foreign foresters.

HKS: They have somebody to arrange for foreign tourists.

MBD: There's been a lot of discussion within SAF about the place of the World Forestry Committee, and whether it is really needed. One thing that I should mention about the SAF was the establishment of the International Forestry Working Group. That came about directly because of the interest in the activity of the World Forestry Committee while I was serving as chairman. We decided that the international activities were of sufficient importance that we ought to have a working group at the national conventions.

HKS: You got a petition out.

MBD: And got endorsements from SAF members all around the world. The council agreed to go ahead and set up the working group. I believe today it has the largest membership of any of the working groups.

HKS: I'm sure. International forestry is really where the action is.

MBD: And they've had some pretty good programs. That's a good example of what I think the World Forestry Committee should be looking out for, where are the new organization opportunities to strengthen the society. This will not come through International Forestry Working Group. This is too big a group to focus on the policy concerns of the society. This is what the World Forestry Committee can and should do.

HKS: We lack hard data for some of the questions that are being debated now.

MBD: You know, right now the interrelationships with various international organizations don't pose any particular problems, but you look back in the years following World War II when FAO was getting organized. Henry Clepper, representing the society, used to go to the Committee on Forestry meeting at FAO and contributed to the FAO in those formative years. Same thing now with IUSF—International Union of Societies of Foresters. It has not developed as a particularly strong organization yet. But this is where the World Forestry Committee can function and effectively assist in the development of the organization. There are, I'm sure, other examples.

HKS: I knew Henry Clepper pretty well, a lot better than I knew Hardy, because Henry was so active in forest history.

MBD: Yes.

HKS: I was riding on a bus someplace with Henry. He was critical of Hardy in that Hardy spent all of his time on the SAF problems and not on outreach.

MBD: Oh.

HKS: Moving to Wildacres was a mistake. You want to be where the action is in Washington, D.C. Like you say, Henry used to go to the meetings. He saw SAF isolating itself. Would that be a fair characterization?

MBD: It's something you have to guard against. I don't think it's a good characterization, particularly now with the Metro system available. You get back and forth so easy from out there that I don't think it's isolation. With other resource organizations locating there at RNRF, there's getting to be more and more of an interchange of people in organizations. Hardy was driven by the rapid growth of the society and new responsibilities. Certainly there was more than enough to do.

International Society of Tropical Foresters (1980–)

HKS: Let's move on to the International Society of Tropical Foresters.

MBD: This was one of Tom Gill's favorites which he started back in the fifties. He organized informally a group of forestry-interested people in Latin America into what he identified then as ISTF—the International Society of Tropical Foresters. He was getting along quite well. He developed a membership of four or five hundred.

HKS: Was most of that membership from developing countries?

MBD: There were many U.S. members in the initial organization. Then as Tom traveled in Latin America, he picked up more and more members. That was the make-up of the group. It didn't reach out to Asia particularly. There were a few members scattered elsewhere, but it was primarily an American organization. Tom put out a quarterly newsletter. When he passed away, the organization just kind of folded. Charlie Larsen had a continuing interest. He agreed to take over the leadership and made considerable effort to do so, especially with the French. But the French never moved ahead.

HKS: Is Frank Wadsworth the permanent editor of the newsletter?

MBD: No. He wasn't in on the original. I think Tom and his secretary probably did the original newsletter. Frank took over in the late 1970s. During the 1970s, after Tom passed away, the ISTF organization was inactive. In the late seventies Gordon Fox and I attended a meeting at the State Department in which it seemed the main emphasis was on criticizing foresters and the forest industry for what was happening in the tropical forest. I thought this was unfair and unrealistic. A large part of tropical deforestation was coming from the natives in the type of agriculture they were practicing which resulted in deforestation.

HKS: Tell me about Gordon Fox. His name comes up throughout all the interviews. Did he have a specific job? He was involved in everything.

MBD: I've known Gordon since he was a ranger out in the Huron National Forest in Michigan, some fifty years ago. Gordon had a capacity to reach out into a lot of things. His interests were broad. As I mentioned at the beginning of this interview, he was involved in a forest cooperative on the Huron National Forest. Then he shows up during World War II down in Latin America working on, I believe, balsa wood supplies for the War Production Board. Then later he became the associate deputy to Clare Hendee. At the time we became interested in ISTF, both Gordon and I were retired and had time on our hands. This looked like something that might be worthwhile to reactivate. So Gordon and I spent quite a bit of time trying to get ISTF going again.

HKS: So he had some tropical background.

MBD: Yes. He worked in the tropics during World War II.

HKS: Was this a natural outgrowth for you, since you had an overall interest in international affairs?

MBD: Yes. I did not have and do not have any particular skills or knowledge about tropical forests in Latin America. I've been in the tropics in southeast Asia and some parts of rural Africa.

HKS: What was the primary rationalization? Why is the group necessary? What does it accomplish?

MBD: The main purpose as we saw it was to get helpful forestry information out to the various countries and to their forestry departments. Sort of a one-on-one basis in contrast to the typical international organization which has a hierarchy of several layers before you reach out to the working forester. Gordon was always very socially minded in the sense that he always wanted to help the man on the ground.

That's what we set out to do, to get organized so that we could send our periodic newsletters and other publications. By about 1979, we had our first reactivation meeting and quite a number of the charter members of the original organization were there. Charlie Larsen became president as a natural successor to Tom. He was still up in Syracuse and very much interested. The SAF agreed to give us some office space where we could have our office and use their facilities. In the first few years I spent as much time as a volunteer on SAF international activities as I did on ISTF. One thing that we decided early on was that since we didn't have any grant money, we would operate with volunteers completely. That has continued. Other than the secretarial help, it's been a volunteer organization with a staff of four or so.

HKS: So the dues pay for the secretary and the newsletters.

MBD: That's the way it goes. It's been that way thirteen years now.

From that '79-'80 period, the five hundred membership has now built up to about two thousand. There's been much interest and many people involved. We have over seventy country vice presidents and members in about one hundred ten countries. Every country where we have members, we try to have a vice president designated. That's the core of the organization. There's the board of directors with three directors representing major tropical regions and several directors at large.

HKS: But you've gone beyond the American focus of Tom Gill days to worldwide.

MBD: We have. We're worldwide. This is what Tom had in mind, I believe.

HKS: Roughly, how many tropical countries are there in the world?

MBD: I'd guess about one hundred fifty, the number keeps increasing.

HKS: Are there any really important tropical forestry countries that aren't involved in the organization?

MBD: Not that I know of. There may be a few in Africa. Africa is a difficult country to work in organizationally because of instability of governments.

HKS: Countries come and go.

MBD: That's right. Countries come and go.

HKS: When you say "tropics," is that any country between the Tropic of Cancer and the Tropic of Capricorn or is it humid tropics?

MBD: It's not limited to the humid tropics. We take in the arid areas also. We don't have any real firm lines. We have quite a few members in Europe, the Middle East, and Scandinavian countries. Many of them like to get the newsletter, because this is one way that they can get the word on activities that are going on around the world.

One of the things we've done with our two thousand membership is to publish a list every year of the members. That is one of the most widely used things we put out. It gives people contacts. If I'm going to go to Nigeria or to Indonesia, I look over that list to see who are some of the people you can write to and ask them where, what, who should I see, that sort of a thing. We started the list as a means of recruiting members, to let them know who were members and to attract others. It's worked that way, too.

HKS: It's limited to English in terms of publication?

MBD: No. Spanish also. We tried a French issue. We put out two or three issues in French, but we soon decided that the cost of a French edition was beyond our means. We would like to get back to the French edition, because the French speaking countries in Africa are some of the more important tropical forestry countries.

HKS: What is the language for Southeast Asia? English?

MBD: English is general. In Southeast Asia most anyone that's in forestry has gone to school elsewhere. Indonesia and Australia have good schools.

HKS: I don't have any idea what the other common language might be.

MBD: I don't think there is any other. The nearest I think you'd get to it would be probably Chinese, Thai, Malaysian, and Indonesian.

But I think ISTF has a place. It never will be a big organization. I don't visualize it as such unless somebody finds a way to finance how-to-do-it type of activity.

HKS: Would it be appropriate with the apparent major growth of international forestry in the Forest Service to link up with ISTF?

MBD: Maybe so, time will tell.

HKS: In a sense, you have it already with Frank Wadsworth directing the newsletter.

MBD: Right now for example ISTF has a contract for distributing publications which are available from any organization or conference. I don't know how many hundreds of publications are mailed out by ISTF. We send out a list of what we have. If you want a publication, check it off on the list and send the list back back to ISTF. Just a few weeks ago, I spent all day assembling packets of material to be mailed out. Some had checked off five or ten publications to be mailed to them.

HKS: We get some pretty heart rending letters from Nigeria. Send us anything. We don't care what it is. We need it. Just send it to us.

MBD: We get many letters "Send us everything you have on tropical forests." [both laugh] Some of the schools must have told students to write to us.

HKS: I didn't follow up on his comment, but when I was talking to Wadsworth, he said, you know, maybe I ought to retire. I can spend all my time on the newsletter. I don't know if it was doing research to get information or if he saw a need for a larger and more elaborate newsletter.

MBD: I can't speak for Frank, but I would say this, that the amount of material that is coming in for the newsletter is increasing rapidly. The more we put in the newsletter, the more we get coming in. We've talked several times about trying to put it into a more formal type of publication, a *Journal for Tropical Forestry* or some other format. I don't think we're anywhere near that in funding or organizationally. I'm sure Frank would have no trouble at all getting material to double the newsletter.

HKS: Is there anything more on this organization? You mentioned some of the key players.

MBD: I am almost inactive now. The progress that has been made in recent years in ISTF, about 90 percent of it is due to Warren Doolittle and several other volunteers.

HKS: What's his interest in the tropics?

MBD: I don't know. He became interested while in the Forest Service. He's a good organizer, a good leader, and well versed. When Gordon and I were starting the reactivation, we were looking around for some more volunteers. I mentioned to Warren one day that this might be something that he'd find of interest when he retired. He started showing up, and the more he came, the more he was interested. [laughs] And he does a tremendous job. He puts in a tremendous amount of time all on his own.

HKS: He lives in this area, so it's convenient for him to go to Wildacres.

MBD: It's convenient for him. He's been active in the SAF as president and on various committees. So he has more than the ISTF organization to interest him. But I don't want to pass up the opportunity to acknowledge the part Frank Wadsworth has in editing the newsletter. He has a tremendous background, knowing people and things that are going on, what's important and what isn't. It's been interesting. It's been a lot of fun. We've got up to what I call the second level. We made the thousand level and held that. Now we're up to the two thousand level.

HKS: With all the financial problems and so forth at SAF, are these sorts of programs in jeopardy?

MBD: I expect it's of some concern, all right.

HKS: As a matter fact, with the shrinking staff, you may have more space.

MBD: More space. We have about four or five volunteers now.

HKS: Are any of the groups that are in the Natural Resources Foundation involved in international affairs the way SAF is?

MBD: As far as I know, no. But I don't speak with authority on that.

HKS: Are they all headquartered at Wildacres?

MBD: No, they're not. Several are downtown. Most of the space is taken up now. RNRF has two buildings.

INTERNATIONAL FORESTRY (1945–92)

HKS: Okay. The final major section here is international forestry.

MBD: Yes.

HKS: You had assignments in thirty nations. It's an impressive list of countries you've traveled to.

MBD: For various and sundry reasons. One of the things that we haven't touched on in this respect is what we refer to as the Food for Peace Program (PL-480). The counterpart funds that become available from having shipped food and other supplies to various countries have financed a number of forest research projects in various countries. That started in the early 1960s. By the time that I became deputy, many of the research projects were being completed and the question was: are we going to renew them? Also some countries once eligible were no longer eligible, new countries were interested.

Many countries were interested in forestry research funded through the PL-480 Program. So one of the things early on that we decided to do was to contact various countries as we were traveling on other activities and see what the possibilities were for additional research projects. That's the kind of interest I had when I went to Poland. We contacted a number of the institutions there to review on-going projects and some new ones. Poland had several forestry research centers and did some excellent research. We were anxious to have more contacts. So I spent several weeks in Poland. Stan Krugman was with me, and he had a good background in some of the project activities. We did this same look-see activity in a number of countries in Asia and Africa.

I remember one time going down to northern Africa when I was attending an FAO forestry meeting in Rome. Later I went on the Russian science and technology exchange. I spent time in the Philippines, and then went on into Indonesia and several other Southeast Asian countries. In all, either reporting on what had been accomplished on the projects that were underway or trying to size up people and possibilities for additional work. Just as much time as one could spare was an excellent way to get some good talent involved in research subjects in which there was a mutual interest. Travel costs being what they were, just tying PL-480 exploration in with other activities was an effective way to carry on. That accounts for a considerable number of countries I listed. Then the international organizations like IUFRO, IUSF, FAO and UNESCO required frequent travel.

HKS: How about the Peace Corps? Is that of an interest to you officially?

MBD: Not especially, we didn't get involved with the Peace Corps program. Being on the executive board of IUFRO required travel and was a way to get well-acquainted with research activities by others. The reason I stopped in the Philippines, for example, was the Man-In-Biosphere Program.

MAN AND THE BIOSPHERE

HKS: Talk about Man and the Biosphere.

MBD: It was developing about the time I retired. In the Philippines they were setting up areas as biosphere reserves and wanted to get the news of others on areas they had under consideration.

HKS: Sort of natural areas?

MBD: Somewhat. There were a number of countries involved in MAB. I believe it was when I was coming back from the Russian trip. I had to stop in Thailand, so I went on over to the Philippines. One of my contacts in the State Department wanted to have me go down and look over a proposed area on Mindanao, which is one of the large islands off from Luzon. When I arrived in Manila the next morning, the forestry group took me down to a navy ship. They had a cruiser ready for us to go on. It was a very elaborate overnight affair.

HKS: Sounds that way.

MBD: I was embarrassed with the arrangements. I spent several hours looking over the proposed reserve and encouraging them to go ahead.

UNESCO

HKS: What about UNESCO?

MBD: UNESCO was centered in Paris. They were more interested in the social aspects of rural life in the various countries. And they had encouraged the MAB program. Coming back from COFAL in Rome, I stopped in Paris and spent some time with people in UNESCO, but we never really did get anyplace working with UNESCO.

HKS: Is that part of the problem when we pulled out of UNESCO, what was it, ten years ago or something?

MBD: This was before we pulled out of UNESCO. But the problem was there. Their operation just didn't fit into the research we in forestry were interested in.

HKS: We call it social forestry today?

MBD: I think probably that's about the closest you can come to what they had in mind. That was the only contact that I had with the UNESCO. Also I went to the IUFRO congress in Munich, and some years later to the congress in Oslo.

When you go on trips, you try to tie into various countries along the way. One of the big things, and you consciously have to work at it, is getting acquainted with key people and give them a feeling of support and interest. That is so important, just to be supportive of others who are struggling to get ahead with their organization and programs. I don't think we realize this enough. I've had several people from various countries say, you know, just the fact that you stopped here and showed an interest in some of our problems gave us a little extra edge in our government. It doesn't take a large amount of time or expense if you can do this while you are on other travel.

HKS: They knew you were coming, but you dropped in.

MBD: You just dropped in. They knew you were coming, and they might want to take a day or two on a field trip. But you don't have any big proposal to make. Just show an interest.

NEPAL

MBD: Another interesting session that I participated in in 1977 was in Nepal. I went there with a group of five. The group leader was Roger Revell from Harvard. There was one fellow from Sweden who was involved in small hydroelectric developments. And another one in small business. I represented forestry, and someone represented agriculture. We held a seminar for the best part of a week on the campus of the university in Katmandu discussing development possibilities.

We had half a day or more to discuss forestry objectives and programs and how to get some vegetation reestablished on the barren slopes of Nepal. It was a real interesting session. I had a chance as I always tried to do, either before or after, to get out in the country and see some of the forests. I got down into long plains— the Terai—where most of their timber is. In Nepal there was once a tremendous supply of large timber there, much of it has been cut in recent years.

HKS: That's where the tigers and stuff are.

MBD: Yes, that's where the tigers are. It was rugged going, poor roads, no places to stay or eat other than in a local village.

HKS: I'm always impressed by Nepal, it's so prominent. It seems like everyone has gone through Nepal at some point.

MBD: Part of it is just the geography of its location. Nepal sits there as a buffer zone between India and China. Both sides are interested in keeping Nepal going as a country.

HKS: But in other countries like Bhutan, you don't hear about people going on forestry assignments.

MBD: No.

INTERNATIONAL FORESTRY RESEARCH INSTITUTES

HKS: Nepal is where they go. The last topic, international forestry research institutes. We've already mentioned some. What did you have in mind, other than the one in Puerto Rico?

MBD: In agriculture there are institutes like the one in Mexico City—Cimmyt, the Wheat and Corn program, and the Rice Institute in the Philippines.

HKS: Borlaug.

MBD: Borlaug was a long time member of the staff in Cimmyt.

Viewing the substantial needs for forestry research worldwide, it's obvious that you shouldn't have a research program in every country. How do you try to put together a research effort that meets the needs of a region, several countries? It seemed to me that regional research institutes might have something to offer if we could find a mechanism for organizing and funding. Through my contact with Norm Borlaug, I went to Mexico City to see how Cimmyt was organized and funded and what might be the possibilities for something like this in forestry.

I had mentioned this at one of the IUFRO executive board meetings, something that IUFRO might get interested in and take on as a project. They didn't pick it up. It was just an idea. That's about as far as I carried it or could carry it, in the time I had left. Then Bob Buckman followed me in IUFRO and was most successful in getting interest and support. I understand several regional forestry institutes are now being developed.

CONCLUSION

HKS: Okay. Let's conclude with, any regrets, other than there wasn't enough time to do everything? Were there some failures? You tried but it didn't work?

MBD: I suppose there were. I don't know as I could identify them—or want to! You don't get all the things done that you want. Probably the most gratifying thing is to see how some of the younger men you recruited have come along and see how well they've done professionally. As far as my personal concerns, all I can say is that I feel privileged to have had the opportunity to participate over the years in the development and success of the Forest Service research program.

HKS: Of all the awards you have listed, somehow it strikes me that the Schlick Medal offers the best recognition for the breadth of the things you've done. Would you agree with that?

MBD: Yes. I was very proud when that came to me. I didn't expect it, because I had not been in top positions very long, usually I was the second or third man along the way. When you get a chance to do a few significant things, that's what you look back upon as most gratifying.

HKS: I suppose one of things you might regret in a way is language. I didn't force myself to remember the languages I learned.

MBD: Sure enough, that is something you just have to keep at all the time.

HKS: It's pretty embarrassing in this international stuff.

MBD: But there is a sign language that is universal. You learn not to let language differences be an impediment. When I worked in Italy, I learned a little Italian; and in school I learned a little German; elsewhere, I picked up a little French; and never was fluent in any language, but always enough to get around and enjoy the company of others.

HKS: Thank you for your time.

MBD: I do appreciate the opportunity to touch on some of the highlights of my many years in forestry. I only hope that what I've done has in a small way helped others and helped to advance the practice of forestry. Thank you.

❧

Robert E. Buckman

Introduction

I knew Bob Buckman only slightly when I gave a lecture at the University of Oregon College of Forestry, where he holds a faculty appointment. A longtime history buff, he invited me to his office afterward for a chat. We agreed that

a history of Forest Service research would be useful, and he subsequently helped clear the way in the Washington Office for the proposal that would yield this interview and that of two other former deputy chiefs for research, Dick Dickerman and Keith Arnold.

The interview took place in Corvallis in July 1992, in a small conference room in the Forest Service experiment station adjacent to the forestry school. We had worked together to construct the interview outline, and additionally Bob had prepared intensively. He came armed with an impressive stack of 4 x 5 cards that contained facts and figures in the same sequence as the outline. Frequently, he asked that the recorder be shut off while he reviewed his notes. Meticulous by nature, he provided carefully crafted responses to my questions. Later, as he reviewed the transcript, Bob reworked the text, line by line, until he was satisfied enough to send it back for final polish. Thus, what follows is his written narrative based upon the interview.

❧

THE EARLY YEARS

WHY FORESTRY

HAROLD K. STEEN: Bob, let's start with why forestry. I don't know what other options you thought about. I wanted to be an engineer or a geologist or a forester when I was in high school. What did you want to be?

ROBERT E. BUCKMAN: I wanted to be a forester and I knew it for a long time. I grew up in northwestern Wisconsin, a very rural part of that state, and I loved the out-of-doors. I had two role models, both uncles. One of them was a farmer, my father's brother, who would do anything to go fishing or hunting, including neglecting the farm. I was eager to be out with him because he brought a sense of excitement to the out-of-doors. The other uncle, on the maternal side of the family, had some college training at West Point. He dropped out of West Point because of the death of his father in a mining accident in northern Wisconsin. He joined the Forest Service in the CCC days. He was a technician and a general district assistant. He used to take me out on the Chequamegan National Forest to look at plantations and forestry activities. My interests in forestry crystallized when I was thirteen years old. I had to write a paper for an English class, "What Do You Want to be When You Grow Up." I still have that paper. It was then I knew I wanted to be a forester. And you know, despite all of the things that followed, I have never wavered in that interest. I see students today who can't decide what they want to do, and I reflect on how fortunate I was. However this was about 1940, and World War II was coming on, which caused some deviations in my career.

HKS: University of Illinois. What did you study there?

REB: I was in high school in the early 1940s, and sometime during my junior year I read about ASTRP, Army Specialized Training Reserve Program. It was for seventeen-year-olds and was intended to give you college training. It was quasi-military because I wore a uniform but did not receive army pay. I was so eager to get on with life that I compressed my junior and senior years and actually graduated from high school in three years. Just a few days after my seventeenth birthday I was in the ASTRP in Champagne, Illinois, in civil engineering. This was an accelerated program designed to give college training to potential military people; accelerated in the sense that we were taking twenty-two to twenty-four credits each quarter. By the time I was eighteen years old I already had two years of college behind me. The war in Europe ended in May 1945 and in Japan in August and the ASTRP program folded. I went on then into basic training at Fort McClellan, Alabama. I was then just eighteen years old. I finished basic training toward the end of 1945. I applied for officer candidate school, was accepted and completed officer training in May 1946. So I was an eighteen-year-old second lieutenant in the Corps of Engineers. I was called Junior. I went on for a little more engineering training, and then went to Germany in August 1946. I came back to the U.S. in July 1947, after spending nearly a year with the Army of Occupation, headquartered in Frankfurt.

HKS: Did you have a chance when you were in Europe to look at any of the forests?

REB: A little, yes. I used to hunt in the forests surrounding Frankfurt, usually by myself because no one else was available who liked to hunt. I had a jeep and so could prowl around the countryside and hunt and fish, but never very successfully. Europe was a very grim and gray place in those days.

HKS: I remember the newsreels, all the bombed-out cities.

REB: Reconstruction hadn't started. Late in my tour the Nuremburg trials ended and the executions of Nazi war criminals took place. Ludwig Erhart became the economic minister of Germany and devalued the currency. The next day the shops were filled with goods. That was the beginning of the resurrection of Germany.

HKS: You were in Frankfurt, you were close to Carl Schenck. There were other GIs that saw him after the war. But you may not have known about him...

REB: I didn't know enough about forestry at that time. I saw the forests but didn't understand forestry principles or the foresters who contributed to the practices.

HKS: He was in Darmstat, which is not very far south of Frankfurt.

FORESTRY EDUCATION

REB: The irony is that it was twenty-five years before I returned to Europe. I've probably been back a dozen or more times in the last ten years. I've seen the forests in much greater detail in these later visits. In 1947 I was discharged from the Army and promptly went back to the University of Illinois to convert my preliminary engineering training into forestry. Illinois had a two-year forestry program at that time. I completed that in one semester. Since Illinois didn't have a four-year program, I had to go somewhere else. I had my heart set on the University of Idaho in Moscow. I got there about two weeks too late to register for the courses required to advance my degree. It was with a lot of disappointment that I just couldn't afford to spend more time at the University of Idaho. In March 1948, I went to the University of Minnesota, which is close to my home in northern Wisconsin. As it turned out, this was a most salubrious choice.

HKS: Was Henry Schmitz dean there?

REB: No, Frank Kaufert was.

HKS: Okay.

REB: Frank, I think, had become dean just a short time before. But that first discussion with Frank Kaufert was only one of many with him over the next thirty years.

HKS: Frank was president of the Forest History Society.

REB: Yes. And you know what a warm person he was, and how supportive he could be. I experienced that relationship with him up to and including the time I was deputy chief.

HKS: Yes. We always thought that Frank and George Garratt were the last two deans that had authority over faculty.

REB: I'm not sure that Frank thought himself as an authoritarian. [laughter] Maybe the faculty did, although I doubt it. Frank Kaufert was very supportive of me. He must have thought that I had some small capacity to do things. He arranged for me to get the Minnesota and Ontario Paper Company Scholarship in 1950. He gave me a teaching appointment and offered my name later for job references. I think that I had two or three job offers from various universities because of his interests.

HKS: At that time were you thinking academic as opposed to the Forest Service?

REB: It was still an option. We'll get to that in a minute. In any event I entered Minnesota in 1948 and completed my bachelors degree in March of 1950. That was a competitive environment because the GIs were back, mature and eager to get on with life. It was a great but arduous time. I neglected one point, a mistake in my career. When I came back to the U.S. from Germany in 1947 I thought to myself if we ever have another war we're all going to be in it again anyway, so I signed up for the inactive military reserve in order to maintain my commission.

The Korean War

HKS: Okay, that's a mistake.

REB: Back to the University of Minnesota. I completed my bachelors degree in March 1950. Kaufert arranged a Minnesota and Ontario Paper Company Scholarship for me. I was embarking on that program when the Korean War started in June. Can you see where this story is leading? I was recalled as a filler officer in an understaffed Mississippi National Guard battalion stationed in Camp McCoy, Wisconsin.

HKS: Wow.

REB: Shortly after I was recalled, Marie (née Eidenshinck) and I were married. I met Marie in Minneapolis, where we both worked, she full-time and I part-time while at the university. Marie's home was Detroit Lakes, Minnesota, at the border between the Great Plains to the west and the forests and lakes to the east. We return there often, where Marie's mother is now in her nineties.

Back to Camp McCoy. The core Mississippi battalion was a redneck, racist group. In fact, the battalion commander was a brother of the then governor of Mississippi, who had been the vice-presidential candidate on the States Rights ticket in 1948. Now the racial implications for me are a separate story.

HKS: Truman had desegregated the armed forces, officially.

REB: But that didn't touch the heart and soul of those Mississippi boys.

HKS: I'm sure.

REB: Anyway, there were actually some race disturbances at Camp McCoy. By this time, the battalion commander was sufficiently aware of my liberal racial views and he peddled me to a newly forming regular army engineer combat battalion. Another happy event at an otherwise unhappy time. My new regular army battalion commander was aware that my wife and I were expecting our first child. He said he would arrange for me to stay in the U.S. for one more month. At the end of that month I no longer had enough time remaining (two years was the maximum time for recalled military) to go to Korea, so I avoided it but just by a whisker. Thus I finished my second military assignment in the spring of 1952; both tours combined took nearly five years out of my forestry career. I went back to Minnesota and finished that master's degree over the next twelve months. During that time Steve Spurr was at Minnesota, and he was to have a significant impact on my career.

RETURN TO THE UNIVERSITY OF MINNESOTA

HKS: I'm sure.

REB: I worked for Steve in Itasca State Park, Minnesota, that summer of 1952. My work included fire ecology, but Steve, you know, was also something of a mensurationist and a photogrammetrist.

HKS: When I was an undergraduate I used his photogrammetry textbook and I thought that was what he was. I was surprised when I saw other books come out later.

REB: His interests were eclectic.

HKS: Yes.

REB: But he had a quantitative bent and an ecological one, and he was a stimulating guy to be around. In any event, I worked under Steve that summer at Itasca State Park, and he stimulated my interest and curiosity. I completed my master's degree in 1953, and by that time Steve had gone to the University of Michigan. Then in the spring of 1953 it was a matter of deciding where I'd go. I was exploring some academic appointments with Frank Kaufert's help.

HKS: Dana was still dean at Michigan at that time.

REB: Yes.

HKS: Close to retirement but...

Northern Rocky Mountain Experiment Station

REB: Very close to retirement, and Fontana was Sam Dana's replacement. So in the spring of 1953 there weren't all that many jobs. I talked with both the Forest Service and the academic community. Universities really weren't interested in a young person with only a master's degree. but the Forest Service was. My appointment in 1953 was with the Northern Rocky Mountain Forest and Range Experiment Station in Missoula, Montana. I joined Forest Survey. George Jemison was the director of the station.

HKS: Alright.

REB: And the assistant director I worked under was Harry W. Camp. Dick Dickerman had just vacated the position that Harry Camp occupied. In many respects my later career was entwined with all three (Jemison, Camp, Dickerman). My work with the Northern Rocky Mountain Station was in Forest Survey and that meant exposure to a lot of country; we were taking inventory plots at four-mile intervals across the country. First, in the panhandle of Idaho and northwestern Montana and later in the Stanley Basin of southern and central Idaho. It was a marvelous experience, but it is one that you don't want to repeat for too many years. Then another thing occurred. There were rumors floating around, this was during the Eisenhower administration, that the Northern Rocky Mountain Station might soon be closed. The rumors became more and more persistent, and one day George Jemison and Reed Bailey, who was then director of the Intermountain Station in Ogden, Utah, called the whole station staff into an office and announced the termination of the Northern Rocky Mountain Station.

HKS: Is that because of the economy, budget cuts, was that why it was terminated?

REB: No, I think it was part of Eisenhower's streamlining of government. I don't know all of the details. In any event, George Jemison announced that he was going to become director of the Pacific Southwest Station, Reed Bailey would remain the director of the consolidated Intermountain Station. For me personally it would have meant a transfer to Ogden, Utah, where I would continue with Forest Survey.

I still had a year or so left on the GI Bill. I wrote Frank Kaufert and said that I had decided to go on for a Ph.D. at the University of Michigan under Steve Spurr. However, there were about six months remaining before the fall semester started at Ann Arbor, and I asked was there something I could do at the University of Minnesota in St. Paul. I got a letter back from Frank giving me a teaching assistantship. While at Minnesota, I also completed French, one of the two languages required for a Ph.D. A tutor located near the campus at the University of Minnesota guaranteed 95 percent success for his French language students after only fifteen days of instruction. But it required absolutely total immersion; the tutor would badger, harass, and intimidate people. I passed the exam.

UNIVERSITY OF MICHIGAN

REB: So, I went back to the University of Minnesota in March 1954, and spent six months then moved on to Ann Arbor, Michigan, where I joined Steve Spurr. With Spurr's help I was awarded a Rackham School Scholarship. The GI Bill and the scholarship permitted me to be a full-time student. That, plus our two children, made 1954-1955 a marvelous time. My graduate committee consisted of Steve Spurr, Sam Graham, Ken Davis, Bob Gregory, and John Caron, all distinguished teachers and scholars. Steve Spurr was outstanding in several fields. Ken Davis was an authority on forest management, Bob Gregory was an economist with strong international connections, Sam Graham was an outstanding entomologist, and John Caron a mensurationist. It was a marvelous committee. I completed my residence requirements for a Ph.D. at Ann Arbor rather quickly, in nine months.

HKS: What was your specialty?

REB: Ecology and silviculture with a minor in the quantitative sciences, statistics, and mathematics.

HKS: Because your work in Forest Survey was economic or statistical.

REB: Actually, Forest Survey for the most part was grunt work, it was climbing mountains and measuring trees. During the winter months when the survey job was mainly office work, I did some of that mensurational work. But, ecology, silviculture, and quantitative sciences background were my specialities at the University of Michigan. I completed German in a couple of months while there. With the family we decided that we couldn't stay in Ann Arbor to finish the Ph.D. dissertation. About that time I received two offers from the Forest Service to come back. One of them was at Grand Rapids, Minnesota, and the other one was at Cordele, Georgia. F. H. "Windy" Eyre was the guy who was orchestrating this, that grand old man of forestry. For the family it was an easy, easy decision. Marie's family comes from western Minnesota and I came from northwestern Wisconsin. By then there were two grandchildren and four grandparents and, you know, it was just going back home again. And so we moved to Grand Rapids, Minnesota, with the understanding that I would do my dissertation research as part of my Forest Service assignment.

HKS: That was not that uncommon then, that you could do your dissertation as part of your assignment.

LAKE STATES FOREST EXPERIMENT STATION

GRAND RAPIDS, MINNESOTA

REB: Yes. And many weekends, holidays, and non-work hours were included, which was part of the job. We arrived in Grand Rapids, Minnesota, in the summer of 1955, and my boss at that time was Zigmund A. Zasada. He was one of the several people who had a significant impact on my career, and I'll talk about him and the others a bit later. In any event, Zig was the research center leader at Grand Rapids. He came out of the National Forest System—a quiet, low-keyed, unpretentious guy who could be just as stimulating and challenging as any person I worked for.

Zig gave me one major assignment. I was to look after the upland forest research. We had wetlands and swamp research and we had entomology and economics, but he gave me the responsibilities for looking after mainly pine and aspen research. And that's what turned into my Ph.D. research. I worked on growth and yield of red pine. I collected all of the permanent plot information I could find, primarily from Minnesota, but also Wisconsin and Michigan. I wanted to put that information together in a new approach to growth and yield forecasting for red pine. I also used some of Spurr's mensuration work. He developed a concept called the volume line which accurately predicted the volume of a tree or stand if you knew the basal area and height.

HKS: Was the concept of a normal stand still acceptable at that time?

REB: Yes, they were used, but they assumed fully stocked stands, which was not often a realistic assumption. I wanted to bring into the research the concept of variable densities. And, I wanted to bring some other things in too, like thinning methods; that is the removal of the largest or the smallest trees or some variation thereof. Furthermore, I wanted to treat growth as a differential equation. Now I'm into the mathematics part of my background. I wanted to treat growth as a differential equation and yield as integration of that equation.

HKS: What was that going to show you that we didn't know already? That was obviously a new way of looking at it.

REB: The growth equations did in fact deal with variable densities, and that was, I think, fairly new. That was the next generation of work after the normal yield tables. But the idea of integrating, that is, summing up those growth increments mathematically meant that you could track any one of a thousand varieties of management regimes. Frequent thinnings, light thinnings, heavy thinnings, variable thinnings, and so forth, and you could track them through time.

HKS: Was this based on real stands or hypothetical stands?

REB: It was based on growth plots in real stands. However, a good deal of the information that I had was imperfect, and that led to a philosophical difference with Zig about which I will comment shortly. In any event, that research turned into my doctoral dissertation and it was I think, far and away the most significant bit of research that I did. At that time it received a fair amount of attention. It was published as a USDA technical bulletin. But it was the methodology, not the growth forecasting, that many researchers followed. Interestingly enough, a colleague, Al Lundgren, who stayed in the Lake States, tracked red pine growth and yield for another twenty years. Al fed independent sets of information into those forecasting equations. They turned out to be remarkably good predictors, which was as much luck as good science.

HKS: And you did all that with an adding machine or a calculator.

REB: Essentially all of it except for development of the prediction equations themselves with the first computers, an IBM-650. But the point I wanted to make here is that those equations turned out to be just remarkably good predictors of independent sets of data. But I was lucky. Statistics and math don't serve all that well with highly variable field plot data.

HKS: Did that mathematical model work for something other than red pine?

REB: People used it for other tree species, but that was only a stop-gap measure until individual species equations could be developed. I ran into red pine equations being used for Sitka spruce in southeast Alaska. But we can come to that when we talk about the Pacific Northwest Station. In any event, the growth and yield research worked out reasonably well, and it attracted some attention from others such as Carl Ostrom and Dick Dickerman and maybe even Les Harper. I think it was that work that really tilted me toward the Washington office of the Forest Service, although I didn't realize it at the time.

HKS: I see.

STATISTICS IN THE FOREST SERVICE

REB: Now I want to touch upon what I consider an intergenerational question in science, and it involves Zig Zasada. Zig came out of the National Forest System, and he was (and is) a remarkably insightful guy but with little formal training in science. His insights involved good judgment and intuition. Many of those growth plots that I used were just terribly inadequate in terms of statistical design. My concerns all came to a head with a famous old red pine plantation near Ely, Minnesota, called the Birch Lake plantation, about sixty acres of red pine that had been planted in 1918 or thereabouts. The trees in 1957 were eight or ten inches in diameter and sixty to seventy feet tall. The question at that time was what were we going to do with the Birch Lake plantation? I said let's install a well-designed stand density and thinning methods study. Zig said fine. So I laid out the experiment. I wanted it to be a very contrasty experiment; that is, very low densities and very high densities and several densities in between. I insisted that the study be replicated and that the treatments be assigned randomly. This is where the conflict occurred. Zig said, and representatives of the Superior National Forest agreed with him, that's all well and good except we don't want those low densities next to a road because we know that they're going to blow down or collapse in snow storms. I insisted that we observe all the principles of the experimental design. I did so because I was so uncomfortable with some of the permanent plot information I used in the red pine work that didn't observe those principles.

HKS: Did you ever deal with Les Harper? In our interview he talked about the introduction of regular statistical analysis in the '50s, and you were part of that, apparently.

REB: Yes, I think I was. But let me come back to your question.

HKS: But it wasn't typical of Forest Service research, it was more measuring and describing.

REB: My concern about this issue goes back to my undergraduate and graduate studies in statistical methods. I insisted that experiments be contrasty, that the treatments be assigned randomly, that every treatment have an equal chance of selection. The conflict in the Birch Lake plantation was one of visual effects of low-density treatment along a road. I knew that those heavily thinned plots were vulnerable to wind and snow, and as it turned out one of two of them were severely damaged by snow. But I insisted. This was a matter of principle for me. Art Greeley and Dick Dickerman, I think, knew about this boiling point, because they came to Ely and they visited the plantation. They didn't really talk to me about it but it was...

HKS: What was Art doing there?

REB: He was the regional forester in Milwaukee.

HKS: Okay.

REB: And Dick was the station director.

HKS: Right.

REB: I don't recall that I talked to Dick about it, but for me it was a matter of principle. I wasn't sleeping at night, this was such a major issue for me. But do you know that at the end of that time Zig said, Okay, we're going to do it your way. It wasn't a hostile response; it was that you made your case. Zig Zasada drew out of a hat the random assignments of the treatments. Zig is now in his eighties and still lives in northern Minnesota. His only child, his son John, chose forestry research as his career and has an office with the Pacific Northwest Station here in Corvallis.

HKS: How about that. Okay, Zig always challenged you.

REB: Zig had the capacity to challenge the dickens out of you. Often times he was right, but he also had the capacity, after extended discussion, to yield, and to yield gracefully. Zig and I have visited often over the intervening thirty-five years, sometimes rehashing the Birch Lake plantation issue. It was an important matter for me personally. I would not have stayed with the Forest Service had the outcome been different.

HKS: How did he get into research? I mean, he represents an earlier generation.

REB: Yes he did. The time that we're describing, which was in the late '50s and early '60s represented the arrival of the next generation of researchers. People with graduate training, Ph.D. training, meeting up with people who were recruited by and large out of the National Forest System to start a research program, to organize it, and to establish community relationships. It was exactly because research center leaders did those jobs that younger scientists could turn to modern research. I don't know whether you have ever heard that old shibboleth, if you need statistical design to prove a point, it probably isn't worth proving. But, that statement in my mind characterized those intergenerational problems.

I want to insert a point here. My recollection is that sometime in the 1930s there was a very small group of Forest Service researchers who recruited R. A. Fisher, Sir Ronald Fisher, of the Rothamstead Experiment Station in the U.K., to come to the U.S. Fisher, and that small Forest Service group, apparently created an interest in experimental design and statistics. I say that because while most people in the Lake States Station had only an indifferent appreciation of statistics, one of the experiments I worked on, a jack pine thinning study installed in 1940 near Aurora, Minnesota, portrayed all the principles of experimental design. I know it goes back to that Sir Ronald Fisher's visit and several Forest Service people: Ted Osborne of the WO, Tommy Evans, SE, Roy Chapman, SO, and Al Bickford of NE. There's a small chapter in Forest Service research that really needs exploring, concerning the origins of statistical sensitivities and experimental design. It had much to do with improving the quality of research in the Forest Service.

HKS: If I remember correctly, when I interviewed Dave Smith at Yale a couple of years ago he talked about Fisher's visit. Fisher on a one to one basis or in front of a group was terrible. He literally turned his back to the audience and wrote on the blackboard, and he was boring; but his writing was so influential. I used Fisher as a textbook in the '60s; he was still influential then.

REB: Ted Osborne, who was in the WO when my career started, gave national leadership to the program. He was followed by Tom Evans and then George Furnivall, now at Yale. The point that I want to make here is that there were small beginnings of good solid statistical design even in the 1930s and early 1940s. Back to those Lake States days, there were conflicts between the old and the new, and Zasada was good at challenging young scientists and very gracious about yielding. I later put in several additional density experiments in red pine, white pine, jack pine and aspen, during my Grand Rapids years, that I think they would stand the test of modern day statistical and experimental design.

STRENGTHENING FORESTRY RESEARCH

REB: Les Harper had a profound impact on research programs at that time. Remember that I started in the middle '50s and Les was beginning to implement some of his ideas, and we felt that in the field. Dickerman and Zasada wanted to build a laboratory at Grand Rapids, Minnesota. At that time, we were working in an old beauty parlor above a hardware store. That's the way it was over much of the country. Harper and Jemison began to equate a laboratory construction program as a necessary adjunct to a growing research program. George Jemison deserves a great deal of credit for this. He wrote a paper called "Get Scientists Out of the Woodshed," which was the beginning. So one of the early laboratories in the Lake States Station, it may have been the very first one, was at Grand Rapids. It was dedicated in 1960. Zig was very active, I'm sure with Dickerman's encouragement, to mobilize political support for that laboratory. It was done and it was successful.

HKS: Do you think this is compatible with the Eisenhower philosophy that the role of the government is to assist and to help industry and so forth? Research would fit into that, but regulation and so forth would not. I mean the Forest Service made some big shifts in the '50s in terms of ways it viewed its forestry role.

REB: We need to keep in mind that there was a congruence of events that really favored research at that time. This was the time of Sputnik, when the U.S. felt terribly inadequate scientifically. It was also a time when the Forest Service, with a lot of help from people like Briegleb and Harold Mitchell in the South, were beginning to work directly with constituent groups regionally, to enlist congressional support.

HKS: Which was technically illegal...

REB: If you follow some of these histories, both the written ones and the oral ones, you'll see a lot of euphemisms, and I'll probably use some as well. But it was lobbying, and it was lobbying sometimes in violation of the Hatch Act, which says that federal funds will not be used to influence legislation, but it was done. And it was oftentimes done with a great deal of encouragement from members of congress and from constituent groups.

HKS: Would the Hatch Act have allowed Senator Humphrey to invite foresters out on a show-me trip?

REB: Absolutely.

HKS: If it's initiated from Congress, it's okay.

REB: Yes. If Congress or the administration gave even a pretense of legitimacy, if they requested information, it was not in violation. Frequently things were, by mutual consent, manipulated to do just that. That movement toward research centers and working with local constituent groups really began shortly after World War II. It came out of the Southern and the Southeastern stations. So, Zasada and Dickerman were using that same approach to help build programs, and I'm sure that Harper was very much encouraging it.

Back to my original point about strengthening research programs. There was a congruence of events in the 1950s. I don't think Eisenhower had anything to do with it. It was a sensitivity to interests of local groups, a decentralization of the research programs into regional experiment stations and satellite laboratories. There's still one more thing that made it all come together and that was the fact that Les Harper was an extremely astute mobilizer of programs, people, and events to make things come together. Harper was very comfortable with a number of key senators such as Stennis of Mississippi, Hayden of Arizona, and Russell of Georgia. The climate for accelerated research was favorable. Harper developed a program for forestry research that gave a background and legitimacy to the expanding research program. But Harper himself was also a key.

DEDICATION OF GRAND RAPIDS LABORATORY

REB: In any event, the Grand Rapids laboratory was dedicated in 1960, and the dedication ceremony included Chief Richard McArdle. Senator Hubert Humphrey was also invited. Humphrey was then campaigning around the state. It was my job at the dedication to be away from the platform where the ceremonies were taking place and to welcome guests. Humphrey was late. I looked down the road and I saw a big black Buick that I recognized; the driver was George Parshall, a technician in blue bibbed overalls then with the state of Minnesota forest service. It turns out that Humphrey had radioed the local airport, and the state forest service had sent George Parshall out to pick him up. As the car came up the road I met them, and said to Senator Humphrey that I would escort him to the platform. Humphrey was in a vigorous conversation with George Parshall, a warm and animated conversation. The reason I mention this is that my admiration for Senator Humphrey soared, because it was apparent that he was a warm and thoughtful person, and as you know, had a profound influence on forestry.

When the dedication ceremony was completed, McArdle and Humphrey walked through the laboratory, and Humphrey asked McArdle, "How are things going, Mac?" McArdle said, "Senator, I've got troubles. The Multiple Use-Sustained Yield Bill is locked up in committee." Apparently they talked about it a bit more, and Humphrey said, "I'll see what I can do." Dickerman and McArdle then left the ceremony and drove back to the Twin Cities. McArdle stopped along the way to call his office; I think it was Ed Crafts that he was calling to ask where the Multiple Use-Sustained Yield Bill stood. Finally, after a number of calls, Ed Crafts replied that it had broken out of committee.

HKS: Didn't Humphrey introduce the Multiple Use Bill?

REB: He may have. But it became stalled in committee, and it was that dedication event that broke it free, one of the many ways some of these forestry and political issues get resolved. I was to observe many variations on this theme in the years ahead.

Appointment as Project Leader

HKS: Given the way your career developed, do you think you were a typical research scientist at the time? Did you have broader interests, were you looking left and right rather than just going to your laboratory?

REB: Maybe that's right, but I didn't plan it that way, nor do I trust my objectivity on that question. The ten years at Grand Rapids were most rewarding and satisfying to me. But maybe, just maybe, I had some small instincts for interorganization and interpersonal relationships that not all scientists have. I didn't plan it that way, but maybe there were a few things that took place that reinforced that impression on those who influence career pathways.

The dedication of the Grand Rapids laboratory took place in 1960, and Harper and Jemison were moving ahead on that construction program. Zig was a proven commodity because he had mobilized the support for the Grand Rapids laboratory. So Zig was invited to join Harper's staff, and it was a mystery about who was going to take his place. We were at a farewell party for Zig Zasada about two days before he was to fly to Washington, and none of us, the nine or ten researchers at the laboratory, was bold enough to ask him who was going to be his successor.

Frankly, I did at the going-away party. I said who's going to be your successor, Zig? He replied that I was. That was two days before he departed for Washington. That came as a great surprise to me because I was doing research and I was very happy.

HKS: All your time at the Lake States was in Grand Rapids?

REB: All of it, except that I was in proximity to the station in St. Paul as a student at the University of Minnesota in the late 1940s and early 1950s. So I knew a little bit about the station.

FROM RESEARCH CENTERS TO PROJECT RESEARCH

REB: I want to use that juncture, the dedication of the laboratory and the departure of Zig, to mention the first of three internal reorganizations that I witnessed in Forest Service research. This one goes back to the McKenzie Report in the mid-1950s. It was a study of the organization of forestry research. At that time we were organized into research centers, where the center leader would be located in a satellite laboratory to the main station. The center leader was responsible for everything—community relations, science, everything. The McKenzie Report, as I recall, recommended a continuation of the research centers. But Les Harper was not comfortable with the report findings. Les and I have exchanged correspondence about the event. He felt that there was too much overhead, that we weren't really emphasizing research quality. He wanted to go to a concept called the project organization. His views prevailed. The departure of Zig from Grand Rapids coincided with the organizational change at Grand Rapids from a center to a project concept. I became one of four project leaders at Grand Rapids. My project was the largest—six scientists, including myself. My responsibilities were silviculture of upland pines and aspen. I was also the director's representative, which meant I was the unofficial chairman among the four project leaders to resolve internal and external problems for the Grand Rapids laboratory.

There was lots of discussion in those late '50s and early '60s among scientists about how we were going to reorganize. I was in favor of the reorganization because my background and training leaned heavily toward science. I could embrace that project leader's job enthusiastically, because I could continue my research. That concept was adopted all over the country, and I consider it to be a significant juncture affecting the quality of Forest Service research nationwide. We went from a research center to a project concept. It streamlined administration and put a lot more emphasis on science. Coincident with that was the development of the Man-in-Job concept in the Agricultural Research Service. Les Harper saw the significance of the concept and adopted it, which meant that a researcher's career was dependent on what he produced, not on his organizational position. That just has to be another one of those major milestones that upgraded the quality of Forest Service research.

HKS: Getting that through civil service and all the other bureaucracy must have been quite a battle.

REB: Oh, it probably was. The Agricultural Research Service deserves a great deal of credit for generating what is now called the Person-in-Job concept.

HKS: I understand.

REB: I came to realize later that the Person-in-Job complex complicated the life of an administrator because it provided a two-track career ladder for scientists. Sometimes that situation made it difficult to recruit people into research administration who were also doing well as scientists. If the two-track career system had come along a few years earlier it might have posed a dilemma for me too, because I liked doing science.

HKS: Was it used—I'm not sure how to characterize this—as a place to put some of the master's level senior scientists at that time, because they really weren't very good scientists by the new standards?

REB: Some of that happened. However, still another innovation of the Harper/Jemison era was the Government Employees Training Act (GETA). Harper and Jemison and all of their successors, including me, very much encouraged the stations and the projects to take advantage of GETA. So if people came in with master's degrees and displayed an interest and a capacity for science, it was very easy to encourage them to go on for a Ph.D. Many scientists did that in the '60s and '70s. This is another one of those Harper/Jemison areas of emphases. I was so interested in these policy developments that I wrote a paper on them while at Harvard in 1968-69. I sent you a copy of that paper.

HKS: Yes, I was going to ask you about that later.

REB: In it I recount a good many of those things that occurred at that time. Back to the early 1960s. I became a project leader as we moved into that new organization concept; those were some of the best years of my life. I had more control over budgets, equipment, facilities, and technicians. My research productivity wasn't all that bad. I was writing four to six papers a year, and I had that USDA technical bulletin on growth and yield. I was also doing fire ecology and prescribed burning research. I never sought to leave Grand Rapids, but I said that if I were to leave, I wanted to do something very much differently, I didn't want to go someplace else and do growth and yield and fire and fire ecology. I wanted to do something entirely different. And I was beginning to get inquiries about a change in jobs.

FIRE RESEARCH

HKS: Fire. Do you want to comment at all on Ashley Schiff's allegation—his book came out about 1964—that the Forest Service administration was, if not censoring, at least controlling release of research data during the '30s that showed that fire was good. What was your feeling at the time? Was it controversial? Were people mad or did they shrug it off?

REB: I visited with Schiff who, in my estimation, was doing research in the tradition of Harvard, kind of a polemical, iconoclastic approach with sensationalism built into it. I think there were some ingredients of truth in what Schiff was saying, but I really think he made a caricature out of what was really a relatively minor problem. My view on this is that the first step in forest conservation, beyond the establishment of the national forests, was to protect forests from fire. Forest fires were the major cause of forest loss nationwide. We had to get that under control, especially in the southern United States, where arson and agriculture and all of those things were a way of life. It was the Weeks Act of 1911 and the Clarke-McNary Act of 1924, and the creation of state forestry organizations, that were absolutely essential. I tend to be more charitable about the role of state foresters and Forest Service people at that time than as described by Schiff. They had an enormous educational problem, and they were terribly concerned about sending mixed signals to people. It didn't take very long before fire, prescribed fire and controlled fires, became a way of life in the South. Schiff never gave any credit to that.

HKS: He was tempted, I am sure, by having someone as quotable as H. H. Chapman. Chapman let it all hang out. He was one of the people who believed in prescribed burning in the '30s, so he was one of the antagonists in the book.

REB: H. H. Chapman could be pungent. He had a lot to do with silvicultural practices on the newly created Chippewa National Forest, which did not come out of Indian lands. Pinchot was a visitor to that area, by the way. The Chippewa National Forest was different than most, very special. Chapman's contribution had to do with leaving first, 5 percent of the old growth pine; later, 10 percent, as seed trees. Some of those reserved trees are still standing.

HKS: I distracted you from Ashley Schiff.

REB: I think that Ashley Schiff overemphasized, somewhat unfairly, his point. I came to realize that later as I worked under the same professor as did Schiff (Professor Arthur Maass) at Harvard, Maass created an aura of sensationalism that often made a caricature of an issue.

HKS: *Muddy Waters?*

REB: That's exactly right, the same Arthur Maass.

HKS: A very dull book to read, I thought.

REB: I didn't read *Muddy Waters* but I took a course under him. I don't think he ever made the case for water as he attempted to do for fire. One of the interesting points about that book and your question, however, is that the Smokey Bear syndrome continues to come back over and over again. Smokey Bear becomes the enemy because he is perceived to stand in the way of controlled or prescribed use of fire.

HKS: The Yellowstone certainly was on the nightly news.

REB: The problem is not Smokey Bear; the problem is the operational difficulty that goes with the use of fire. If we would quit tying a can to Smokey Bear's tail, we might get at some of those operational questions. For example, the risks and the rewards system, the narrow weather windows that we have for the use of fire. The sanctions that go with the maladroit use of fire are far greater than the rewards that go with the proper use of it.

Back to those formative years at Grand Rapids. If I changed jobs I wanted to make a big change. In the meantime (I'm just guessing that Dickerman engineered some of this), Dick encouraged Harper to visit the Lake States Station. Dick very carefully arranged for Harper to meet what he considered to be some of the more promising people in that station, especially during a canoe trip in the Boundary Waters Canoe Area. Bob Lucas, who became a wilderness researcher, was one of those people; Roger Bay, who became the director of the Intermountain and of the Pacific Southwest Station, was another and I was involved also. And Carl Ostrom, director of timber management research in the WO, came out and was also to have a major influence on my career. Carl visited field experiments and I'd talk about experimental design, contrasty treatments, the creation of response surfaces and research methodology. I think maybe that caught his attention as did the publication on red pine growth and yield. In any event the invitation to Washington did occur, and I went there as branch chief of mensuration under Carl Ostrom.

McIntire-Stennis Act

HKS: McIntire-Stennis was enacted while you were at the Lake States. Was this controversial, were you waiting for it to happen, or did it sort of ease in and you learned about it and started making use of it?

REB: The McIntire-Stennis Bill was peripheral to my interests at that time. But my contacts with Frank Kaufert were sufficiently close that I knew that he was one of the major progenitors of that act. He was working with Professor Westvelt, who was at Missouri. There were two Westvelts, one in Missouri and the other in the Northeast.

Frank Kaufert was a major shaker and mover in the enactment of McIntire-Stennis Act. Frank was also working with Bill Cummings, formerly with TVA, on a study on forestry research needs in the United States. Sponsored by the Society of American Foresters, Kaufert and Cummings came out with a book in the mid-'50s having to do with forestry research. I'm reasonably sure that the SAF study had much to do with Kaufert's interest in what became the McIntire-Stennis Act of 1962.

If you read the Harper comments on the formation of the McIntire-Stennis Bill in some of that correspondence that I sent to you, you may recall that the final hang-up in the McIntire-Stennis Bill was whether the program would be administered by the Forest Service or by an independent agency. That agency today is the Cooperative State Research Service (CSRS). In other words, McIntire-Stennis would be administrated separately from the Forest Service. An industry group very late in the congressional deliberations insisted that McIntire-Stennis be administered outside the Forest Service.

HKS: Why was that?

REB: I suspect that it had a lot to do with distrust of government. But you also know better than I that there was a lot of hostility to the Pinchot philosophy that carried way up until the 1950s about federal regulation of private lands. I suspect that hostility was also involved in the McIntire-Stennis Act that was in the *Journal of Forestry*. Harper mentioned to me in one of our informal exchanges that he had been mentioned in the article, but he insisted that his name be removed. Harper indicated to me that he was a far greater contributor to the passage of McIntire-Stennis than he'd given credit for because he insisted that his role be downplayed. I don't know whether Harper wanted it in the Forest Service or as an independent program. My own view, developed in later years, was that the Forest Service was fortunate indeed that McIntire-Stennis was administered independently of the agency. It made for much more productive and fruitful working relationships between the Forest Service and the forestry schools.

HKS: That's something that I wanted to go on into, that is the need to coordinate Forest Service and university research. This laboratory of the experiment station (Corvallis, Oregon) was built in 1960. That's two years before McIntire-Stennis, so there's already obviously cooperation with universities.

REB: Yes.

HKS: But this law made a mechanism for what? For funding? Or for projects?

REB: It authorized funding for forestry research and encouraged cooperation.

RESEARCH AND POLICY CONFLICTS

REB: One of the issues of the late '50s and '60s was the Boundary Waters Canoe Area (BWCA). That was one of the first hotbeds of the wilderness issue. The BWCA was set up under special legislation. One of my colleagues at that time was Myron L. "Bud" Heinselman.

HKS: The name is somehow familiar.

REB: Bud Heinselman was a talented ecologist, did some marvelous work in fire ecology and in peatlands ecology.

HKS: Is this the area that Truman by executive order prohibited over-flights?

REB: Exactly.

HKS: That's pretty early.

REB: Yes, and this was in the late '50s and early '60s. Bud Heinselman was a lifelong user of the BWCA and was very much environmentally oriented. The BWCA was set up in such a way that the canoeing, the water based recreation, was partially screened from timber harvesting that was going on beyond the buffer zones. There were lots of people, including me, who really thought the BWCA was a treasure that ought to be protected. The timber harvest there was heavily subsidized. It was a time when the Forest Service might get fifty dollars stumpage per acre for the jack pine but would pay one hundred dollars per acre to reproduce the forest.

HKS: So reproduction was the problem, it wasn't the harvesting costs, road building and so forth.

REB: Oh, they played a role too, but the main problem was that regeneration didn't come easily.

HKS: Aspen would take over or what?

REB: Aspen and shrubs would take over. Red pine was the preferred species for reforestation. Jack pine was acceptable but it was very short-lived and commanded lower stumpage prices. And in many respects it was an argument about below-cost timber sales not different from the ones I heard thirty years later. The point that I want to make is that instinctively, philosophically, intuitively, Heinselman wanted the BWCA enlarged. That was my first encounter with conflicts between Research and the National Forest System.

HKS: I see.

REB: Heinselman was considered the villain in this relationship. The supervisor of the Superior National Forest and the regional forester and other Forest Service and industry people really had an antipathy toward Bud because he was violating Forest Service policy.

HKS: But they weren't opposed to the idea of enlarged...

REB: Yes they were.

HKS: Oh, that too.

REB: At that time, by and large, they felt that the Forest Service was on a reasonable course of multiple use where recreationists could use the water and industry could use the timber. That situation made things very difficult for me, but even more difficult for Heinselman because he was using all of his free time to lobby for the BWCA but was using his working time as a very productive scientist. I used to talk with Bud about the dilemmas. Instinctively I shared this view. Why should we spend one hundred dollars an acre when we only get fifty dollars back.

HKS: That's right.

REB: That was a major conflict in Minnesota, and for the Forest Service I think time has vindicated Heinselman. I encountered similar conflicts between policy and research in later years, but I'm not sure that any of them were more acrimonious than this one. A few years after I left the Lake States Station, Heinselman arranged with the station director on a change of assignment which meant that he could decline and take early retirement. Bud's internal conflicts were so great that he felt that he had to leave the Forest Service. I thought that was an appropriate and an honorable thing for him to do. Of course the BWCA was established as a special wilderness area and significantly enlarged.

HKS: Does this go across the Canadian border?

REB: Yes it does. It used to be called, as I recall, Quetico-Superior Wilderness Area. The Canadian side of the BWCA might be even larger than the U.S. side, and it is also a national treasure.

An Exciting Research Environment

REB: Another point that I wanted to make about the Grand Rapids years was the excitement that went with the synergism that goes with working with unusually stimulating and able people. I used to inquire about this in later years as I visited various laboratories. Where are the centers of excitement and ferment? I think we had some of that at Grand Rapids. There were ten scientists during my time in Grand Rapids and several that were unusually stimulating and able. Roger Bay, who later became director of Intermountain and the Pacific Southwest stations, was one, Bud Heinselman who was an outstanding ecological researcher was another. So was Al Lundgren, an economist, and Bob Wombach, who later went to the University of Montana. One of the most stimulating persons I have known.

The point I want to make is that quite by accident there was a great deal of intellectual ferment among that group of ten. I think it's more a random event than anything else. Somehow a group of scientists got together who interacted extremely well. Several of those folks went on to have distinguished careers in their own right. I saw the ferment in other laboratories in later years and tried to offer administrative and financial support where I could.

HKS: Does that reflect on Dickerman? Does the station director select the people like that? Were they all there because he invited you guys to be there or what?

REB: I'm not sure. Dick, in my estimation, was an astute judge of people. He had two centers in those Lake States days that I thought were exciting places. One was at Rhinelander, Wisconsin, and the other was Grand Rapids, Minnesota. There was some chemistry and synergism among people that made them exciting. I saw that happening elsewhere in the country. For example, Lake City, Florida, produced just a great number of unusually able people; Harper was one of the people who went through Lake City. I think now it's called Olustee. Carl Ostrom was there as were Karl Wenger, Francois Mergen, and many others. One sees those creative centers and you wonder why. What is it that makes one place more creative and exciting than another? This Forest Service lab in Corvallis, which is a fairly large one, also has some of those ingredients. It's big enough that it may have two or three subcenters, for example, in ecology, entomology, and genetics. It was something I've asked myself about through the years, why is it that some places are so much more productive and stimulating than others? I think it is people and their ability to stimulate and reinforce one another. It's also distressing to see once productive centers revert to a lower level of performance. It is a major administrative responsibility.

HKS: Sure.

REB: I think it is people more than the work environment. But how you attract people like that, I don't know. Some leaders attract good people or otherwise are more able recruiting them. Still there is a luck-of-the-draw element also.

SILENT SPRING

HKS: You may want to deal with this concept a little later in your career, but in 1962 *Silent Spring* came out. That must have had an impact on research. I don't know how immediate it was. Did biological research become more fashionable because of *Silent Spring*?

REB: *Silent Spring* didn't really impact me directly because I wasn't in that area of research at that time. I certainly knew about the book, and I know about its consequences. I probably became more involved with *Silent Spring* after I became director of the Pacific Northwest Station.

HKS: It became an icon eventually.

REB: Yes.

HKS: Congress supposedly would have been more amenable to increasing budget requests toward certain kinds of research because of *Silent Spring*.

REB: Yes, and that did have a profound impact on Forest Service budgets. I don't recall whether Jemison or Harper mentions that in their papers. Dickerman and Arnold can comment on that better because they were in more senior positions at that time.

HKS: Maybe because you didn't have severe bug problems, right? The use of DDT for insect control.

REB: That was the first issue that I was confronted with, DDT and the Douglas-fir tussock moth, when I became director of the Pacific Northwest Station

RAPHAEL ZON

REB: I want to make one more reference to my Minnesota years. As a student at the University of Minnesota I was very much aware of the Lake States Station, and I certainly knew about Raphael Zon. And I probably saw him on the campus in the late '40s and early '50s, but it didn't register. After I joined the Lake States Station at Grand Rapids in the mid-1950s, I occasionally went to St. Paul. On one of those visits at a Christmas party, a very old man was introduced to the group. His name was Raphael Zon, very frail and old. That was the only time I ever recall seeing him. Less than a year later, Zig Zasada, who was still center leader at Grand Rapids, said that he had been asked by Dick to scatter Raphael Zon's ashes on a set of plots that Zon had helped establish in 1926 on the Cutfoot Experimental Forest. Zig scattered the ashes and later remarked to several of us that scattering the ashes was okay; it really didn't bother him but discarding Raphael Zon's glasses was more troublesome. You may recall from photographs that Raphael Zon wore those little round glasses. I knew at that time, and of course Zig and Dick did too, that Raphael Zon's influence on forestry research went back nearly to the turn of the century.

HKS: Sure.

REB: But Zon spent the last half of his professional life as director of the station. Soon after his death I became the project leader, and Dick Dickerman and I talked about some kind of a memorial for Zon. What should we do? One day Dick sent me a longhand note on a half-sized sheet of paper, written in ink, and, paraphrased, it said: "GP. I have the pleasure of transferring to you a plan for experiment stations." At that time both Dick and I realized that we had some ingredients for a monument to Raphael Zon. With help from the Chippewa National Forest, we arranged for a very large field stone, more than six feet high, to be placed next to the plots where Zon's ashes were placed. A bronze plaque was cast on which the words of the Raphael Zon letter to Gifford Pinchot were inscribed. Those events surrounding Zon's life triggered my interest in the roots and origins of Forest Service research. It's a story that both Dick and I use when we give talks on history of the origins of Forest Service research. Zon, among others, deserves a great deal of credit for creating what are now the regional experiment stations of the Forest Service.

HKS: Pinchot created an image of himself that really was false—he was totally practical, he was a field person. He would brush the cow chips out of the pond and drink the water. But under his administration the Forest Products Lab was established. The Office of Silvics, it was called, with Zon and Sam Dana, and it did a great deal of research. But Pinchot didn't want to call it research because he didn't want to look like he was a professor or something. He wanted...

REB: One of the accounts that I have read is that 25 percent of Pinchot's work force in the first year or two after he became bureau chief was in investigations in silvics, timber physics, and so forth. I'm asking you, why did Pinchot want to disassociate himself from research?

HKS: Part of it was he wanted to show he was hands-on. When you look at the political cartoons of the western papers like the *Denver Post* when they were opposed to the conservation movement, they always called him Professor Pinchot. That was a pejorative to call him Professor Pinchot. He just didn't want to say, "I'm a scientist, I'm doing research," because that added fuel to the western fires in opposition to the conservation movement, so he downplayed research.

REB: And yet he created an environment for research...

HKS: Absolutely.

REB: ...and the history of the Northeastern Station gives Sam Dana credit for the creation of Fort Valley in Arizona.

HKS: That could be.

REB: Now apparently Sam Dana and Raphael Zon worked together in the Office of Silvics. Sam Dana was actually at Fort Valley when Raphael Zon arrived there in 1908. The Northeastern Station history gives Dana credit for the creation of Fort Valley. It probably was a joint undertaking between Zon and Dana.

HKS: Yes.

REB: I've wondered about the same point that you've just made. Why is it that so many early scientists and progenitors of research found a home in the Gifford Pinchot years.

HKS: Silcox, all those guys were active under Pinchot. Pinchot wanted a vastly different image, like Albert Potter in range. He wanted people who actually knew how to ride a horse and that stuff.

REB: Okay. That was the final story of the life of Raphael Zon, and it came to an end on the Cutfoot Experimental Forest for which I had some responsibilities.

HKS: He wrote a letter to FDR about shelterbelt. Zon did a marvelous number of things.

REB: You'll want to ask Dick about that because Dickerman worked as an assistant under Raphael Zon when Zon was drafting parts of *Breaking New Ground.*

HKS: Okay.

REB: Dick will tell you about some of Zon's left-leaning tendencies and his lack of protocol in dealing with high-level administration officials.

HKS: So what happened? You liked Grand Rapids, you're doing good research and everything is great for ten years. What was the incident that caused you to go to Washington?

THE WASHINGTON OFFICE, 1965

REB: I don't think I ever made an overt gesture to change jobs. I only had one personal requirement, that if I changed jobs I wanted to make a major change, not a minor one, not the same work somewhere else. Several inquiries came along. The one that said Washington, D. C. was from Carl Ostrom, director of timber management research.

HKS: You went back there to do research, not to be in management?

REB: No, I went back there to head up the branch called Forest Mensuration, and it was really a very small program. I came to realize in later years that those invitations were oftentimes not to deal with the specific tasks that you were assigned. They were for you to do a variety of things, for you to size up that work environment, and for others to take stock of what and how well you performed in a variety of jobs.

HKS: You didn't see it as a lessening of your interests as a scientist?

REB: No, at that point I didn't. Later on, in retrospect, I did. In any event, three of us from the Lake States Station went to Washington roughly at the same time. Bob McCulley in 1964, and Dick Dickerman and I in 1965. Bob McCulley was another one of those people that had a strong influence on my career. He had been the assistant director responsible for my program in the Lake States Station. He went into a staff position under Harper. He later went on to the Pacific Southwest Station as director. But Bob was a good counsel both in Minnesota and in Washington. He was another one of those crusty guys that would challenge the hell out of you and then it was all over say go to it. Dick Dickerman, very much senior to me, was also most helpful in the new environment. My new boss was Carl Ostrom. Carl had been assistant director of the Southeastern Station. He was a great developer of people; he had that ability to tutor, to help steer and develop you but not frustrate.

HKS: Wasn't he an economist?

Washington Office Environment

REB: No, he was a silviculturist, and a good one. He was a great person to work for. Quickly my limited assignment as branch chief of mensuration research enlarged into things like assistant director for timber management research and a person available for a variety of assignments. At that time timber management research included genetics, silviculture, timber related crops, forest mensuration, and so forth. A number of people who later had distinguished careers came through those offices, people like Tom Nelson, John Barber, Steve Boyce, Karl Wenger, Bob Callaham, and others. In any event, I went to work for Ostrom. My first year there was extremely troubling. I kept asking myself what have I done, because I didn't understand that work environment. It was a jump from a very rural location in Minnesota to Washington, D.C.

HKS: Lyndon Johnson is now president, we have civil rights, we have Vietnam, and then Washington, D.C. is a whole different environment, you're suddenly in the middle of a whirl rather than out there in Grand Rapids.

REB: Washington, D.C. wasn't as foreign to me as it was to other Forest Service people because I had been in Officer Candidate School at nearby Fort Belvoir in '45 and '46, so I knew the city, and it wasn't intimidating to me. It was the work environment about which I was extremely uncomfortable. I remember that one or two occasions I was so unsure of myself that I would take long walks in the Mall just to ask myself whether I had done the right thing. I saw similar concerns among WO recruits during my time as deputy chief; most, but not all, eventually adjusted as did I.

HKS: These were administrative challenges rather than scientific challenges.

REB: Yes. One of the impressions I brought with me to the WO was that it was D.C., it's going to be very procedurally oriented, people will know what they're doing, what the protocols are and so on. It took me a year or so to discover that exactly the opposite was true. That the Washington office of the Forest Service is far more unstructured than any field organization.

HKS: Why is that? Too much power?

REB: No, I think it's because of the fluidity, the fluidness of the environment in which one works. Fast moving situations. Issues can come into the agency from anywhere, from Congress, from the White House, from constituent groups. The most important business in Washington is conducted with longhand notes and personal conversations. The memos, the memoranda, and the published stuff are not the important decision points. It took me a year or so to recognize that, and it was a point that I built on a lot in later years, that what you're looking for are those people who can cope with that kind of unstructured environment. Not everyone can. There's a cyclical quality to the Washington office; you do some of the things year after year geared to the rhythms of Congress and the White House. I became increasingly comfortable with that environment, and Ostrom and Jemison who was deputy chief gave me some jobs that turned out to be reasonably satisfying. For example I worked with the personnel office on reclassifying Carl Ostrom into a supergrade job. The effect was successful and brought all WO research staff directors into supergrade positions. Carl Ostrom's promotion was a labor of love because I had very warm regard for him and still do.

    The mensuration branch chief's job was only a small part of my WO work. I became involved in the International Biological Program (IBP), which came out of the National Science Foundation as I recall. It turned out the IBP was an important program for the Forest Service, including the Pacific Northwest Station.

HKS: Is that when they rated or evaluated government research, is that what you're talking about? The quality of research?

REB: No, it was an effort on the part of those involved in the more basic sciences, the National Science Foundation and others, to address natural resource issues internationally and more comprehensively than we were doing with our narrowly focused research. The reason I mention that is IPB was a major supporter of three watersheds, Coweeta, Hubbard Brook, and H. J. Andrews, and that had a lot to do with the influence of those experimental forests in later policy issues.

HKS: I didn't realize that H. J. Andrews was one of three; I thought there were dozens...

REB: There were eighty-four in 1992 under Forest Service jurisdiction, but the three that I named have had an unusually strong impact on ecological research. The scientists who led those programs are major shakers and movers in policy issues today.

I had another job in the WO concerning natural areas. I was chairman of an interagency committee monitored by the Office of Science and Technology (OST), concerning natural areas. I presented some ideas that said that we ought to set aside still more examples of all the natural environments on federal lands. These suggestions were not original with me; I was building on the old SAF program called natural areas. But the Forest Service was also an active participant in the SAF natural areas program and had a series of natural areas all over the country. We were trying to be sure that we had examples of all natural forest and range ecosystems. I also became chairman of the SAF Natural Areas Committee while I was in Washington. The natural areas in the Society of American Foresters, and various names by other public agencies and professional societies, is in my estimation a very important program that's little understood. It deserves a lot more attention and support.

HKS: We have the SAF records in our archives; there's a big chunk on natural areas. What is in those records?

REB: I don't know what's in the SAF records, but I know what is in the field.

HKS: Okay, let's consider the field.

REB: There are now two hundred fifty natural areas in the Forest Service system, prime examples of naturally functioning ecosystems, and in my estimation a tremendously important adjunct to the preservation of biological diversity.

HKS: It strikes that they're very practical these days. Almost more valuable than they were when they were set up.

REB: Yes. I became involved in natural areas in Washington, D.C. from a national perspective, and then when I went to the Pacific Northwest, it gave me an opportunity to work, particularly with Jerry Franklin, on greatly accelerating a program on natural areas here in the Northwest. Not original with me, the earliest natural areas go back to 1927, in fact there's a major one at Wind River, if you have ever been on the Wind River Experimental Forest.

HKS: Oh I have, yes.

REB: It's now called the Thornton Munger Research Natural Area, emphasizing the Douglas-fir ecosystem. But there must be a hundred natural areas in the Pacific Northwest, counting those on all ownerships.

HKS: Some of the rationale for the establishment of wilderness and their uses in the '20s and '30s was...

REB: Research.

HKS: ...research. To set aside these benchmarks.

REB: Wilderness areas only served part of that purpose. One of my arguments in Washington, D.C. was to create natural areas within a wilderness. The feeling was, from the National Forest System, no, we don't want anymore classifications within wilderness, but finally I talked to Larry Neff, who was the deputy director of recreation. Larry signed a policy statement that said yes we can create natural areas within wilderness. What this meant is that we wouldn't build a campground or a trail through the natural area; they had to have some additional protection from human interference. Wilderness areas have been important for research, but I have the impression more sociological research rather than biological studies. Also we need to recognize that wilderness tends to represent only a portion of natural ecosystems, generally the high elevation or otherwise attractive scenic areas.

ZERO-BASE BUDGETING

HKS: It was Kennedy or McNamara...

REB: Zero-base budgeting was McNamara. Lyndon Johnson apparently was so taken with McNamara and his work at the Department of Defense having to do with zero-base budgeting, called PPBS (Program Planning and Budgeting Systems) that he wanted all agencies of government to use it. That part of research budgeting became another one of my assignments. I worked with several economists including John Fedkiw and Bob Marty. I entered into the task with great enthusiasm, because it seemed so sensible to develop an analytic procedure to weigh various research alternatives. I left that exercise thoroughly disenchanted, because it was extremely demanding of data, extremely sensitive to assumptions, and in the end not used at all by administrators. One of the lessons that came out of the old PPBS work for me was that incremental changes are very much more realistic in government than is a zero-based review.

HKS: How do you do zero-based budgeting philosophically when you're dealing with the concept of applied research and basic research?

REB: You can't do it. When you think about it, budgets are invariably presented in incremental terms. If you've ever looked at a Forest Service budget, it gives the base year and it displays departures from the base year. In any event, I was so disenchanted with that PPBS system that it took me several years to develop any enthusiasm for another look at an analytic as contrasted to an incremental approach to research budgeting. But we did come back to it, and I think with some positive outcomes.

HKS: Theoretically you're in timber management research still.

REB: Yes.

HKS: But obviously you're experiencing much more.

REB: I was beginning to move off in other directions. I don't remember whether it was because I displayed some interest in other activities or somebody was pushing me in that direction. I certainly was underemployed as branch chief for mensuration. By the way the Branch of Mensuration folded when I left that job. I also served as assistant director to Ostrom; when he was gone I assumed some of the leadership in timber management research. I became more comfortable with the Washington office as time passed. My second year was a lot more pleasant than the first. My third year was also stimulating and pleasant, but a certain repetition was beginning to show up, because the work is, as I mentioned, geared to the annual rhythms of Congress and the White House.

THE HARVARD YEAR

REB: I had always had a goal that at an opportune time I was going to do the equivalent of a sabbatical. I wanted to reinforce my skills for whatever the next job was to be. If I had continued to be an active scientist I would have sought training related to my next generation of research. But I was already heavily involved in research administration, so I chose to spend a year studying public administration. I received an okay from Carl Ostrom, Dickerman, and others. And I was awarded a Bullard Fellowship at Harvard that paid my tuition. The Government Employees Training Act covered other costs. The Forest Service paid my salary, I spent 1968–1969 at Harvard. I left our family in Washington but came back about once a month. It was an arduous but productive year.

HKS: Max went through that program.

REB: Yes he did. You can choose almost any combination of things—early, mid or late-career, just almost anything. Intellectually it was the richest academic environment I've ever experienced. The Lyndon B. Johnson crew was coming back and the Richard Nixon appointees were leaving. The people who were departing included Henry Kissinger and Patrick Moynihan. At the moment I don't recall the names of all those returning. But, the J. F. Kennedy School was rich in faculty that had served in senior government positions. I had an option, I could spend a year there in residence with or without a degree. I chose to go for a degree. So I earned a second master's degree. It turned out to be a useful and valuable experience. Some courses were lousy but the ones that were good were super good.

HKS: What kind of classmates did you have? I mean were they like you or private sector or what were they?

REB: They were mainly from public agencies, two or three from Soil Conservation Service, a number from the military and Department of Defense and from other agencies in government. Yes, and there were several from state and local governments and there may have been a few who came out of the private sector.

One or two of the officers were thoroughly disenchanted with the Vietnam war. I suppose my class had fifty or sixty people. Courses that were most useful to me had to do with economics and congressional and executive supervision of public policy. Arthur Maass was an instructor on congressional supervision of public policy. Despite my concerns about his role in *Muddy Waters*, his teaching was among the most useful.

HKS: You observed that this was really very significant, congressional oversight, and so when you went to Harvard you had this in mind. This is something you really wanted to understand better, the role of Congress in the way the Forest Service functions.

REB: Yes. I knew that before I went there because I knew that Russell, Stennis, and Byrd were influential, but I never understood some of the more subtle relationships. I might have discovered this all by myself but the year at Harvard really helped. And a person like Richard Nuestadt who wrote *The Power of the President*, a classical book that came out of the Kennedy years. He talked about the White House, how the White House sees the various departments and bureaus of government. It was most helpful to see the Forest Service and other agencies from vantage points of those who created policy. So it was a rich year. I also took a course in science and public policy, and it was at that juncture that I chose to look into the origins of Forest Service research. That gave rise to that...

HKS: History paper on origins of research policy...

REB: Yes, an unpublished paper titled "Evolution of a Science Policy in the Forest Service." There are some errors in the paper, but it served me well at that time and in the intervening years.

HKS: When you wrote that you obviously learned something, you learned some details and some cause and effect and some specifics. Were there any surprises? Or was it you just understood better what...

REB: Both. I understood very much better. Yes there were some surprises too. I came to realize how important Earle Clapp was to the origins of the Forest Service research. Earle Clapp's role in research is not well understood. He was influential in research in the pre-World War II period as Les Harper was in the post-World War II period. If you skim that paper, you'll also see that I recount some things that happened in the Harper/Jemison period that did so much to enhance the quality of the research in the Forest Service. There were no blinding revelations, but the insights were very useful to me in later years. And the preparation of that paper then permitted me to begin a dialogue with some people like Les Harper and George Jemison and Dickerman, and others and you've seen some of those letters that I've exchanged with them.

Harvard, 1968-1969, I've never worked harder in my life. Marie was back in Washington with our four youngsters and I would come back about once a month. By then there was a looming question as to what job I would have when I came back to Washington. When I left I was anticipating coming back to the Division of Timber Management Research. But in that period, 1968-1969, George Jemison retired for family reasons. He came to this school, the College of Forestry at Oregon State University. But the point here is that changes were taking place.

STAFF ASSISTANT TO THE DEPUTY CHIEF

REB: Keith Arnold was recruited to the deputy chief job from the University of Michigan. He went to Michigan as the dean of the School of Natural Resources about two years earlier and was invited to return to the Forest Service to head the research branch. Dickerman would have been a very strong candidate to be the next deputy chief of research, but his wife Marge was seriously ill with what became a terminal illness. Dick, I'm told, declined to be candidate but continued as associate deputy chief. So Arnold returned to the Forest Service in 1969. Keith previously was the director of the Pacific Southwest Station and later director of forest protection research in the Washington office.

He doesn't get very high marks from Michigan alumni because he dismantled that school and put it back together, from a hard science and professional school that Sam Dana, Spurr, Davis, Gregory, Graham, and others put together. The school became swept up in the environmental movement. So at least some of the older alumni think that he took it from a hard, discipline-oriented program into some fuzzy environmental stuff. However, Michigan was in serious trouble at that time with three forestry schools in the state and a rapidly changing environment for forestry. Duke and Yale went through some of the same travail during those years.

HKS: So he got them more into policy rather than...

REB: I'm not sure it was hard policy or soft policy. Keith changed the School of Natural Resources at the University of Michigan. He was a shaker and a mover and an innovative guy. I liked working for him, but he needed a Dickerman or an Ostrom or a Herb Storey around him to discipline and challenge that wide-ranging mind, and I mean that in a very positive way. In any event, Keith Arnold came back after two years in Michigan to serve as deputy chief. Someone decided that I should come into Keith Arnold's office as a staff assistant responsible for budget. That's where I came back after Harvard. This was one of Harper's innovations—the position of staff assistants to the deputy chief. These are people who do things like budgets, personnel matters, and program formulation for the deputy area. I think an organizational expert would look at a chart and ask why the hell do you need all of those positions?

The important things about them were that they were training slots for the next job. If you look at some of those staff assistants you'll see the leadership of the Forest Service. It started in research and was later adopted by other deputy chief areas. John McGuire, Tom Nelson, John Barber, Bob McCulley, Bob Callaham, Roger Bay, and many other leaders were at one time or another staff assistants in research. Those were so valuable as training slots that the pick of the litter went through them. Again, important for what you did but equally important as a precursor for the next assignment. I moved into the budget slot. I must have done two or three things really well in that job. Perhaps I could illustrate with a couple of anecdotes.

HKS: Good.

REB: About what you look for in leadership, people who occupy those positions, because this is exactly what I looked for when I occupied the deputy chief's job several years later. I got a phone call from the administrative assistant to Senator Hiram Fong of Hawaii. Senator Fong was on the Interior Appropriations Committee chaired by Senator Alan Bible. His assistant was Earl Nishamora. He said, my senator is running for reelection in Hawaii, and he helped to get an appropriation of one hundred or one hundred fifty thousand dollars for your research laboratory in Hawaii. He would like to know what was accomplished with those dollars. I said I'll get the information for you. Before Nishamora hung up, however, I asked "Is there any possibility that the senator might like to accelerate the programs in Hawaii?" I assured him that we were doing good research. He said, you know I hadn't thought about that, I'll ask the senator. He called back and said yes the senator would consider another increment to the funding. I said, okay, when I report to you on what we've already done, maybe I could give you some suggestions about what the new programs might be. So Nishamora and I worked on a set of questions that Fong would ask the Forest Service during appropriations hearings. Both he and Fong were really naive about this sort of thing.

Alan Bible was chairing the appropriations hearings when Fong and Nishamora arrived late. Of course Keith Arnold had all of the questions and all of the answers, and he knew Hawaii from his Pacific Southwest days. When Fong came in I signalled to Nishamora. We talked on the side and I said, suggest to Senator Fong that he ask Senator Bible to yield. Fong then went through the set of questions. He held a press conference right afterwards. Keith went to the press conference and talked about all the things that the Forest Service was doing in Hawaii. One of the things that Keith did, though, was talk about the importance of timber in Hawaii. The then current issues in Hawaii weren't timber at all, they were about the environment. I tried to get Keith to steer away from the timber subject. In any event, Fong added another one hundred fifty to two hundred thousand dollars to the Hawaiian budget of the Pacific Southwest Station. It all came because of that chance phone call from Senator Fong's assistant.

HKS: The introduction of exotic species was and still is a serious problem. Is that one of those environmental issues?

REB: Yes, and decline of the native forests called Ohia decline. Those were the things that I suggested to Keith that he talk about—not the need to grow more exotic timber. In any event the press conference worked fine. The point that I'm making here is the opportunistic nature of that unstructured, fast moving environment. I responded in a way that got a couple of hundred thousand dollars out to Hawaii, and I'm sure that didn't escape Arnold or Dickerman. Let me give you one other example.

HKS: All right.

REB: The laboratory construction program was in full swing in those days, in the very late '60s and the early '70s, still a carry over from earlier years. There were two laboratories under consideration in the Northeastern Station, one in Burlington, Vermont, where Senator George Aiken was the benefactor. And, the other was at Durham, New Hampshire, and I can't remember the name of the senator. But Warren Doolittle was the Northeastern Station director. He was preceded by Dick Lane, who was a very aggressive guy. They'd been working on getting those two laboratories. Warren said he was going to pay a visit to the Hill. He asked what were the Washington office priorities.

Warren wanted to be in step with the Washington office, and I told him what Keith and Dick Dickerman thought—that Durham, New Hampshire, was the most important and that Burlington, Vermont, was second. As I reflected later, my response was accurate but not astute. Warren paid a visit to Senator Aiken and told him that Burlington, Vermont, was the Forest Service's number two priority for funding after Durham, New Hampshire. George Aiken just exploded. It is important to know that Aiken and Richard McArdle were good friends. Ed Cliff was then the chief, and George Aiken got all over Ed Cliff and Warren Doolittle. If I had been more perceptive I would have anticipated that conflict. In any event, George Aiken arranged the money for the Burlington lab even though he wasn't on the Appropriations Committee. He had so much influence in the Senate that he could easily do it.

HKS: And that lab went on the campus.

REB: Yes. It's now called the George Aiken Laboratory on the campus of the University of Vermont. Durham, which was our number one priority, was also funded.

HKS: How did you get the ranking of the laboratories, the timber species are very similar.

REB: The rankings are based much more on physical need, such as office and laboratory space and what Keith and Dick perceived to be political realities.

HKS: The quality of the forestry schools or the faculty at that moment.

REB: No, both forestry schools were important and of course the Forest Service wanted both laboratories. But there were other political realities, budget constraints, and other things that were the major determinates about which was first or second.

Let's go to the Durham lab after the Doolittle visit. The administrative assistant for the New Hampshire senator called and said my senator has agreed to support the Durham lab in the appropriations but he can only get one million dollars. I knew enough about the design of the Durham lab to know that we could construct only about half the building for a million dollars. It was a two million dollar job, it was a square building and you just couldn't build half of it. I said that the design doesn't lend itself to partial construction. And she understood what I said and she called me back an hour or so later and said my senator has agreed to go for all the funding. The point that I'm making here is that there are all kinds of junctures and opportunities to advance an agency's program, and those were two, Hawaii and New Hampshire/Vermont that I remember vividly. Please understand that it took other participants—the station director, the deputy and associate deputy chief, for example—to make the action complete.

HKS: But also you have to have this environment that you were authorized to broker a deal in response to an opportunity without checking back with the boss, because you don't have time.

REB: Well, I knew...

HKS: But you knew it was going to be okay. There is a certain unstated delegation of authority to go ahead and go for it when an opportunity presents itself.

THE CHIEF/DEPUTY CHIEF OFFICE

REB: That's part of the unstructured nature of the Washington office. Now with that kind of delegation of authority there are also chances to make some fairly substantial mistakes.

HKS: A mistake can be something that didn't turn out okay.

REB: That's right.

HKS: If it turned out okay then it was smart.

REB: Those were a couple of examples, and I'm just guessing what supervisors were looking for. Certainly it was a quality that I was looking for in my later years—the capacity to take risks, but with reasonable judgment.

Dick Dickerman and Keith Arnold were interesting people to work for. And Dick was another one of these mentors that meant so much to me through the years, another great developer of people. Arnold was the deputy chief and Dickerman was associate deputy chief. Keith Arnold was a swinger, and I don't mean that in a pejorative way, a person with a wide ranging imagination, spinning off ideas but sometimes not screening them very well. Dick was much more deliberate but also imaginative and impressive in his own quiet way. I was working with Keith, who was bouncing off ideas all the time, but I would also work with Dickerman. I had easy access to both of them because of that budget position. Dick, always very loyal and supportive of Keith, would tell me, you know we don't do things the way we did when Jemison was here. [laughter] And, I damned well knew it. Keith really was an imaginative guy, but my assessment is that he also needed somebody like a Dickerman and two or three other people around him who would caution him from time to time.

HKS: Ed Cliff, in terms of the folklore of the Forest Service, ran a pretty damn tight ship. Why would he have selected Keith Arnold? It strikes me as inconsistent, where you have sort of a free agent. The way you characterize Keith doesn't fit the stereotype of Ed wanting to keep the lid on or making sure that what happened is what he wanted to happen.

REB: I don't know the answer to that question; it's a good one. I had the impression that Ed was more innovative than that, and more willing to take chances than you might have suggested. If you look at Ed Cliff, open flannel shirt, crusty, probably more development than environmentally-oriented. John McGuire, more intellectual, more urbane, an entirely different demeanor. I'm sure that Ed Cliff had a lot to do with bringing John McGuire in. McGuire came in as deputy for programs and legislation followed by associate chief, then chief. I think Ed deserves a lot more credit for tolerating different points of view, in fact encouraging different points of view than he might get traditionally.

HKS: Others in the Washington office at that time have commented that at chief and staff, Ed would make a series of announcements and then walk out. That may have been a perception of someone who didn't like what was going on.

REB: That could be although I wasn't aware of his early departures from confrontation. I was a staff assistant, and I sat in on some of those meetings. Boy, if you ever wanted to see a smoke-polluted environment it was an Ed Cliff staff meeting.

HKS: With his pipe going.

REB: Oh, it was just incredible. [laughter] My relations with Ed were more distant, but I thought that he especially appreciated his science arm. Have you encountered the Byrd hearings, the Saturday appropriations hearings about the Monongahela case?

HKS: No, I've heard a little bit about the Monongahela stuff, which we can certainly talk about.

REB: Well it's not germane to this central topic, but Senator Byrd was fronting for Senator Jennings, both of West Virginia. Randolph was deeply involved in the Monongahela situation. Ed Cliff was still chief, and the Forest Service was unwilling to change some silvicultural practices. I'm sure it had to do with even-age versus all-age management and especially clearcutting. Senator Robert Byrd was chairman of the Appropriations Sub-committee, and he was very forceful; he's strong in his own right concerning the Forest Services appropriations, especially those concerning West Virginia. He was holding the budget hearings and Jennings Randolph was trying to get to him to perform some oversight functions about the Monongahela. Byrd said that he wanted to hold Saturday hearings on the issue. Jennings Randolph was in the room during the entire hearings, but he's not nearly as swift as Byrd, so Byrd brought in about six or eight witnesses from West Virginia, and all of them were absolutely opposed to the Forest Service timber harvesting activities on the Monongahela. Only one person was sympathetic and that was Dr. White, then head of the forestry school at Morgantown. Byrd spent the whole Saturday badgering Ed Cliff, much of it for public consumption in West Virginia, about the Monongahela. Ed Cliff was accompanied by Carl Ostrom and two or three other scientists.

This is one of the reasons that I say that Ed had high regard for his science group. He was staunchly defending the science behind the silviculture used on the Monongahela. I was sitting behind Ed, and I could see him getting red and really upset. There was much posturing on the part of Byrd, as nearly as I could tell, on behalf of Jennings Randolph. It was one of the most significant days for me in Washington. After the hearing, Ed mentioned to a couple of us, that shortly thereafter he visited with Byrd. According to Ed, Byrd said, "No hard feelings, Ed, what do you need in the budget next year?" This does illustrate a little bit about my perception of Ed. Ed Cliff had been chief for upwards of ten years at that time. My impression was that Ed represented a different era, and it was time for change. I've always admired the change that took place. McGuire would never have come in if Ed hadn't supported it. And John McGuire brought an entirely different personality and a different perspective to the job of chief. I thought he was just the right person for the time.

RESEARCH PLANNING

REB: I want to pick up one point that goes back to the Harper/Jemison period. Now this program for research, if my memory serves me correctly, was really a Richard McArdle exercise. McArdle was trying to strengthen the program for the national forests, and he needed to display to the Congress that he had a well thought-out plan. That really was the basic purpose. A program developed about three or four years before was called Program for the National Forests, or something like that. In any event, that Program for the National Forests did outline what they needed. There was one paragraph in it saying that there was also a program for research. Harper had in his hip pocket a very much abbreviated plan about where he wanted research to go. It was that plan that Harper was using to build the research program of the Forest Service in the late '50s and early '60s. It was only a one paragraph entry. I'm sure that Jemison was assigned the job of fleshing that plan out. The irony about that situation is that most of Harper's accomplishments were made before the formal plan of 1964 came out. Harper doubled or tripled the research budget of the Forest Service in those years. And Jemison also built an accompanying construction program. I don't know whether construction is mentioned or not but that had a lot to do with new laboratories.

ORGANIZATIONAL CHANGES IN RESEARCH

HKS: Were you making a lot of trips out into the field when you were doing budgets, did you do a lot of travel to places?

REB: Not very much as staff assistant. I did more of it when I worked for Ostrom from 1965-1968. That brings me to the next event. I was in my second year with Arnold and Dickerman. A major research review was planned in 1970 for the Pacific Northwest Station. Dickerman led the review team, including visits to Oregon, Washington, and Alaska. Because of budget activities in Washington I couldn't be in on parts of the Oregon and Washington review but I was on the Alaskan part. Bob Harris, then an assistant director of PNW, was also there. I welcomed the opportunity to get out of the Washington office and to see Alaska. With a bit of hindsight I recognized that this review was a prelude to a whole series of major retirements. In essence it was the recruitment of the late 1920s and the 1930s that were retiring including Phil Briegleb, director of PNW; Joe Pechanec, director of Intermountain Station; Charlie Connaughton, and many others. The post-World War II age class was coming in.

HKS: Sure.

REB: On June 1, 1971, Charlie Connaughton and Phil Briegleb retired. I was named director of the Pacific Northwest Station. Bob Harris was named the director of the Intermountain Station, because his predecessor, Joe Pechanec retired. There was a whole series of changes. Rex Resler became regional forester of R-6 on that very same day. In many respects the late '60s and early '70s were the transition from pre-World War II to the post-World War II leadership. I became the director of the Pacific Northwest Station after six years of a variety of assignments in Washington.

HKS: Les spent a lot of time in his interview talking about the administrative structure, the Man-on-the-Job program. Were you involved in that kind of business?

REB: No, those were all done before I got there. Harper retired a few months after I came to the Washington office in 1965. Harper was surely the most influential person affecting Forest Service research in the post-World War II period.

HKS: How about the reorganization of the stations?

REB: I was involved in three significant organizational changes during my years with Forest Service research. The first were those Harper initiated in the mid and late 1950s. At that time, Forest Service research went from division chiefs generally with unidisciplinary portfolios to assistant directors with a number of projects involving several disciplines. The second was really an Arnold/Dickerman innovation. I was confronted with that as soon as I came to the Pacific Northwest Station. The third was also an Arnold/Dickerman innovation, but had to do with organizational concepts to do team research.

PIONEER UNITS

HKS: How about pioneer scientists. That strikes me as a very profound innovation.

REB: That's another change borrowed from the Agricultural Research Service, and Harper bought into it. The concept was sound but the implementation was difficult.

HKS: So you weren't involved in the creation?

REB: No, but I was involved in the pioneering research questions later as deputy chief. The concept was to take the most productive, creative, and imaginative scientists and set them up in separate units where they would receive essentially no supervision. The pay would be determined by the Person-in-Job provisions, but would be high level. The first pioneering scientist in the Forest Service was Lou Grosenbaugh, who was the inventor of 3P sampling or Probability Proportional to Prediction sampling. Phil Larson at Rhinelander, Wisconsin, was another. A great idea. ARS became disenchanted with the concept in later years, and I did too, because one would build a team of scientists and technicians around a pioneering researcher. The problems occurred when the pioneering scientist moved on or retired. What do you do with the laboratory and the people built around that scientist? We tried to create that kind of environment in other ways but with a little less formality and more flexibility. I think maybe we accomplished that. Conceptually, the pioneering scientist was a good idea, but operationally it was difficult to use.

HKS: Who's the senator that gave the Golden Fleece award?

REB: That was Senator Proxmire of Wisconsin. I view the Golden Fleece award as a mixed bag—some legitimate issues, some that were phony and unfair. It's a price one paid for government service. None came to Forest Service research during my time.

HKS: I could see Proxmire really tearing into that. Seven million dollars to study something, and he'd describe how silly this is. But it didn't suffer at the hands of Congress, apparently. Congress put up the money.

REB: Scientists and administrators have a capacity for self-inflicted wounds—pompous and exotic titles for grant proposals, and subject matter difficult for the public to relate to. Society needs a Senator Proxmire, but not too many.

Back to the Pacific Northwest Station. I came here on June 1st, 1971 and departed almost exactly four years to the day to return to Washington. One of the most important things confronting the PNW Station was internal reorganization. This was a Keith Arnold innovation. Keith wanted to decentralize research administration and increase technology transfer. That meant that assistant directors would be stationed at field locations when it was logical. He also was responding to a General Accounting Office (GAO) study of a year or two earlier having to do with technology transfer. The GAO study was critical of how some of the findings of Forest Service Research were used. The GAO based its criticism on ten case histories. The GAO report said (for example) you did all this research on crop tree release and nobody is using it.

HKS: What kind of people do the studies for GAO?

REB: Ambitious young people who get no rewards out of saying that something is okay.

HKS: What are their skills? Are they trained as scientists?

REB: Generally not. They come from a variety of backgrounds such as law, economics, and political science, but relatively few from science. The GAO report was seriously flawed in many respects, but it touched on an underlying issue about getting scientific findings into practice. That was and is an important issue. So, in order to address that question, part of the Arnold organization created in the experiment stations the position of planning and applications assistant director. You've got three things—creation of deputy director and application assistant director and the moving of assistant directors to field locations with interdisciplinary portfolios. The Pacific Northwest Station hadn't adopted that scheme when I came here.

THE PACIFIC NORTHWEST STATION

HKS: You succeeded who at the station?

REB: Phil Briegleb. Reorganization was right here in front of me. Dick Dickerman came out and simply told me in his quiet way to implement this organization. It didn't fit the Pacific Northwest Station very well, but we did it. Bob Tarrant became deputy director, headquartered in Portland. We moved Bob Romancier to Corvallis, which was our largest field location. But, it just didn't fit well to move either of the other two assistant directors, Ken Wright or Don Flora, away from station headquarters although we seriously considered moving Ken Wright to Alaska. George Garrison, who was then at Le Grande, became the planning and applications AD in Portland. The impacts for us were not all that great. Nationwide, I think, the most important thing was that it put emphasis on technology transfer, mobilizing research information so that it was more useful. Other aspects of this reorganization, such as field location of ADs, didn't work as well, although the basic concept is still in place at some stations.

TECHNOLOGY TRANSFER

HKS: Isn't the other half of the equation the quality of staff over in the regional office? They have to absorb and transmit. How do you affect the technical quality of those people? Most of them really weren't trained as specialists, to be honest.

REB: It certainly is a shared responsibility among scientists and users. However, as a generalization, research was expected to be far more active and aggressive at conveying technology to the users. The emphasis on technology transfer was appropriate for that time, and it's appropriate yet today. The tools for technology transfer are many. Personalities also come into play. One of the most effective forms of technology transfer, reinforced in my mind since I've been here at Oregon State, is continuing education. This college must have twenty-five or thirty workshops a year for mid-career training. Some are one day long; some six weeks long. That's where technology can be mobilized appropriately and conveyed in large bunches to people who just don't have time to read a publication or listen to a talk when they're on their regular job.

HKS: So the Forest Service would authorize or direct staff people to attend these courses?

REB: Yes.

HKS: That's the mechanism by which they are updated from time to time.

REB: Yes. Continuing education is one of the most effective forms of technology transfer that I've ever seen. Of course, there are others. An additional comment: the P&A AD's job was also becoming important to handle the planning requirements of RPA. That was the Arnold organization. Dick Dickerman in some of his personal remarks to me has said this wasn't popular with the stations. Dick may have been talking about me because I didn't display much enthusiasm. But it really worked okay, and was an organizational innovation in the Keith Arnold style.

REORGANIZATION

HKS: Is this an extension of the Person-in-Job?

REB: No. It was a principal response to that GAO report on technology transfer, and also a need to get top level station administration closer to the field and to the problems. Now, let me comment a little further on that aspect of the job. With that reorganization, assistant directors were moved away from station headquarters in many places in the country. For example, Riverside, California, in the Pacific Southwest Station. The Northeastern Station put its assistant directors at three field locations, Delaware, Ohio, Morgantown, West Virginia, and Durham, New Hampshire. The Intermountain Station positioned an assistant director at Missoula, Montana. It put the stations in better contact with client groups. It was important to have an administrator in the field. That decentralized organization was in place when I came back to Washington as deputy chief. It was a mixed bag right from the start. Some of the station directors were complaining. They wanted to pull their AD's back to station headquarters where they could work with them on a day-to-day basis. Gradually they began to pull back the AD's to station headquarters. I think now essentially all of the stations have brought their AD's back. In later years, especially after 1980, increasingly tight budgets also caused streamlining of this organization.

HKS: Are all the stations generally the same?

REB: In principle but not in detail. There are some broad policy guides such as Person-in-Job, research reporting, budget formulation, and overall station organization that are standard, but every station has a personality and an environment of its own. Here research content, relations with cooperators and user groups, historical development of forestry and research, land ownership patterns—all affect the individual station. This is as it should be in a decentralized organization.

    The Forest Products Laboratory, as I recall, was last to adopt the overall organization of the eight regional stations. In fact, up to and including some of the

Harper years, FPL tended to play a highly independent role from the remainder of Forest Service research. Interchanges among stations and FPL by such leaders as Bob Youngs, Bob E. Thington, and John Erickson helped. Today, in my estimation, FPL enjoys close working relationships with stations but still works closely with its traditional constituents in the forest industries.

HKS: I suppose there's less transferring in research, so people are there for a long time, as opposed to National Forest Administration; if you're successful, you're moving around.

REB: That's right. National forest systems have a hierarchical career ladder—district ranger to supervisor's office then on to the regional and Washington office. Research has a dual track. They have the Person-in-Job concept, that is scientists can stay in place and still be promoted depending on the quality of research they do. Or, they can move into an administrative position. It complicates life a little bit in research. It permits scientists to spend their career at one location. However, there is much more emphasis on mobility for those who choose research administration.

HKS: I am trying to understand the various kinds of management reorganizations like the Person-in-Job and the creation of assistant directors and all the rest. It's all part of a change in the way research is structured. There wasn't a master plan that in ten years we'll have implemented these things. Each new deputy chief came along and turned the ratchet another notch on reorganization. The stations must have been involved and reacted under Nixon and under Carter to reorganize the Forest Service.

REB: I said that there were three internal research reorganizations during my time in the Forest Service. The first one was Harper's, the second one was Arnold/ Dickerman, and the third one, which had to do with new mechanisms for conducting team and larger scale occurred research, also in the Arnold/Dickerman period. In addition to internal reorganization there were external organizational questions, of course, that did or would have affected research. The Eisenhower consolidation of stations and regions, the ten standard regions proposed in the Nixon era, and Jimmy Carter's failed efforts to create a department of natural resources. These were external to research, and while they had an impact on research in the end, they didn't influence it very much.

STATION BUDGET TECHNIQUES

HKS: Alright, you're in Portland. You've reorganized the station.

REB: I want to make a comment on the station that I inherited. Phil Briegleb was a long-term station director. He'd been at Central States Station, then at the Southern Station, and he came to the Pacific Northwest as station director about 1960.

He spent at least half his career as a station director. Phil was, in my estimation, among if not the most capable fund raiser of any of the station directors. In all three stations, there were substantial increases in funding after he arrived.

HKS: Is that right?

REB: ...with members of Congress and other key cooperators.

HKS: You don't ever deal with OMB or that side of the budget?

REB: Oh, yes! But Congress has had more influence on the budgets and the programs of the research branch of the Forest Service than does the administration, including OMB, ever did.

HKS: So here he would have had Wayne Morse, is it too early for Hatfield?

REB: No. He had Wayne Morse, Mark Hatfield, Julia Butler Hansen, Wendell Wyatt, and Al Ullman, who was chairman of the Ways and Means Committee and was very influential.

HKS: Scoop Jackson.

REB: Scoop Jackson was never much of a supporter of the Forest Service or Forest Service Research. Neither he nor Maggie Magnuson, both powerful senators from Washington, displayed much interest.

HKS: Is that right?

REB: The Pacific Northwest also always had somebody on the Appropriations Committee in the House—Julia Butler Hansen, Wendell Wyatt, Bob Duncan, Norm Dicks, and Les AuCoin. In terms of research, it is the Appropriations Committees that really makes funding decisions. On the Senate side, there were people like Wayne Morse and Mark Hatfield that influenced appropriations. By and large, congressional appropriations were influenced far more by the interests of individual members than by political party whether a senator or congressman.

STATION PERSONALITIES

REB: Phil Briegleb was a great fund raiser. However, he was viewed as an austere, distant, reserved guy by station people. That was an unfair characterization. Phil rarely visited individuals and wasn't comfortable dropping in on offices or with confrontation issues. Still Phil had his finger on the pulse of virtually every aspect of station life and left a strong station upon retirement. When he learned that I was going to be the director of the station, he arranged a joint visit to all key congressional contacts in Washington. Phil and I have stayed in contact ever since. He's writing his recollections of the PNW Station. Phil also had a top staff that was, in my estimation, one of the best in the country. It included Don Flora, Ken

Wright, Bob Harris, and Chuck Peterson, solid performers all. Bob Tarrant and Bob Romancier joined the station staff soon after I became director. Bob Romancier took Dave Tackle's place and Bob Tarrant followed Bob Harris, also solid performers. Bob Tarrant became deputy director a year later.

HKS: I knew several of them from my days with the PNW Station.

REB: They were just marvelous people to work with, and several of them, Wright, Flora, and Tarrant had chances to move into supergrade positions. Flora wouldn't leave the Northwest.

HKS: He had his son and daughter in that ice skating business.

REB: That was before they became ice skaters, but he wouldn't or couldn't move. McGuire wanted him to succeed Joe Josephson as head of economics and forest inventory research.

HKS: Is that right?

REB: Ken Wright could have been the director of forest insect and disease research, also a supergrade position in Washington; family considerations intervened. Bob Tarrant did become director of the station when I left. The Pacific Northwest Station also had something else going for it. I knew this from my TMR days in Washington. It had a group of researchers who were shakers and movers, and you know some of them, Jerry Franklin with the H. J. Andrews Experiment Station

HKS: Yes.

REB: Jim Trappe is one of the world's authorities on mychorrizae. He's retired, but has an adjunct appointment with OSU. Les Viereck's ecological work in Alaska. Jack Ward Thomas, who came in during my time, at Le Grande, Oregon. Val Carolin in insect research coauthored a book on insects of the West. Gary Daterman, Mauro Martignoni, and Hank Thompson for their Douglas-fir tussock moth research and there were several other outstanding researchers. But the point I'm attempting to make here is, perhaps serendipitously or perhaps accidentally, there was just a great deal of ferment in this station and one wonders where the credits go for it.

HKS: It's as though we were talking about earlier in the Lakes States.

REB: Yes, but the Pacific Northwest Station had more than its share of shakers and movers. All stations, however, had and have centers of excellence.

HKS: Do you think the problems here are more exciting? That it attracts?

REB: All of that plus the fact it was a larger station than most. Of course there were average and less than average people, but with that kind of leadership you could place the people who weren't as able or weren't as aggressive under the shakers and the movers. We did a lot of that. So I inherited the station here that was, in my estimation, one of the most robust and healthy of any in the country. Phil Briegleb certainly deserves credit. Those three or four assistant directors who were here or came shortly after were among the most capable people that I've ever worked with. I've never lost sight of that.

HKS: So Bob Tarrant succeeded you?

REB: Yes. Permit me to back up a bit. I had the Bob Harris vacancy here when I came. Did you know Walt Hopkins?

HKS: By name. He's recreation, right?

REB: He was recreation. Walt was a good friend in Washington, D.C. He was frustrated as hell with Washington, and when he learned that I was coming here he asked if he could be a candidate for the Bob Harris job. I talked it over with Dickerman and Arnold and they said yes, and in fact they were eager to have me take Walt. They didn't want the station to fill from within.

Walt Hopkins came to visit the station but first he stopped to see that damn pirate Carl Stoltenberg, dean of the College of Forestry at OSU. Stoltenberg offered Walt a job to teach introductory forestry. Walt had already accepted my job, but changed his mind without telling and accepted the Stoltenberg one at OSU. That turned out to be one of the happiest circumstances both for Walt Hopkins and for the Pacific Northwest Station. Walt taught introduction to forestry and was a super teacher of young people. The faculty still talk about him. In any event, Hopkins' retirement permitted me to ask Bob Tarrant to fill the Bob Harris vacancy. I'd heard many good things about Bob Tarrant. Bob was living here in Corvallis, and he accepted the job provided he could stay in Corvallis until his youngest daughter finished high school two years later.

HKS: He was assistant to George Meagher in timber management research when I was at the station, that's how I know Bob.

REB: Bob talked frequently about his work with Meagher. It was Bob Cowlin who encouraged Bob Tarrant to go to Corvallis and get away from station headquarters. So Bob came here. He had trained himself as a soil scientist. Bob intended to get a Ph.D., except that like me, he was recalled to the Korean War.

HKS: Oh.

REB: And by the time the war ended family responsibilities were heavy. He may not have had a master's degree, I don't know, but intellectually he was...

HKS: No question about that, but still...

REB: Bob was a capable person. He had that ability to work well with people. He was also a strong AD on external relations, especially political ones. He understood the Pacific Northwest; understood Washington, D.C. He and Jean moved up to Portland after two years in Corvallis. Soon after Bob was appointed deputy director of the station. Bob Tarrant had the ability to handle external relations with great sensitivity and warmth. He was doing the jobs that I wasn't doing.

HKS: In a sense deputy directors function like an associate chief?

REB: Yes.

HKS: Freed up the chief to do the chiefly job, but they run the organization on day-to-day basis.

REB: When it came my time to leave the station, I participated in the search for my successor. I was going back to the Washington office as associate deputy chief, and so was consulted about potential successors. I very much supported Bob, because I had so much confidence in his abilities. Bob was selected as station director and stayed in that job for four years. Bob would have been an even stronger director if he had previously had a tour in the Washington office in order to better understand that Byzantine world. Still with his quick mind and previous Forest Service and military experience he was well prepared for the job. Again the point I'm making is that I inherited a really solid program here.

THE DOUGLAS-FIR TUSSOCK MOTH

REB: I want to discuss now two more things in the Pacific Northwest. One of them was the Douglas-fir tussock moth outbreak. The second was innovative organizational concepts to deal with the Douglas-fir tussock moth problem and other large-scale research programs. About the tussock moth. When I came to PNW, a Douglas-fir tussock moth outbreak was brewing in eastern Oregon. Eight hundred thousand acres of forestland were threatened. DDT previously had been used effectively to control tussock moth outbreaks. This time there was a public outcry against using DDT. EPA, which had gained a great deal of authority in intervening years, would not grant authority to use DDT unless the Forest Service met some rather stringent conditions, including an accelerated research program that included alternatives to DDT. Ken Wright, Gary Daterman, Hank Thomas, Mauro Martignoni, and Boyd Wickman, who had been working on tussock moth research, were major players before the outbreak and even more important after. This brought about the third set of organizational changes internal to research, as I saw them. Keith Arnold and Dick Dickerman with strong assistance from others put together something called the combined Forest Pest R&D Program, commonly called the 3-Bug Program.

HKS: Right.

REB: Okay. There was some Machiavellian behavior here too on the part of the Forest Service. The focal point was Douglas-fir tussock moth and DDT, and the strong promise that there would be accelerated research funding to go with that Douglas-fir tussock moth program. But Dickerman and Arnold said, you know there are other major insect problems in the country too, why don't we piggyback them on the tussock moth program. And, that's where the 3-Bug came from, the second bug being gypsy moth in the Northeast and the third bug being the southern pine bark beetle. Bob Long was the assistant secretary in charge of the Forest Service at that time. And he, Ned Bailey of his office, and a long-time colleague Keith Shea...

HKS: I know the name. Did he work for Weyerhaeuser?

REB: Yes he did.

HKS: When I was on the senior field trip at the University of Washington we went on a tour and he was there. I always thought of him as a pathologist.

REB: He is a pathologist. Keith has a large capacity for organization of questions. In any event, that group, Keith Arnold and Dickerman, Bob Long, Ned Bailey, and Keith Shea, put together an organizational concept, an RD&A concept, to address problems associated with those three insect problems. Let us have tightly drawn plans, they said, let us bring together an appropriately large team to address the problem and provide milestones and markers and things of that sort. In addition to the Forest Service, Cooperative State Research Service, Agricultural Research Service, and Animal and Plant Inspection Service were partners in the program with additional funding allocated in the Forest Service budget. The 3-Bug Program was funded as I recall at about six million dollars per year, a large budget increase for forestry research by early 1970 standards.

HKS: RD&A, that's Research, Development and A?

REB: Application.

HKS: Application. Part of the technology transfer process.

REB: Yes, a research and development mechanism to bring together larger teams to solve more complex problems more quickly. My part in the scheme was as the station director administering the tussock moth program. In many respects, my role here was peripheral, because the strengths were in the scientists that I named earlier. But I think maybe my understanding of how the Washington office worked and what Keith and Dick were trying to do may have helped organizationally in bringing people and organizations together to solve a complex problem. And if Keith Shea had a special talent for organization in the Washington office, Ken Wright had that same skill in the PNW Station.

HKS: Ken Wright told me, it must have been the tussock moth, but it was so controversial that the secretary of agriculture created a special team, and Ken was pulled out of the station to serve as director of that team? It was some insect problem.

REB: Ken was selected to be the program manager in Portland for the tussock moth program.

HKS: Maybe I misunderstood what he was saying, but it was a such high level concern that to make sure that the Forest Service politics didn't get involved, the secretary's office actually handled this.

REB: Yes. The program was administered by the secretary's office under the direction of the assistant secretary, Bob Long. Keith Shea was transferred from the Forest Service to work as a staff officer reporting to Bob Long on that national program. Each of the three program managers and their immediate staff also were assigned to the Office of the Secretary, USDA. The overall policy direction came out of the secretary's office as well as approvals for annual plans and budgets. But as you can imagine the administration of it was decentralized even more. My job was to provide overall direction to the tussock moth program, but with people like Ken Wright and a few others around...

HKS: Is the tussock moth still an issue?

REB: Yes.

HKS: Would you say the research was effective?

REB: Yes.

HKS: How do you solve the problem then? I'm a member of Congress, I've been giving you all this money, and we've still got the bug.

REB: There are two issues involved here. One of them is organization and the other one is scientific. These research, development, and application programs had their start at that time. Administratively it's a model that we've copied a number of times since then. It was also a way to mobilize funding to achieve coordination and accountability. That was the organizational question. In terms of science, the Douglas-fir tussock moth program turned out to be far and away the most successful of any of the 3-Bug programs. The reasons were only partly related to the organization. There were two technological developments underway at the time of that outbreak, both of them here at the Forestry Sciences Laboratory. One of them was the work with a virus, nuclear polyhedrosis virus (NPV), by Mauro Martignoni and Hank Thomas. They knew that tussock moth outbreaks collapse because a naturally occurring NPV begins to sweep through the insect population. By then, however, the damage to the forest is already done. Those two scientists developed the techniques for reproducing the virus in the laboratory and testing its efficacy and safety. That technology was just beginning to emerge when the 1971 tussock moth outbreaks began. The second major scientific break-through came from Gary Daterman.

Gary worked on the sex attractant for the Douglas-fir tussock moth in cooperation with the Oregon Graduate Center located on the west side of Portland, which had strong organic chemistry skills. It took several years to identify the sex attractant, pheromone, but the team finally was successful. That technology came to fruition about the same time that the outbreak began. So there were two major technologies to feed into that tussock moth RD&A program — a way of detecting and monitoring the insect population, with sex attractant, and a means of controlling it with early insertion of the virus into the outbreak area. The RD&A programs tested both technologies and all the signals were positive. Boyd Wickman and his associates understood the ecology of the tussock moth and deserve a great deal of credit for the field testing. Then the epidemic collapsed. Ironically we haven't had a major outbreak in the U.S. since that time, but Canada has. And so the Pacific Northwest Station and Canada have collaborated on those two technologies in the intervening years and the results were positive. Simply stated, we can monitor the population with the sex attractant and manage it with the virus.

HKS: So, did EPA give the Forest Service permission to use DDT?

REB: Yes. Reluctantly, hesitantly, but they did with many constraints. And Jack Ward Thomas, a wildlife biologist, had just joined PNW from Massachusetts. He monitored birds in the DDT-sprayed areas. He couldn't find any difference between bird survival in areas with and without DDT. So far as I know that's the last time that the agency ever used DDT—or any other agency in the U.S. for that matter.

Biological Control of Forest Insects

HKS: Introduction of a virus to control an insect outbreak. Are environmentalists, however you want to define it, concerned about that kind of contamination? It's unnatural? It's quasi-natural. Obviously it's better than DDT, but you're still messing with the environment.

REB: Some of the environmentalists are so hostile to forest spraying of any kind that they react viscerally. You know, BT, *Bacillus thuringensis*, is widely used a biological control agent for insects, and even then there's hostility to it. So there are some people who are opposed to any spraying. But it didn't take long to win this NPV case which is actually a naturally occurring virus specific to the tussock moth. We went through all of the protocols for testing it for safety and efficacy and it was approved. Russell Train, then heading EPA, presented in person an award to the research team which developed the technologies.

HKS: What does a station director do? You've got this difficult problem, the mechanism is in place, you've got super scientists working on it, how much time do you spend? You've got a lot of things on your plate.

REB: The biggest part of my job at that time was liaison with the National Forest System, because Region 6 had the administrative responsibilities for spraying that eight hundred thousand acre area. They were deeply involved in the consequences and interpretation of the research because they had responsibility for preparation of environmental impact statements and the operational aspects of the control program. Other parts of the job included frequent contacts with the Washington office and various user groups. It was a matter of being a good listener and a cheerleader.

HKS: Right.

REB: As I said before, this was probably the most successful RD&A program of the 3-Bug effort. It was successful because there were two technologies emerging that only needed to be adapted to field application.

This brings up another aspect of RD&As. They are good where a good body of knowledge exists that can be sealed up and otherwise adapted to the problem. The difficulty is that they are not good at doing basic research and creating new knowledge. You've got to have technologies emerging or available. To some extent, that was the problem with gypsy moth and especially the southern pine bark beetle. We did not have major breakthroughs to back up the adaptive work. To be sure there were marginal breakthroughs but not major ones. All those RD&A programs were completed within specified time frames and have now been discontinued. We've gone back to more bench science and more basic research.

HKS: I always thought the gypsy moth would be controlled by finding out whatever keeps it under control in Europe, but apparently it's not that simple.

REB: A major strategy for all insect pests is to find parasites and predators that control them in their natural habitats. There have been expeditions to Europe and Asia searching for parasites and predators of the gypsy moth but up to this point they haven't been very successful.

We had a marvelous related story here in the Pacific Northwest Station that was highly successful. It had to do with a larch case bearer, a defoliating insect introduced from China. It affected western larch, an important tree common on the east side of the Cascades and in most of the Intermountain West. Tom McClintock, then WO director of forest protection research, was so concerned by the case bearer defoliation that he said we're just going to drop almost everything and go to work on it. In the meantime, Roger Ryan of the PNW Station was assigned to work on the case bearer. He collected parasites and predators of larch case bearer in China. Within a relatively short time in the 1970s and early 1980s that parasite/predator complex had increased to such an extent that the larch case bearer is no longer a problem. A simple story of successful biological control.

THE SPOTTED OWL

HKS: I'm not sure how this fits into your chronology, but it may have been about fifteen years ago. I told you earlier I knew Don Flora well. I was passing through Portland. We chatted for a while and he was saying, tomorrow I'm going over to Bend. He said I don't know if it's going to be a snipe hunt or not. They're studying something called the spotted owl, and I'm going to go out and observe this. Anyway, the station was studying the spotted owl long before it became famous. Was it predicted how serious...

REB: What year was that?

HKS: Well that would have been in the early '80s.

REB: The spotted owl was not a significant issue when I was the station director. I left PNW in 1975. The spotted owl became a concern a year or two later. Eric Forsman, now with Forest Service research here in Corvallis, wrote his masters and Ph.D. thesis in 1976 and 1980, respectively, here at OSU. The spotted owl moved up in visibility year by year.

RD&As

REB: What I wanted to do is introduce the idea of RD&A programs which came out of the Arnold/Dickerman period. And the fact that I saw this from the administrative side. It set a pattern for innovation in research organization that persists to this day. The 3-Bug programs were the start, but then there was the CANUSA, the Canadian-U.S. RD&A program in spruce budworms and SEAM, Surface Environment and Management, which was coal mine reclamation, that came along in the energy crisis years.

HKS: It's in my list of things because in the annual report of the chief, year after year under the section under research, coal mine reclamation was something that for whatever motivates something to get into the chief's report, that was one that made it every year.

REB: That was another RD&A. It followed the ones that I'm talking about here.

HKS: And that's on the eastern national forests.

REB: Western as well.

HKS: Is that right?

REB: The coal fields of Montana and Wyoming were being opened in the late '70s, because western coal had lower sulfur content than the eastern coal. Much western coal continues to move east. It was in the 1970s that the public became concerned about the effects of surface mining on both eastern and western coal fields. The Forest Service began the SEAM program in the mid-1970s with research both in the East and West. We even created a budget line item—SEAM. The RD&A program continued for several years and in my estimation was another one of these success stories, second only to tussock moth and maybe even equal to tussock moth. The research demonstrated both in the East and West that surface mining rehabilitation techniques were available and reliable. Have you noticed how little public concern there is today on environmental consequences of surface mining. The program ultimately was closed and there is little Forest Service research on the subject today.

The upshot of this is that the Forest Service developed a whole family of ways to organize research. When I became deputy chief I had almost no interest in generic organization questions, because we had so much flexibility in how we could put programs together. All I wanted to do was take advantage of the tools that we already had. Institutional reorganization of Forest Service research was of no interest to me. Now case-by-case groupings of scientists and resources around specific problems were of keen interest. SEAM, COPE, and FIR and the Eisenhower and Pinchot consortia were cases in point. We didn't need more authorities or precedents to put research together in almost any way that we needed or wanted to.

A STATION VIEW OF WO LEADERSHIP

HKS: When you were at PNW, that's when Jemison retired and Dickerman became deputy?

REB: No. Arnold was deputy chief when I came to the PNW in '71. He retired from the Forest Service in 1973 to go to the University of Texas. Steve Spurr, the president of the University, was tugging him down to Texas. Steve and Keith had worked together at the University of Michigan, Keith as dean of the School of Natural Resources and Steve, as I recall, dean of the Graduate School.

HKS: That's right.

REB: By that time, Dick Dickerman had lost his wife. John McGuire thought a great deal of Dick. John has given lots of credit lines to Dick and properly so. McGuire appointed Dickerman as the deputy. And Dick comments on that in his letter.

HKS: That letter of Dickerman's, that summary of his that you sent me, said when he became deputy chief, John said he wanted a steady hand, he wanted to settle things down. What needed settling?

REB: I think we're talking about Keith Arnold's...

HKS: You were here at the station. Was that a Washington office settling down?

REB: I think that McGuire and Dickerman were getting feedback about how uncomfortable the stations were with Keith's wide-ranging ideas. I didn't feel that way about Keith because I worked with him. I understood his manner of doing things as you only do when you work day-to-day with a person. I liked working with Keith. But I think other stations were not so comfortable. They saw Dick as a steady hand, and that's the way I perceived him then, and still do today.

HKS: Maybe it was Harper commenting that when Keith went to Texas, he really wasn't well suited for the hurly-burly of Forest Service Research. The perception that a civilian would have is civil service is pretty calm, you have job security and so forth, but it's characterized as a pretty rock and sock 'em operation. I thought it was kind of an interesting characterization.

REB: I reread that Dickerman memo yesterday. There are subtleties in there; you've captured some of them. But Dick was still a remarkably good and loyal lieutenant to Keith Arnold as well as a steadying hand.

The Harper/Jemison Years

HKS: In a sense I suppose the way Jemison was to Harper.

REB: Yes, a good parallel.

HKS: A very good lieutenant but also a good...

REB: But a solid performer independently. By the way, Marie and I had dinner with George Jemison a short time ago, and we were reminiscing about these days. George was never critical of Harper, but his wife, Bea, was. [laughter] Oh she didn't like Harper, because he was so demanding.

HKS: In other words you didn't work a forty-hour week when you worked for Harper.

REB: Oh no you didn't. The exchanges I've had with Les Harper would suggest that George was a very capable lieutenant. George would take on such tasks as developing a national program for research, but really the genesis of the idea was with Harper. I have the impression that George was an innovator in his own right, he never got credit for it. For example, that speech, "Let's Get Scientists Out of the Woodshed," which gave rise to the laboratory construction program, I think was his undertaking.

We also need to keep in mind that Jemison succeeded Harper as deputy chief. Also, Harper resigned from IUFRO upon his retirement, and Jemison succeeded him, becoming president in 1968. These leadership positions were unlikely to have occurred if Harper had thought less of George Jemison and his abilities.

HKS: In Jemison's interview, he was a little bit critical of Harper in terms of recreation research. Les apparently didn't support the sociological studies, maybe the soft science, he didn't say why. That was one of the things that they disagreed about enough that Jemison wanted it on his record.

REB: Yes, but that was the only critical thing that I've ever seen him do on paper. Les just didn't give visible credit.

AERIAL YARDING SYSTEMS

HKS: FALCON, spruce bud worm, prescribed fire, Alaska, you had some specific topics you wanted to talk about.

REB: Yes. The environmental movement was underway in a major way when I came to PNW. Much of the ire at that time was directed at mountain logging systems. Another one of those innovative organizations was an attempt to speed up research on environmentally acceptable logging and transport systems for the mountains. That was FALCON; Forest Advanced Logging and Conservation.

HKS: Where was this done? I didn't know there was logging engineering in this area.

REB: The Forest Service had a research center in Seattle, engineering and transport systems, headed by Hilton Lysons.

HKS: Okay.

REB: Hilton was an imaginative guy. We outlined a concept called FALCON which was also an RD&A program. Don Flora was giving intellectual leadership to it; we brought Ed Clark out of the Washington office to assist Don. It was intended to develop logging methods on a broad front. There were three principal approaches, balloons, helicopters, and cable systems. That program was moving along and Congressman Wendell Wyatt was on the Appropriations Committee. Constituent groups here, with some collaboration from the Forest Service, approached Wendell Wyatt and he got a very substantial add-on to the budget about 1973. It was three million dollars, a huge add-on for that time. Julia Butler Hansen of Washington was chairperson of the committee, so she and Wyatt had a great deal of influence. That was also during Nixon's second term, and Nixon was beginning to impound congressional add-ons. I don't know if you remember those days or not.

HKS: I do.

REB: Nixon latched onto those FALCON dollars (and many other congressional add-ons) and wouldn't release them. Congress was outraged, but the Congress also knew that it had to do something about budget discipline.

HKS: There was the heliostat too.

REB: That was a modification of balloons and that came along a little bit later. It was the politics of FALCON that I want to call attention to here. Anyway, the PNW was set to get the largest budget add-on that a single Forest Service experiment station ever received. But Nixon wouldn't turn the money loose. Congress knew it had to do something and enacted the Budget Impoundment Act of 1974. Up until the early post-World War II period, budgets were largely built, regionally and locally, by interactions among the Forest Service and constituent groups. I was confronted with that problem when I went into Washington as deputy chief. So the FALCON program never really blossomed into the full-blown one that we imagined. We continued to pursue the technologies, but much more slowly. Balloons, cable systems, the helicopters—I think we have a pretty good appreciation of what each system can do today as a result of that work.

HKS: Bohemia Lumber Company did that balloon logging, was that in conjunction with this? Was that how it was field tested?

REB: Yes. The balloon work already underway was largely done by Bohemia and Fay Stewart. The industry really did the balloon logging work and Bohemia was responsible for it. The other two approaches were done in part by the public agencies, with a lot of cooperation from the industry and the National Forest System.

HKS: How did Bohemia get selected as opposed to Weyerhaeuser or a larger company?

REB: It wasn't a question of being selected. Fay Stewart was already in the balloon business when FALCON came into being. As it turned out, balloons didn't work out very well. There was an amusing story about balloon logging. We always had the question, what would happen to these huge balloons in a snow storm? [laughter] The balloons are seventy-five feet or more in diameter and they have fairly flat top surfaces. We knew that the storms in the mountains could be violent. One day a storm did occur. I don't remember who the logging contractor was, but he took the balloon off the cable system and hooked it up to a very big tractor, a D-7 or D-8, and he was marching the balloon down the road attached by cable to the tractor. Can you imagine a balloon tethered to the back of a tractor? Overhead? And it was snowing heavily. Suddenly the balloon became unstable. Understand, we didn't know how a balloon would behave in snow. So the balloon is loading with snow and all of a sudden it tipped. Tons and tons of snow, so the story goes, came cascading down and then the balloon snapped upright. The rear end of the tractor came off the ground and then settled back down. You can speculate on how that tractor driver felt. [laughter]

HKS: I was skeptical. It's the amount of time, if it takes you twenty minutes to pull a turn of logs in, you can't make any money at it. The cables are so long.

REB: Conceptually a great idea. But slow. Also helium was expensive and nobody wanted to use cheaper hydrogen. Also, they are easily damaged. Stories about balloons getting pulled into yarding drums and things like that. You can't put up with many of those problems before the system is just not useful. Alright, FALCON was another one of the RD&A programs, and it had some political and budgetary implications for research in general that stayed with me for years afterward.

PRESCRIBED BURNING RESEARCH

HKS: Prescribed fire, that's very fashionable. Congress must love that, during the environmental times. Go out and do what God always intended.

REB: The research on prescribed fire was started, or perhaps it would be better to say reactivated, during my time here. When you were here at PNW we were doing slash burning and broadcast burning and that certainly is prescribed burning. The model that I was looking for was understory prescribed burning, patterned after research in the South and my later work in the Lake States. The Douglas-fir tussock moth also called attention to the significant successional changes in the forests of eastern Oregon where shade-tolerant trees were coming on in abundance. The issue of fire and forest succession is still with us; today we label it the forest health problem.

   Bob Martin was part of a cooperative Forest Service unit at the University of Washington. The purpose of the unit, training fire researchers, was drawing to a close. We transferred Bob to Bend, Oregon, where he started prescribed burning research. He had a special talent for that work, and as a result of that research I am told that eastside Region 6 is burning as much as fifty to sixty thousand acres a year. It started several years ago. Bob left the Forest Service and went to the faculty of the University of California, Berkeley, where he is now conducting fire research and teaching. I think that it was the research program that we started in about 1973 in Bend, Oregon, which had a lot to do with the legitimacy of fire research in eastern Oregon. We've only touched the surface—much more fire research needs to be done.

HKS: Harold Weaver did thirty or forty years worth of research with the Bureau of Indian Affairs. Was that useful in a scientific sense?

REB: Yes, Harold Weaver was, as nearly as I can tell, more a promoter than he was a researcher. His favorite theme was to use fire for thinning in ponderosa pine. And that's an extremely difficult thing. It certainly had a lot to do with stimulating interest in fire. Who was the person who headed the institute in Florida?

HKS: Komarek at Tall Timbers.

REB: Tall Timbers. He was, as nearly as I could tell, in the Harold Weaver mode, an evangelist and a promoter.

HKS: Okay.

REB: I don't want to suggest that the initiation of prescribed burning at Bend was new, there were many antecedent efforts. My concern was that the potential benefits were high and the efforts so minimal.

HKS: I was tutored by Dave Bruce, who was extraordinarily cynical of the people working prescribed burning because of the lack of science, the lack of hypothesis, the lack of structure. They went out and touched it off, and they made notes. Purely descriptive stuff. I couldn't judge, but Dave said that and so I've always remembered that. It became so fashionable. The Park Service was going to burn Yosemite and the photographers were there.

REB: Dave is right on. In my days at Grand Rapids, Minnesota, the year was 1958, Zig Zasada and I visited Charleston, South Carolina, to look at the work that Lee Chaiken had started on the Francis Marion National Forest. Have you ever seen those plots? They are world class, unfortunately destroyed by Hurricane Hugo. We found something that I had always been looking for, and that was an acceptable experimental design for prescribed burning. The study had a series of burning schedules, all replicated. We carried that experimental design back to Grand Rapids in red pine. I think it would meet Dave Bruce's concerns about the quality of research. A number of variations of those study designs have been used elsewhere, but so far as I know not here in the Northwest.

HKS: He's very statistical in his approach to things.

REB: There are appropriate experimental designs to go with prescribed burning. Some of the fire research was descriptive stuff, too, but I always tried to have credible scientific backstopping as well.

HKS: Dave probably had some fire research experience in the South.

REB: Yes, Alexandria, Louisiana, for one. I think we're already revisiting the whole prescribed burning issue in the West. The Bend fire research has been underway for fifteen years, but my instincts tell me that with the forest health issue so dominant now in eastern Oregon and Washington, that we're going to see a dramatic upsurge in the use of fire.

RESEARCH COORDINATION

HKS: As I understand it, the Wallowa timber type slops across the boundaries into Region 1. How do you coordinate that as a practical matter? You've got two different experiment station areas, plus two different regional offices, different national forests.

REB: That's in part what the staff groups in Washington, D.C., are supposed to do, but there are less formal systems as well that contribute to coordination.

HKS: Does that work?

REB: Yes, reasonably well. This in part was also a product of Harper's organization of the late 1950s. For example at Grand Rapids, Minnesota, Zasada's—as center leader—responsibilities stopped at the Minnesota border. My responsibilities as the project leader were Lake States wide. In other words, I was responsible for red pine growth and yield studies in Michigan, Wisconsin, and Minnesota. On the question of prescribed burning and wildlife research or any other topic of common interest, some accommodation would be made between say the Intermountain Station and the Pacific Northwest Station. Not always perfectly so, but there is a very high order of understanding about what each other is doing and who's going to do what.

HKS: I can see if you have a really high-class crew...

REB: The problem of research coordination is far more exaggerated in the public's mind than in reality. It is not a major problem. Informal networking goes on among scientists that's just incredible. And the same thing applies internationally. IUFRO provides a marvelous international networking system among forest researchers.

HKS: Maybe the touchy part of that question I asked is in application, not research.

REB: I think you're right. But in defense I would argue that adaptive research really tends to be site specific, client specific, and you have to go through some of that duplication in order to get that work into practice. But you're right, it's at that end of the RD&A spectrum where you see redundancies.

ALASKA

HKS: Alaska was in your domain here. Alaska has been an enormously controversial natural resources issue. Some of that controversy must have affected the kinds of research decisions. What to study and when to study and how much money you've got. Is that true?

REB: Yes. Let me back up here just a little bit. Alaska, after World War II, was not part of the Pacific Northwest Station. And I can't remember when it was combined with PNW, but it was in the '50s or the '60s. Some Alaskans resented that state merging with the Pacific Northwest Station. But when I became director it was a part of the station and the Alaska D-2 lands issue was in full flight. You know that the issue was resolution of the native claims. The urgency was triggered by the Prudhoe Bay oil discoveries. I spent lots of time in Alaska with Ken Wright who was the assistant director responsible for Alaska programs. The issues were mainly national forest oriented, but research was at least peripherally involved. The regional forester was Charlie Yates, one of the most difficult people I have ever worked with. Charlie Yates was, of course, ambitious, as were all of us, that there be additional national forests created in Alaska. As it turned out that didn't happen, it became wildlife refuges and national parks. Still, interest in Alaska called attention to a lot of technical issues.

HKS: I would think silviculture would be controversial.

REB: Yes, except that silviculture was mainly a problem for heavily forested southeast Alaska, much of which was already in the Chugach and Tongass national forests. Tundra and tiaga vegetation associations in the interior was the focus of concern during the D-2 debates. There was a feeling that research would be important after the D-2 issues were completed, not so much for timber research but for basic ecological studies related to all resources. The Forest Service already had a laboratory in Fairbanks and a research group in Juneau but no laboratory. We wanted a laboratory there and came close to getting it. There was so much money available in Alaska that the state would have built it. Unfortunately, the financial situation in Alaska tightened up just before the money was to have been appropriated.

As I recall, our research team at Fairbanks was substantially strengthened during that time, and there were people like Les Viereck and John Zasada (Zig Zasada's son who is now in Corvallis with PNW) who were in Alaska and developed a lot of the creative silvicultural techniques for managing spruce along river flood plains, the only place there was much timber. So we contributed substantially to the scientific technical issues at that time. Science had little impact on the land allocation questions which were political. In the meantime, at Juneau, Admiralty Island and other places were coming up for consideration as wilderness areas, and timber cutting for the Sitka and Ketchikan paper mills was becoming increasingly controversial. Our research program in southeast Alaska also was substantially strengthened during that time. Lots of emphasis in southeast Alaska was placed on silviculture and on anadamous fish.

HKS: Sure.

REB: Partition Alaska into two parts. The interior with the D-2 lands, and southeast Alaska with the Tongass and the Chugach national forests, which has been in the Forest Service hands for a long time and the two research programs were somewhat different.

One more comment. There were some vocal people in Alaska who wanted Alaska to have a separate experiment station. I was not enthusiastic about that, and neither was the chief's office, because it would have added greatly to the costs—the whole administrative structure, and so forth. I think we fairly well fended that off. And that issue so far as I know is now no longer active.

HKS: Let's talk abut natural areas.

RESEARCH NATURAL AREAS

REB: I mentioned earlier that I was involved in the Natural Areas Program, both with the Society of American Foresters and the Forest Service when I was staff assistant to Carl Ostrom in Washington, D.C. The most active Natural Areas Program among any of the Forest Service regions was here in the Pacific Northwest, going back to 1927. Thornton Munger was an active supporter. Jerry Franklin was the main shaker and the mover on natural areas when I came here. It was a pleasure for me to team up with Jerry to see if we could do more with that program. I especially wanted to draw other agencies like the Bureau of Land Management and the National Park Service and Fish and Wildlife Service and the Nature Conservancy into that program.

So Jerry and I developed a program here. We said we need to know what kind of natural areas we have. Up until that time their creation had been more opportunistic than planned. We conducted a workshop to identify overall natural area needs. It had a lot of high level participation from the Forest Service and from other public and private agencies. We developed an inventory of what we had and what we needed so that we could guide the program. It was called the Yellow Book—printed with a yellow cover. It accelerated the creation of natural areas among all agencies. There are two research natural areas, for example, on the Findley Wildlife Refuge about five miles south of here, containing vegetation not found on the national forests.

HKS: Is this tied at least intellectually to the forest biome business?

REB: Related. In any event, the Natural Areas Program had some acceleration at the time, although it wasn't viewed with great enthusiasm by land managers who were under pressure for more grazing areas and more timber areas. A sequel to that story was the emphasis on biodiversity which was written into several pieces of federal legislation. As the land management planners began to work here, they said that this gave them encouragement to be sure that we had vignettes of various kinds of ecosystems. The upshot was that, when I had the Natural Areas Program in Washington, D.C., in the 1960s, we had a hundred research natural areas on the national forests. Today there are over two hundred fifty. In fact it was last month that the ceremony was held for the two hundred fiftieth natural area in the Forest Service, and there are already two or three hundred more in the pipeline. Now imagine, we have two hundred fifty areas in all kinds of vegetation associations that are relatively undisturbed.

HKS: Yes.

REB: As I said earlier, an important program, but one that isn't fully appreciated.

HKS: So the local manager, the district ranger, the forest supervisor, sees this as a complication.

REB: They did earlier. They see it now as an adjunct to the biodiversity requirement.

DEPUTY CHIEF FOR RESEARCH

REB: A concluding point on my Pacific Northwest days—I began to get inquiries about when I might be willing to come back to Washington, D.C., sometimes not so subtle. My reactions at that time were several. We had four children. A daughter who was in high school, and daughters are particularly sensitive on the moving question. I asked John McGuire and Dick Dickerman, "can you wait until she finishes?" The answer I got was yes. In 1975, after I'd been here four years, the inquiry came again. I wasn't terribly eager to go back. But I had a commitment to do so.

HKS: I guess he was the first chief to do this (Max said he did it too), requiring each regional forester to sign an agreement that they would come to Washington when they were instructed to. Did they make the station directors sign something like that?

REB: No, they did not. We had somewhat more success at getting station directors to move than John or Max did getting regional foresters to move and for reasons that I only partially understand. We tried to consult frequently with station directors and to honor special personal needs. For example, my successor as deputy chief was John Ohman, formerly at the North Central Station. He was asked to move and he said that we have a son that we would like to have finish high school. John McGuire honored that and so did I. You know John Ohman sent a letter to the chief when his son graduated. He noted that his son was graduating from high school and I'm now at your disposal. I thought that was really upbeat. I know about that arrangement with regional foresters, but it was never imposed on station directors.

In any event, my immediate family commitments were finished in Oregon and I went back to Washington as associate deputy chief. I filled the slot that Carl Ostrom vacated. About that time, Warren Doolittle filled the slot vacated by Bob Youngs who was also an associate deputy chief. Bob Youngs became director of the Forest Products Laboratory. The question of succession to Dickerman was up in the air for almost a year. Warren Doolittle and I were the two principal candidates. By the way, Dickerman actually retired before his successor was appointed. So Warren Doolittle and I rotated as acting deputy chief. Finally, I think it was nine months later, I was formally appointed as deputy chief.

HKS: John does this, but does the secretary officially appoint the deputy chief?

REB: Yes. It's a supergrade or senior executive position, and the formal approvals are required in the secretary's office. They may actually go to the Civil Service Commission. I think they did after we moved over into the Senior Executive Service. But you can be sure that at that level the chief has enormous influence.

REX RESLER

REB: One of the people who probably played a major role in my appointment to the deputy chief's job (I know Warren was disappointed but he was very gracious about it) was Rex Resler the associate chief. Rex is one of the finest people that I've ever worked with. Rex was appointed regional forester R-6 the day I was appointed PNW station director.

HKS: So you worked with him out here closely.

REB: I worked with him here but I also interacted with him through a group of young turks in the Washington office in the late '60s. Rex, as regional forester, was very generous about inviting me to his advisory committee meetings and inviting me to sit in on his R-6 staff meetings. He would come over to visit me at the station headquarters, and regional foresters don't normally do that. It was a warm, warm relationship. Rex stayed in that regional forester job for not more than two years; he went directly back to Washington as associate chief under McGuire. Did McGuire ever comment about Rex as his associate?

HKS: About the political problems he got into. Crested Butte or something like that.

REB: Yes, Crested Butte ski development. I don't know what relationship Rex and John had. Rex was a great idea generator and had some of Keith Arnold's qualities. An idea generator and a guy who could be enthusiastic about a new idea. Rex wanted to be chief, but I'm sure that those political issues about Crested Butte got in the way. Then Bill Towell retired as executive vice president from the American Forestry Association and Rex went to that job.

HKS: I heard somewhere that Rex was one of the five district rangers that were used as models in Herbert Kaufman's book *The Forest Ranger*. Have you heard that story?

REB: Could be but I've never heard that. Permit me to elaborate on the kind of person Rex was. Rex had been a young forester on the Alsea district of the Siuslaw National Forest. Shortly after Rex became regional forester he took all of his forest supervisors, nineteen of them, and invited me to come along as well. We went on the Corvallis watershed where Rex wanted to show two things. One of them is how you challenge authority and the other one was that he wanted to show that we need to be innovative. The issue was road building. Shortly after Rex graduated from Oregon State University, he was to set up a timber sale in the Corvallis watershed on the Siuslaw. The regional policy was to build roads with a certain radius of curvature and with prescribed widths and grades. And, Rex said we can't do it here in the Corvallis watershed. So, he shortened the turning radius and he steepened the grades, including adverse grades. You didn't do that. All of this was in the face of Charlie Connaughton then R-6 regional forester, who was one of the most intimidating people going. Rex thought he was going to get canned by Charlie, but it was exactly the sort of thing that Charlie was looking for. And, Rex wanted to show this to his top staff as an example.

Changing Budget Procedures

HKS: Do you have more comments to make on your time at the Pacific Northwest Station?

REB: I continued to pay annual visits to key staffers and congressional offices in Washington, D.C., during my time at the Pacific Northwest Station. I remember one visit to the staff clerk of Senator Alan Bible's Senate Appropriations Committee. During that time, a longtime colleague, Bob Callaham, was on an internship in Bible's office. I visited first with Callaham, and he told me that he had described to the staff clerk how the Forest Service does its budgeting, including contacts with cooperation. Then I visited with the staff clerk. He looked at me with some irritation. He reached over and picked up a three-ring binder called the Mark-Up Book for the Forest Service. He said these are the additions that have been proposed for the entire Forest Service budget. He held his fingers up (more than two inches apart, and more than half the amendments) and said these are the ones that are proposed for research. This told him that research was being extremely active in trying to modify the budget and, of course, he was right. The clerk waved his finger at me and he said, if you keep this up, we'll give you these amendments but we're going to take it out of your base program. The lesson for me was that we had to bring a lot more discipline to the budget process than we had had before. A significant part of my job in the next ten years as deputy chief was to bring some discipline to that process, in so far as possible, without diminishing opportunities to strengthen research.

HKS: Research was asking for too much, is that what you're saying?

REB: If not too much, at least it was occupying too much of the workload of the committee. That meeting was prophetic in the sense that we had to look once again at the discipline of the budget process.

Ten Standard Regions

REB: I mentioned earlier about a meeting in Missoula, Montana, when George Jemison and Reed Bailey, the station director of the Intermountain Station, met with us and announced the closing of the old Rocky Mountain Station. During my time as director of the Pacific Northwest, the Nixon administration was moving to place all of government into ten standard regions.

HKS: I remember that.

REB: The western station directors were meeting surreptitiously to deal with the implications of that because it would have closed some stations and had great displacement effects, particularly on the administrative side of the regions and stations. There was great momentum for ten standard regions, and the Pacific Northwest Station was to have assumed all of the research programs of Idaho. Bob Harris, then director of the Intermountain Station, and I agreed not to keep people in limbo with anxieties hanging out all over. We received a letter passed on by the chief saying, proceed with ten standard regions. Within a day or so Bob Harris and I visited Moscow and Boise, Idaho, and announced that Idaho was joining the Pacific Northwest Station. Some of the people sitting in those rooms were the very same ones who were in the room twenty years earlier when the old Northern Rocky Mountain Station closed. The parallels and the irony of that situation stayed with me for a long time. Did you know that within three or four days, the three M's—Senators Mansfield, Montoya, and Moss—overturned Nixon's decision and the ten standard regions just collapsed. So in all of my later visits to Moscow and Boise, Idaho, I used to tell them that it was great having you in the PNW Station for three days. [laughter]

HKS: My impression of Nixon, my perception of the man, is that he was fundamentally a good administrator, one of the better administrators of our presidency.

REB: That agrees with my view.

HKS: The ten standard regions didn't make sense for the Forest Service, because the forests aren't spread evenly across the country. Was this tied to the Roy Ash study on reorganizing the government, putting the Forest Service with Interior and that kind of stuff? Or was this wholly separate from that?

REB: I think that was separate. That Roy Ash issue occurred during the Carter administration, if I recall correctly.

HKS: Wasn't there one too under Nixon? And it happened just before Watergate.

REB: That could well be the move toward ten standard regions. It was the Ash proposal. I think it is. Nixon began his second term by accepting letters of resignation from most of his senior people and he was going to embark on a whole new direction for the federal government, but by that time he was so encumbered by Watergate that he could never carry that out. I think if he'd been strong in his second term we would see a substantially different government.

HKS: He would have been strong, because he won by a landslide.

REB: A very able guy, but with some serious moral flaws.

SELECTION OF DEPUTY CHIEF

HKS: To the best of your knowledge and from your vantage point, how was the selection made for you to become deputy chief.

REB: I don't know.

HKS: You don't know.

REB: I don't know but I can guess. I do know that there were conversations among the chief and deputies. I would get little glimpses of vague conversations that I was going to come back to Washington, and the position I was likely to occupy was the associate deputy chief slot or...

HKS: McGuire didn't call you and ask, would you like to come back to Washington? Research needs your kind of leadership, or whatever he would say. Sort of test the water to see if you wanted to come back?

REB: You know if Ed Cliff was obtuse in his dealing with John McGuire, John McGuire was obtuse in dealing with me.

HKS: Hear that John?

REB: John would make some very vague references about coming back to Washington. Never talked about what position I was going into, but I think he thought I knew. I did talk to John about my postponing until our youngest son completed high school and he understood and acquiesced. Actually, as I reflect on my time as deputy chief, there were many internal discussions about succession, often not communicated well with prospective candidates.

HKS: But you weren't interviewed as such for the job. He knew you well enough to know what he was getting.

REB: Late in Ed Cliff's tour of duty, he asked when I was coming back to Washington. Ed was still chief, as I recall, but very close to retirement. I said, I haven't heard anything about it. He kind of put his hand over his mouth and said, oops, I thought you knew. So it strikes me that there's some internal discussions and I kind of knew that I was going back as associate deputy chief.

HKS: Did Dickerman retire a little earlier than he might have?

REB: Yes, I have a hunch that he timed his retirement so that replacement candidates would be at hand.

HKS: Okay.

REB: You'll have to ask Dick how he saw the affair, but in my estimation there were two choices. One was Warren Doolittle, the other one was me, and I'd like to think the Forest Service would do well with either one of us. As I said, Warren was very gracious when I was appointed deputy chief. Bob Youngs was associate deputy chief under Dickerman and became director of the Forest Products Laboratory shortly before Dick retired. Warren Doolittle occupied his associate deputy chief chair. Carl Ostrom, also an associate deputy chief, retired and I occupied that position.

HKS: What I want to be able to find out in this portion of the interview is, what's it like to be a deputy chief. What does a deputy chief really do. You get up in the morning and you shave and have breakfast, you drive into the office and you go in and you sit down and you've got a staff report or the phone rings. What do you do? It's sort of like cruising timber, we're going to take a 5 percent sample, we can't do it all.

REB: Tom Nelson put it this way. He said the three jobs for a deputy chief are budgets, personnel, and programs. That was my perception also and in fact that's the way it turned out. Of the three jobs, the one least forgiving for the deputy chief is leadership on budget matters, because nobody else can do it in all its dimensions. The second is personnel. It's bringing people along to do current jobs and to anticipate future needs. Adverse personnel action, that is addressing personnel problems of various sorts, is also an important and time consuming part of the job. The deputy chief plays an important role here, but some of those things will be done if the deputy chief isn't there. The third area is programs, in my case research, and here I was well fortified. The Washington office research staff consisted of about ninety people, most concerned with research programs, with an additional three thousand in the field, and so you're well served on programs. What should research to do, how to monitor it, how to report it? All of the things that have to do with research itself. I was an active participant in the formulation of research programs, but most of the activities and most of the ideas came from my colleagues in Washington and in the field.

THE RPA AND RESEARCH

HKS: Let's talk in a generic sense here. We want to talk about the budget cuts and the Gramm-Rudman effect and all of that, but generically at what point in construction of the research budget do you talk to the chief?

REB: A highly interactive process. Could I take a slightly different approach to that issue? I would like to describe some steps my colleagues and I took to give additional visibility and substance to research with the aim ultimately of winning more support for it. Most of my predecessors, starting with Earle Clapp, used a variety of devices to accomplish these same ends.

HKS: Sure.

REB: I went back to Washington almost four years to the day after I came to the Northwest. The environment in Washington at that time was heavily oriented toward RPA. Max Peterson had come in from R-8 and was the deputy chief for programs and legislation, which is where RPA was. RPA was a large part of the WO workload. Actually I'd been on Max's advisory committee as a station director even before I went in. You may recall that the first RPA had to be prepared in about sixteen or eighteen months.

HKS: Right.

REB: Humphrey wanted it in a hurry. The first RPA was due in 1975 and the second five years later in 1980. The first RPA was most helpful in increasing funding for the National Forest System. Congress was eager to have a document that would permit them to increase those budgets. As we got into the second RPA period, Max Peterson, John McGuire, and others said the national forests are now financially healthy and whole. Our next installment is going to emphasize research. I knew that when I came into Washington and had a concept to justify an increasing research program in the 1980 RPA. That story has several parts to it.

HKS: In 1978 you get a renewed and enlarged authority to conduct research. Do you want to talk about that law?

REB: Yes I do. I'd like to unfold that story just a little differently.

HKS: In terms of these three categories of what a deputy chief does, legislation comes under programs.

REB: That's a little artificial in terms of research, because legislation really interacts actually with all three—budgets, personnel, and programs.

HKS: Okay, let's talk about that. Now explain again why this law was seen to be necessary from your point of view. You had the 1928 McNary-McSweeny Act. Was there any other statute that authorized research?

REB: Yes. Minor ancillary authorities but McSweeny-McNary was the basic authority for Forest Service research. I recall John McGuire saying to Research and to State and Private Forestry, clean up your act. We've had two major bills, RPA and the National Forest Management Act. He did not want Research or State and Private Forestry to be caught with incomplete or obsolete legislation as we were with the old Organic Act of 1897.

HKS: Okay, so there was a lesson learned.

REB: Yes.

HKS: You really do have to upgrade the legislation from time to time.

REB: And we in Research and State and Private Forestry wanted to make sure that we weren't caught with outdated or inadequate authorities. That I think was McGuire's attitude. My view was that we also wanted to use new legislation to give visibility and momentum to a research program as Earle Clapp had done in the 1920s with passage of the McSweeny-McNary Act.

HKS: Okay.

FOREST SERVICE RESEARCH AND USDA

REB: Are you familiar with the relationship of the Forest Service to the Department of Agriculture in terms of science?

HKS: Only what little bit I've read in my preparation for the interview.

REB: All of my predecessors, at least all of my post-World War II predecessors, were keenly sensitive to the science and education agencies of the rest of Agriculture. That means Agricultural Research Service, Economic Research Service, Cooperative State Research Service, and Cooperative Extension Service.

HKS: Sensitive in what way? Be sure you're compatible with that or what?

REB: Because much of the science policy of the Department of Agriculture including the Forest Service was coming out of that group. For the Forest Service, the principal contact in USDA was McIntire-Stennis, which was administered by the Cooperative Research Service.

HKS: I see.

REB: The top level USDA planning committee was called ARPAC, Agricultural Research Policy Advisory Committee, and that was the pinnacle of the policy-related issues in science in the Department of Agriculture. Forestry research was represented on ARPAC by the deputy chief for research, by a forestry school dean, and by the director of the Cooperative State Research Service. Agricultural interests made up all the rest of the 20-25 person committee. It was chaired by the assistant secretary and one of the deans or vice presidents of a university.

When I first came into Washington, I sat in on those meetings with Dickerman. The assistant secretary at that time was Bob Long, and his university counterpart was Orville Bentley, then dean of agriculture at the University of Illinois, and highly regarded by the agricultural community. I'd gotten to know Orville over the years. In the first ARPAC meeting, Bentley and Long were going over a major agricultural research planning report from a conference in Kansas City. All of the agricultural community felt good about the Kansas City report (ca. 1976) because it sampled a wide range of agricultural users that had commented on priorities for agriculture research. Bentley and Long turned to me and to Don Duncan, who was then representing the forestry schools.

They said we want you to do something like this for forestry. That was exactly the challenge I wanted to hear. The forestry schools, CSRS, and the Forest Service then initiated a major involvement of the forest community in research. That led to four regional workshops attended by about two thousand people where a large cross section of users met to suggest research priorities in 1977. It also involved a national conference in 1978 where top level natural resource people commented on the overall national priorities for research. This advice also provided input for research priorities in the RPA. Understand that we were satisfying an agricultural science and education requirement but at the same time laying out a program for Forest Service research for RPA.

I wanted to do something else as part of this. I wanted to deal with some of the current policy issues. Up to that point our concerns had been the content of the research program, not the conduct of it. We convened a major policy workshop at Airlie House outside Washington, D.C., in 1977. Steve Spurr was the chairman and Carl Stoltenberg was the vice-chairman. Dickerman, now retired, was the secretary to that group, and he was working under the auspices of the newly formed Renewable Natural Resources Foundation (RNRF). We were anxious to have the community of interest that dealt with forestry research to be larger than forestry. We wanted it to be other disciplines, other agencies. Cooperation with RNRF was one way to do this.

HKS: Give me an example or two of a policy or an issue that you wanted to deal with. What are the policies for research?

REB: There were about eighteen or twenty recommendations out of the Airlie House report on forest and rangeland policies. What we wanted out of this review were new ideas and a strong endorsement for several research-related things, including international forestry, competitive grants, enhanced cooperation, and improved delivery of information.

So here we had the content of the program from four regional workshops. And, we had a review of research policy issues from the Airlie House conference. I wanted to use this policy document as background for the drafting new research legislation as suggested earlier by McGuire. Understand that all of these efforts were collaborative with the sixty forestry schools affiliated with McIntire-Stennis, the Cooperative State Research Service, industry and other cooperators. What I was trying to achieve personally was an enlargement of the scientific community that was interested in forestry and forestry research. This wasn't new with the Forest Service, but I'd like to think that I tried to give collaboration and cooperation still more emphasis. As you look at the documents coming out of these conferences, you'll see joint signatures and joint sponsorship of virtually everything that we did. All the time that we were developing this I was participating in the RPA process so that what we did was entirely compatible with RPA.

RESEARCH ON RESEARCH

REB: Let me add a couple of more items to this story. I told you earlier how disenchanted I was with the old PPBS zero-budgeting work that we did during the Johnson era. Up to the late '70s I was reluctant to go through any charades of analytic analyses of forestry research. Then we began to see some research coming out of the agricultural community, where highly credible cost-benefit studies had been done on research on corn, on wheat, on dairy cattle, and so forth. It was apparent that some of the methodologies developing here would be useful to forestry. So we created a project on research evaluation at the University of Minnesota where that group of economists had been especially active in agricultural research policy and research analysis. Al Lundgren of the North Central Station did this research with cooperation from the University of Minnesota and several other stations and universities. Out of this research came a group of studies concerning the benefits and cost-benefit ratios of a whole series of technologies including southern plywood, nursery practices, and so forth. These studies served as analytical background for the Forest Service with its RPA, the forestry schools, and Cooperative State Research Service.

At the same time, Bob Callaham, who was the staff director of Forest Environment Research in Washington, suggested that we develop additional justification for forestry research. We commissioned a study later called Criteria for Deciding about Forestry Research Programs. Bob Callaham with all his energy and his colleagues then canvassed about fifteen or sixteen federal departments and agencies and asked how they plan their research. It turns out that there were few new insights from those agencies compared to what we already knew.

Another step in that research program justification was the commissioning of a study on basic research chaired by Stan Krugman of the Forest Service and Ellis Cowling of N.C. State University. I wanted to increase the basic research content of the Forest Service program. In any event the task force identified a number of basic research needs important to forestry.

NEW LEGISLATION

HKS: And so you were developing background that would serve RPA and new legislative initiatives?

REB: That was the intent. In the meantime, draft bills were requested by Congress. The three major groups—Research, S&PF, and forestry extension representatives met in a hotel in Crystal City near the Pentagon in 1977. They spent a day or two discussing the content of the legislation and actually drafting the bills. Denny LeMaster calls them the baby bills in that active legislative period of the '70s.

I'd like to think the major initiative was coming out of Research but State and Private Forestry was also active and a new bill concerning forestry extension was introduced, coming out of extension forestry in the universities. All three of those bills went to the Hill. John Hendee was on the Hill at that time working, as I recall, for Congressman Weaver of Oregon. He helped to move those three bills through as did several other key staffers. Do you know those three bills went through without hearings and debate? As I reflect on it, we would have been better off if there had been hearings and more visibility. Thus, we obtained the Forest and Rangeland Renewable Research Act of 1978.

HKS: You would have a better legislative history if there's ever a lawsuit?

REB: Right, but there is more to it, including parallels to the McSweeny-McNary Act. Some have said the Forest Service didn't need the McSweeny-McNary Act as we had all of the authorities needed in the old Organic Act of 1897 which authorized as I recall, the Forest Service to conduct such investigations as are necessary.

HKS: Sure.

REB: But Clapp and others, I think, wanted more visibility for forestry research and the SAF study and debates surrounding the McSweeny-McNary Act did that. I just wish we'd had a more catchy title for the research bill. In any event, the research bill (Rangeland Renewable Resources Research Act of 1978) passed but public attention was so heavily focused on national forest activities that there really wasn't very much visibility for research.

HKS: Did the environmental group have a position on this or not? Did you know?

REB: They may have. Jim Giltmier, Bob Wolfe, John Hendee, and Dennis LeMaster were the principal staff people on the Hill at that time. I have no doubt that they touched bases with all the critical constituents. At that time, there simply was little conflict with Research, S&PF, or forestry extension.

HKS: Denny was at SAF at that time.

REB: I think Denny was with Congressman Weaver. In any event, the principal staff got all of the support and endorsement that they needed. So the bill just slipped through. One day it was passed. Then it seemed the rational and desirable thing to do. There are likely to be long-term, positive benefits from a well-conceived forestry research legislation and there are RPA-related benefits as well.

HKS: Who is it that sponsored that bill?

REB: It may have been Talmadge on the Senate side, and perhaps Weaver on the House side. But those three bills just slipped through so quietly that no one knew that they passed.

HKS: The Forest and Rangeland Renewable Resources Research Act of 1978 kind of ties it in to RPA in terms of the cadence of and cumbersomeness of the title and so forth.

REB: The act restated much of the research authorized by the McSweeny-McNary Act. It also emphasized multiple-use research including the addition of outdoor recreation studies, forest protection in all its dimensions, a broad definition of forest utilization, and a careful and wide-ranging restatement of forest assessment and economics research. In short, we attempted to anticipate as many as possible of the areas of inquiry that forestry and renewal natural resources might one day address. It also added authorities for international forestry research, competitive grants, and encouraged cooperation, coordination, and extension. It even has a section concerning acceptance of grants and gifts, a problem then occurring at FPL.

HKS: So it wasn't just housekeeping. Did it broaden the authority or sharpen it?

REB: It broadened the authority and resolved some vagueness in the McSweeny-McNary Act and it filled in some other niches. For example, the authority to do international forestry research was strongly enhanced.

HKS: The Forest Service lost range research early on.

REB: In the Eisenhower years.

HKS: It went over to the Agricultural Research Service.

REB: Most of it did. A little bit stayed.

HKS: Did it come back?

REB: So far as I know it did not. Joe Pechanec was the person who was very active on that question in the early 1950s. The Forest Service lost the Miles City, Montana, location which was primarily range research but I think we retained the Desert Experimental Forest in Utah and the Santa Rita Experimental Range in Arizona, for reasons that I don't understand. It's probably because they had a forested component with them. The Forest Service still does range research, especially in areas associated with national forests and grasslands.

HKS: The Forest Service is a part of Agriculture and you have to be compatible. Were you ever constrained by USDA research policy?

REB: No. In fact I think the opposite was true. Agricultural research policy was very helpful to the Forest Service. Take the Person-in-Job concept and the pioneering research concept that came out of ARS. Budget support was generally positive, although the support was muted because the Forest Service comes under the Appropriation Subcommittee for Interior and Related Agencies, while all the remainder of Agriculture comes under the Agriculture Subcommittee for Appropriations.

HKS: Would it be fair to say that Research operates more closely with USDA than the National Forest System in terms of running across the street and talking to those guys?

REB: Yes. But it's a two-track system, and the deputy chief for research must work with both groups. I don't want to say that we played one off against the other, but reinforcement came from the science and education groups of USDA. In my time as deputy chief, Keith Shea was especially effective at representing Research with the USDA agencies.

HKS: Is this a logical time to talk about how all the activities were integrated into a coherent program for forestry research?

REB: The research program, the one requested by Orville Bentley and Bob Long, and that jointly involved the forestry schools and the Forest Service, was published about 1978. For the Forest Service, the same material was published as part of the RPA in 1980. In the meantime we were satisfying, I think successfully, Rupert Cutler, Howard York, John Fedkiw, and the other budget-makers in the Department of Agriculture about the importance of the forestry research program. The climate for research in USDA, this was during Carter's years, was also supportive of research. Secretary John Block, formerly an Illinois corn farmer, often described how much research meant to him during the corn blight outbreak of about 1970. So, in the first year after the 1980 RPA was approved, Forest Service research received, as I recall, an 8 percent budget increase from OMB and expected to get an 8 percent budget increase for the next several years. I thought that was a successful outcome of the research priority setting, the research policy and legislation, and the various analyses of research accomplishments and needs. The success was to be short lived.

HKS: The rationale behind that increase was the Forest Service can't pull off RPA without some very significant research.

REB: That was part of the rationale.

HKS: So if Congress had accepted the program it had to give more money to research.

REB: That's right. That was all built into it. But the consultations and the economic justification and the new legislation, all of that tended to reinforce that rationale.

HKS: Was there any grumbling from the field that research was getting more money than...

REB: Not that I could detect. The National Forest System wasn't grumbling because they had received hundreds of millions of dollars of increases in those first two or three years of RPA.

HKS: What does OMB say about all this anyway? They're always trying to balance the damn budget, and here's the Forest Service going crazy.

REB: OMB had agreed with it because of the justification presented to them, and because the climate for agricultural science and natural resources was favorable.

HKS: OMB, according to Max and McGuire, opposed RPA.

REB: Yes they did.

HKS: Because it took away the president's prerogatives or something.

REB: More so, much more so in the Reagan years than in the Carter years. I hope Max and John elaborate on how RPA finally received approval from President Ford.

HKS: After it passed, though, then OMB was no longer opposed to it.

REB: No, and Carter was interested in natural resources. So, the increase in research was, as I recall, part of the statement of policy that went on the Hill, which committed the administration to increasing research.

HKS: RPA was '74, you're back here in '75, Jerry Ford's still president for another couple of years.

REB: Yes, yes, but this effort really took place in the Carter years. Ford was involved with that first eighteen months increment of RPA, the one with eighteen months or so to complete. So, what we have is a whole series of steps that were intended to build a rationale and a justification for increased research and we were successful. No problems with Congress on the 7 or 8 percent increase during the first year. The second research budget also went to Congress with a proposed 7 to 8 percent increase, and then Carter lost the 1980 election to Reagan. The disappointment to me was that these budget increases would have represented the first overall growth of Forest Service research since the Harper/Jemison days.

HKS: Carter was defeated. What happened next?

REB: I often use my hands to describe what happened. The budget was going like this for research [gestures upward], and it was in the administration's budget, it wasn't a congressional add-on. That made it much more durable. Soon after the Reagan administration came in, they sent an amended budget to the Hill, so the second 7-8 percent increase went like this [gesturing downward]. In fact it declined more steeply than it went up. This is when John Crowell came into office as assistant secretary. You may want me to comment on the contrast between Cutler and Crowell.

HKS: Oh yes, definitely. I want you want to talk about the changes in administration, how that translates out on the land. So you get a lot of money, I mean you almost have to go out and hire a bunch of people to start spending that kind of increase, right?

REB: Yes, programs were growing.

HKS: And under civil service job tenure, it's a little bit dicey, right? What happens?

REB: No it really wasn't. We could easily handle 7 to 8 percent increases among equipment needs, construction, some recruitment, and extramural or contract research. So we felt pretty good. We, the forestry schools, and the Forest Service felt good about the way that we had built the rationale and justification for the research program. Can I talk about Cutler now?

Rupert Cutler

HKS: Absolutely.

REB: Cutler was a person that I knew slightly from my University of Michigan days. He came into the Department of Agriculture with really only one mission in life, it was the Forest Service. Cutler had a fairly large portfolio. He had the Forest Service, Soil Conservation Service, and the science and education agencies of the Department of Agriculture.

HKS: He was well-equipped by educational experience to handle that, wasn't he? More so than the average assistant secretary?

REB: He was a professional but he was a political...

HKS: But he had credentials...

REB: He had credentials if you mean a Ph.D. and a keen interest in the Forest Service and environmental interests. But, his agenda was so narrowly focused that he really cared only about the Forest Service. The agricultural side of science and education soon became disenchanted with Cutler, and they created something called SEA, Science and Education Administration, which was a super agency within the department. SEA had responsibilities for Agricultural Research Service, Cooperative State Research Service, Cooperative Extension Service, and so forth. That disenchantment was to grow as time went on.

Cutler continued until, as I recall 1980, and it was obvious that he was becoming a liability. I think it was John Block who was the secretary at that time. Cutler left, he resigned and departed, and that was a poignant moment worth describing. Cutler had accepted a senior job with the Department of Natural Resources in Michigan. He had his furniture all packed and it was ready to go, and Michigan withdrew the offer that day. Several of us who had worked with Cutler liked him as a person. We went over to a morning meeting to wish him well and to offer commiserations.

I suspect that Cutler's very strong environmental orientation had a lot to do with who came in as his successor. If Cutler was strongly left-leaning, and environmentally-oriented, Crowell was business and far right-leaning. Cutler's ideology on one side gave rise to an ideology almost as extreme on the other side.

HKS: That's right.

JOHN CROWELL

REB: After some delay and controversy in obtaining Senate approval, John Crowell was appointed assistant secretary in 1981. Max wanted some of the deputies to meet Crowell so we went to his office for our first meeting. For me it was easy to visit with Crowell because he's a steelhead fisherman; he came from the North-west, and we know some of the same people.

HKS: Did you know him from your time there?

REB: I didn't know him personally, but we knew several people in common and we liked the out-of-doors in the Northwest. At that first meeting each of the deputies described highlights of their program. When my time came, I offered a few obser-vations. Crowell abruptly said that research should be a hundred million dollar a year program. He said he had serious reservations at this time when there is a tim-ber shortage and so many other needs whether we can afford a one hundred forty million dollar research program. The abruptness of his observation was startling. It portended things to come.

HKS: So he was looking at the budget priorities, rather than saying you're doing unnecessary research.

REB: He wasn't sympathetic to research at all. He said, I don't understand research and he asked for help. But he never understood research, nor its importance to, for example, Georgia Pacific and Louisiana Pacific, his former employees. I think that he was getting a lot of his input from some of the industry people out here, who didn't associate their business with research.

INDUSTRY AND RESEARCH

HKS: That was my follow-up question. Does industry generally support the research budget?

REB: No, on the average not with enthusiasm. However, there are many positive exceptions to this generalization.

HKS: It seems like they are getting a lot of freebies, and they'd be for that.

REB: One example out here was with a company, which I won't name, that was very proud of its computerized sawing programs. They didn't want the Forest Service doing anything with computer development for sawmill operations.

HKS: I see. A proprietary interest in certain kinds of research.

REB: What they lost sight of is that the original computerizing of the sawmill industry came out of the Forest Service. It was a program called Best Opening Face from the Forest Products Laboratory in Madison, Wisconsin. But they'd forgotten the seminal work. They were already into the second, third, and the fourth generation of computer development. Lots of similar things happened.

HKS: I remember, when I was at the station, that some industrial spokesman said that the Forest Service ought to focus on basic research, let industry do the applied research.

REB: You're opening up a whole additional set of questions. The captains of industry by and large either were indifferent or not supportive of Forest Service research. But the groups within industry, the land managers and product development scientists could be very supportive and often were. The industry was especially supportive of research to backstop lumber grade standards, fire safety procedures, and pesticide studies involving safety and efficacy.

HKS: What company besides Weyerhaeuser has substantial forestry research, as opposed to products development?

REB: Weyerhaeuser was the leading one. The second one was Westvaco and West Virginia Pulp and Paper Company, in the Southeast. Crown Zellerbach had a research program. Did you know Clarence Richen?

HKS: Sure.

REB: Clarence is a grand old guy. He went to Crown Zellerbach from the Forest Service about the time of WW II and became their vice president for lands. He was extremely supportive of research.

The question about the Forest Service relationships with industry is a recurring one. I think all of us at one time or another were required to justify why we were doing research in forest products. The question was so predictable that every one of us had a hip pocket rationale to use in congressional hearings or budget hearings.

During the Carter administration we went through one of those periodic bouts about who should do research, and it was the utilization research that came under the greatest scrutiny. Why are you doing work at the Forest Products Laboratory in Madison? Why shouldn't the industry do it? The budget implications of those questions were clear.

We developed a rationale for that kind of research. One is basic, risky long-term research. The industry isn't going to do it. Another reason was that there are many small entrepreneurs, small sawmill operators, small wood processing plants and so forth, who could never mount a research program on their own. Another was research that backs up consumer interests such as safety and reliability of preservatives, toxic chemicals, and fire retardants. Still another was to backstop regulatory programs including the toxic consequences of wood preservatives and pesticides in forest products and forestry.

The one place that we always said industry ought to do its own research was where there were clear proprietary interests. But every once in a while someone would ask that question either on the Hill or at OMB with budget implications.

CONGRESSIONAL RELATIONS

REB: On the day that Dickerman retired, for example, there was the usual morning chief and staff meeting. John McGuire announced that he had the sad duty of announcing Dick Dickerman's retirement. On that very afternoon we got our first budget marks from the Hill. I think it came out of the House, saying that they were giving us a three million dollar budget reduction in forest products research. I've often kidded Dick that this was the kind of a legacy he handed off.

HKS: Sure.

REB: But it stemmed from the same set of concerns. The industry ought to do the research. Warren Doolittle and I mobilized, with help from FPL, all of the support we could find around the country, much of it from industry, saying we needed that research. They intervened and the three million dollars were restored.

HKS: Generally these cuts would come out of the Republican side?

REB: No, I think it was independent of party. The appropriations process really is relatively party-free. Julia Butler Hansen was a Democrat, Wendell Wyatt was a Republican, and yet they worked hand-in-glove, worked very closely together. There was relatively little partisan politics that I ever saw in the appropriations committees.

HKS: In the media it's always that Teddy Kennedy wants more money for health care and the Republicans want more for something else. Some issues are very politicized, but forestry was not politicized.

REB: There are thirteen appropriations committees, and each one will tend to have a personality all its own; they behave differently. The Interior and related committees were relatively free of partisan politics.

HKS: How much of your budget came out of Interior as opposed to Agriculture, in terms of committees?

REB: All of the research funds came out of the Interior and Related Agencies Committee.

HKS: Okay.

REB: Essentially none out of Agriculture except some tree planting programs, administered by S&PF. One of the significant historical events about Forest Service appropriations is that jurisdiction shifted from the Agriculture Appropriations Committee in the 1950s over to the Interior and Related Agencies Committee. Most people don't comment on that very much, but I have the impression that that was a significant event.

HKS: Was it a good event?

REB: A significant positive event for the Forest Service.

HKS: Why? Because the agricultural people in Congress really only cared about wheat and soybeans?

REB: That's exactly right. This moved all of the Forest Service programs then into a much more compatible budget environment. For example, there were almost no research budgets before in the Interior committee; that meant that Forest Service budgets had more visibility.

HKS: And that's good.

REB: I want to mention another small but important change in the format of research budgets. In all of the budget hearings that I sat in on, and there must have been ten or eleven of them, the ordering of discussion was always National Forest System first, Research next, followed by State and Private Forestry, and then the administrative parts of the budget. This is the way the material was presented, and that really set the agenda for the budget hearings. The National Forest System was, as I said, soaring at that time. I made the suggestion to John McGuire (John and Max were very supportive for more visibility for Research) how about inverting the order of the budget presentation? John and Max agreed, and so we put Research first, State and Private Forestry second, and National Forest System third.

HKS: I noticed that. I didn't know how significant it was, but the annual reports of the chief changed the sequence too. Research was the lead-off a lot of times.

REB: Okay. I think that was also RPA-related. But something so simple led the members of the committee into Research first. That had a significant impact on visibility and the discussions that took place in the Appropriations Committee. Very much to the advantage of Research in my estimation.

HKS: Did you testify directly as deputy chief?

REB: Yes.

HKS: Do other deputy chiefs testify directly?

REB: Yes. McGuire handled this somewhat differently than Max. Max wanted the testimony to flow through him. John would have his deputies at the table. John would sit at the center of the table with the associate chief, Rex Resler, beside him. The assistant secretary and the other deputies, Research, National Forest System, and State and Private Forestry with occasionally others. As the discussion went along, John would turn the questions to the appropriate person.

HKS: So he was always there.

REB: He was always there. Of course his principal preparation was for the National Forest System. That was the big business for John and for most of the others. John was very generous in deferring to me on technical questions, and Rex Resler would occasionally intervene and defer questions to me, so I was getting a fair amount of visibility and airtime during McGuire's tenure.

Max handled it somewhat differently. He sat at the head table, and his associate along with the assistant secretary, as I recall, would sit there with him. Only one deputy would be at that table, and it was the one whose program was under review. Max tried to funnel much of the discussion through himself.

HKS: So under John all of the deputy chiefs were there as part of a learning experience for you too, you got information first hand, so to speak. Did it instill a sense of the mood of Congress?

REB: Much of the learning was done before we went into the room. But yes, the hearings were essential to capturing the mood of Congress.

HKS: I know, but you saw the interchange.

REB: Yes, I sat in on budget hearings as a staff assistant, but often in the back row, and more directly as deputy chief under both John and Max. So I had some feel for how that process worked. The way to characterize appropriations hearings is a little bit like preparing for preliminary exams for a Ph.D. It all came together at that time and there were relatively few second chances.

HKS: You really had to have the data in your head, you couldn't be fumbling...

REB: Much of it in your head. All of us carried notebooks. I always went over the budget item by item with the staff directors and the background for them and their consequences, trying to anticipate questions.

HKS: Did you brief congressional staff ahead of time?

REB: We had some interaction with congressional staff. Generally, at their initiation. They would ask, for example, are there questions that you would like to have asked for the record.

HKS: Alright.

REB: They wanted to build a record. Can you give us background on this issue or that issue? There was often strong interaction between...

HKS: Did they tip you off and say the senator's really looking for things to cut this year, so you better have something to cut.

REB: Occasionally, but rarely directly from the congressional staffers. That usually came through the chief's office and through the Forest Service budget coordinator, who were key people. Did you know Gordon Fox?

HKS: By name only.

REB: Gordon was a consummate contact with the Congress, and he had the confidence of the staffs on the Hill. They would feed information back to him, and Gordon would consult with each of us. They had confidence in him, and so did we. There were others that followed. All were outstanding people crucial to the well-being of budgets and far more important to the agency than their organizational positions would suggest.

HKS: Did you ever sense that other agencies were less or better prepared than the Forest Service when it comes to budget?

REB: I really have no way of knowing. But I have the feeling that the Forest Service relationships with the Congress were the envy of many agencies. Not always flawless, but better than most.

HKS: What you're saying sounds so logical that you wonder why everyone's not doing this. I guess there's a certain sense of esprit de corps in the Forest Service that probably doesn't exist in other agencies.

REB: I think that's right. Maybe they are not as astute in their relationships with the Hill. I remember when John became chief, he said to the RF&Ds, the regional foresters and directors, that he intended to spend 25 percent of his time on congressional relations, 25 percent in administration, and 50 percent with the agency in the field, or something like that. The numbers may be wrong, but John clearly knew he had a major responsibility with the Hill.

HKS: To us civilians, testifying in Congress is kind of mysterious. All you see are the newsclips on Dan Rather or something. It's a controversy, and Congress is beating up on this witness. Most of us don't really know what happens in the hearing, because we never really see it.

REB: One of the things that insures some success in the Forest Service, at least one of the ingredients, is the ability to work with the Congress. You know I gave you some anecdotes earlier about how staffers would call.

Let me relate another story. I came to know Dick McArdle reasonably well after he retired. He was just a delight to visit with. He told me a story. He said, you know I don't want you to repeat this, but when J. F. Kennedy came in, Orville Freeman was named secretary of agriculture. McArdle, who had been chief for eight years and was highly regarded, knew that Freeman, who had been governor of Minnesota, wanted to bring in his own team. Actually it was George Selke, formerly with Freeman in Minnesota and then with him in USDA, who wanted to replace McArdle. Richard McArdle, who enjoyed a very special relationship with Senator George Aiken, called him and described his current situation. Aiken responded that he would call Jack and see what he can do about it. Jack of course was John F. Kennedy. McArdle said, you know I never heard anymore about being replaced. I told you earlier about Warren Doolittle visiting Aiken.

HKS: Right.

REB: When he told George Aiken that the Burlington Laboratory was number two on his priorities, Aiken called Ed Cliff who was then chief and just ate his butt out. McArdle used to have breakfast on Saturday mornings with George Aiken.

HKS: What was the tie there?

REB: It was a very personal relationship. In any event, McArdle would have breakfast with Aiken on Saturday mornings when both were in town, and they developed a very special personal relationship. Those often exist among staffers and among members of Congress. Ed Cliff didn't do that, and that was one of the things that made Aiken so unhappy when the laboratory question came along. The next thing that happened was that Ed Cliff and Warren Doolittle paid a personal visit to Aiken.

HKS: Was Aiken a chairman of the committee?

REB: Aiken was chairman, I think of the Agriculture Committee, not Appropriations but Authorization—the committee that passed the laws. It exercised a lot of oversight, and Aiken was a conservative but highly regarded Republican. He and McArdle got along very well.

HKS: I hear so many anecdotes of this genre that the White House doesn't check with Congress before it makes a public statement. Maybe they do a lot and you hear it only when they don't. They forget that there's a key congressman that's got a vested interest in some idea who's just not going to go along with it. Seems so fundamental.

REB: Those congressional relationships are extremely sensitive, and were a point of discussion in many, many internal chief and staff meetings. Who's going to deal with this congressman or that one. Two or three fairly senior people in the Forest Service did not enjoy good relationships with Congress. One of the tough jobs for the chief is to recognize those problems and keep the person away from the Hill.

I told you that Phil Briegleb took me around to all of the key congressional offices when I became director of the PNW. Bob Tarrant and I did the same thing when it came my time to leave the PNW.

HKS: Right.

REB: As deputy chief, I did the same thing with my successor, John Ohman. I could get appointments with staffers or occasionally with members of Congress fairly easily although it made Max and Lamar Beasley a little nervous at times.

Permit me to carry congressional relations a step further. The House Appropriations Subcommittee was chaired by Congressman Sidney Yates. I have the impression that Yates' number one target in the Reagan years was James Watt, secretary of the interior. There were many stories around about how he badgered and worked on Watt. The story also is that the second man on his hit list was John Crowell. So we would go into appropriations hearings (this was in the early '80s) and now research is the first item on the agenda. Yates would start by calling on me and begin to ask a series of questions. Crowell sat at the hearing table and as the years went on, Yates began to play me off against Crowell, not a comfortable position for me. I don't know whether Max or John ever mentioned that.

HKS: No.

REB: But Yates would ask me some questions and then ask Crowell a question about the value of the research. Crowell would say, well it can be postponed, and the Yates would say, "Dr. Buckman, do you think this research is important?" I would issue some kind of a disclaimer because I was expected to support the administration budget. Yates would keep coming back to me, do you really believe that? Then I would begin to say to Yates what I really believed. This became a pattern in the House Appropriations hearings for a matter of three or four years while Crowell was there. There were times when I think I spent more time testifying in appropriations than Max did.

HKS: I wrote to each station director saying I was going to interview you and asked for questions. One of them came back, ask Bob about his sparring with Congressman Yates.

REB: That could have come from any of them. I don't know whether it was sparring or not.

HKS: Was he being mischievous when he was doing this with you and Crowell, or did he really want to set something up?

REB: I think maybe it was a mixture of both. He didn't think much of Crowell; he was looking for a rationale to support research and he was probably kind of bored with some of the national forest items which had been coming up over and over again.

HKS: Now the assistant secretary would typically sit through all of the budget hearings.

REB: Yes.

HKS: And comment with an opening statement?

REB: Yes.

HKS: And then sort of keep quiet unless called on?

REB: The assistant secretary would intervene on policy related issues. In fact, on policy related issues John or Max would turn to the assistant secretary and let them deal with them. As an example of a policy related issues—do you believe we should have more wilderness or less wilderness or something like that.

HKS: So the chief routinely deferred to the assistant secretary...

REB: On significant policy issues the chief would always do it. On technical issues, of course the chief and his lieutenants...

CROWELL'S AGENDA

REB: Now Crowell brought with him an agenda that was very narrowly focused on timber. He dealt with that issue as a conservative Republican who had been active in the forest industry (legal advisor to Georgia Pacific and Louisiana Pacific). He thought that the Forest Service was wasting resources here in the West. He was searching for every way that he could to increase the timber cut, with special emphasis on the Pacific Northwest. He would give us, for example, a reduced research budget and then even after that was accepted by the Department of Agriculture he would go over and lobby OMB to reduce it still more so he could get another road built or another three or four hundred million board feet of timber cut. He was narrowly focused.

As a person Crowell was a very warm individual. I have visited with him several times since we both returned to the Northwest. As a top level policy maker, he was inept and naive. I've often said to friends he was a well-educated plodder, an ideologue. He could have accomplished some of his agenda if he'd just listened. But he alienated the South, he alienated the supporters of State and Private Forestry, he alienated the supporters of research. He did one good thing; he resigned after four years.

HKS: How come people like that aren't replaced? Why doesn't the secretary see this guy is screwing things up, he's alienating the South. It doesn't help the larger program to do that.

REB: I'm not sure that the secretary really had time to deal with the Crowell agenda. I think that was partly true in Cutler's case too. The assistant secretary is really the key policy person for the Forest Service in USDA.

HKS: In the Reagan years we had Jim Watt and we had the big flap at EPA and Ruckelshaus was brought back. I guess Crowell was sort of below radar in that combat zone.

REB: Crowell wasn't below radar, but he was the lightening rod, he was taking the heat. He wasn't deferring these issues to the secretary or...

HKS: That makes him a good guy from the secretary's perspective.

REB: From the secretary's perspective, no penalty for me, Crowell's taking the heat. But he did some awfully naive things. Did anybody ever talk about his Audubon speech?

HKS: No, Max referred to him a few times. He had lunch with Crowell and Cutler one time, after they were all three out of office. Max said, you know if I could have put you two guys in a sack and shook you up and taken the average, my life would have been a lot easier. And they laughed. That was the only anecdote I've heard.

REB: Crowell was not a vindictive or a bitter person. He was an ideologue. One of the first public appearances for Crowell was a speech before the Audubon Society in Denver. The Forest Service gave him a draft speech which he apparently modified on his way to Denver. Understand that Crowell was a longtime member of the Society. In any event, his speech was poorly received. Max or John, I don't recall which, reported back that Crowell couldn't understand why this happened.

HKS: Sometimes attorneys tend to be that way because law requires an adversary relationship.

REB: But, I didn't see the compromise or the conciliatory role that I think Crowell ought to have played. He knew he needed to bring somebody in from the South, and he was trying to recruit as his deputy assistant secretary the state forester of Mississippi. He didn't get that person who was a former Weyerhaeuser employee. Crowell just had a capacity to alienate people that he needed to make his program go; and, he didn't listen very well.

HKS: We Americans believe in the democratic process. The candidate says, if I'm elected I'm going to do one, two, three, and they get elected and sometimes they actually try to do those things. When you go from a Carter to a Reagan, it's really abrupt. A lot of people think that's good; it's long overdue; we should have done this long ago. When we went from Ford to Carter, we shifted the other way. But when you're in there running a federal agency, and it goes through a shift, it's quite a bit of chaos. How hard is it really to make some of these shifts?

REB: In my case the shift was dramatic and painful. My program was strongly conditioned by budgets. Research requires relatively little new legislation from year-to-year—not like the national forests. Budgets and their justification is the heart of the issue with OMB and the Congress.

HKS: When does Gramm-Rudman kick in?

REB: Gramm-Rudman was actually a post-retirement affair for me.

HKS: Okay.

REB: I think I told you earlier that one of the most useful things that I acquired at Harvard was how the administration and especially the Congress looks at the agencies—congressional supervision of public policy. That served me well, but you have to have instincts for dealing with Congress, and the people and the personalities involved in order to be successful. There were others who were far better at it than me.

HKS: You mentioned during the break that Congress frequently would restore some of Crowell's cuts. How did he feel about that? Did he think you were going around him?

REB: Yes. He was very unhappy. He never talked to me personally about it, but he talked to Max and Dale Robertson. He was unhappy that John Ohman's State and Private Forestry budget and especially the Research budget were being restored, contrary to his wishes.

HKS: He didn't believe that it was all Congress's initiative, that you guys must have been in there working with them.

REB: In some respects he was right.

HKS: Sure.

REB: During testimony I would go through the disclaimers that an agency and an administration witness has to go through, but I'm sure that the nonverbal signals were that I really didn't believe it [laughter]. We were in a desperate situation. Imagine a one hundred forty million dollar budget; overlay inflation, each year it cost more to do business, and Crowell wanting to push that budget down to one hundred million dollars. Those are draconian cuts. I really was secretly pleased when we would get those restorations. Yates seemed to enjoy those exchanges as well. Yates wrote me a nice letter when I retired.

HKS: That's nice.

Budget Cuts

REB: One of the consequences of the Reagan/Crowell budgets is that they quickly got into personnel cuts. We were confronted with budgets that resulted in 5 to 10 percent cutbacks in people. The first year to two after Reagan, I would ask the nine station directors including FPL, and the staff directors in Washington, to come in and we would go over those budget cuts person by person, location by location. If we cut back a program, I wanted it to be consistent and coherent. Kind of a gallows humor—these meetings were called "Christmas parties," because they always occurred just before Christmas. This period was very hard on station directors. They had to identify positions that would be cut. They couldn't talk about it because the president's budget is administratively confidential until it's released to the Congress. But late in January or thereabouts, the station directors would then have to announce the termination of this location and/or the proposed loss of a specific position.

HKS: The position would actually be terminated at the end of that fiscal year?

REB: It had to be terminated during the fiscal year for which appropriations were made.

HKS: So people would get about a six month notice for forced closing.

REB: The budget wouldn't be fully enacted until just before the new fiscal year began. So the person was on notice from about the first of February until the first of October that his position was slated for abandonment, or that location was slated for termination.

HKS: How in the world do you deal with budget cuts and at the same time affirmative action requirements to go out and recruit new employees?

REB: There are bumping rights and things that go with it. We developed a set of criteria about how we would administer that program. I knew that if we, for example, terminated locations in Yates' district or somebody else's district, the Washington Monument syndrome, to use an old cliche, that that would be quickly transparent. So we developed a set of criteria that included such things as—can this research be postponed, is it nearing maturity, can it be terminated? Are there retirements or resignations or other attrition that we can take advantage of? How productive is this research? We had about six or eight criteria. I wanted the justification for those budget cutbacks to be as objective as we possibly could.
    Wouldn't you know it the first year that we did this, Senator McClure of Idaho was chairing the Senate Appropriations Committee and I was testifying. He said, Dr. Buckman, the cutbacks that you are proposing affect five out of the eight senators on this committee, how come? It was apparent that he was ready to badger the hell out of me. I went through those criteria, and I said, "Senator McClure, we did not consider location, we used these criteria." I stayed with that approach for as

long as I was deputy chief. It made the budget much easier to defend. Back to the people impacts and the job that the station directors had, they had to bear that anxiety and that pain. There's just no way to win those situations. A person will say, "why me?"

HKS: Sure.

REB: In fact the thing that happened was that there was sufficient restoration that we really could greatly reduce the personnel impacts. Budget cuts had impacts on other things.

HKS: Were the reductions generally program-related or was this a chance to get rid of the less productive scientists?

REB: Some of that was involved, and we did it within laws and regulations. But, in any event, the restorations generally covered the people problems. I mentioned to you previously prescribed burning research in Bend, Oregon. The Bend location was one slated for termination. We would have moved Bob Martin to another location, but he was so embittered that he left the Forest Service and went to the University of California at Berkeley—a loss for the station.

Those budget cutbacks enabled us to make some program adjustments that we otherwise would might have wanted to make anyway such as termination of unproductive units. The upshot was that cuts were proposed; the Congress would make some adjustments, but we never came back to the same starting point. People would have retired; they would have transferred; they would have done something else. Even in those trying days we were able to generate some discretionary funding that could be used for program redirection. I took advantage of that. For example, I wanted the Forest Service to get into biotechnology research. I took some of those funds and sent them to the Pacific Southwest Station and the North Central Station to work on biotechnology.

HKS: Give me an example of a biotechnology project.

REB: Genetic engineering, moving DNA and genes around. For example, inserting a gene for herbicide resistance into a woody plant, that was one of the successful things that the North Central Station did. I wanted the two smallest stations, North Central and Pacific Southwest, to be the recipients of some of those surplus funds. Do you see what I'm saying, in that yo-yo process—cutback and partial restoration—we generated some surplus funds even in those days to create new programs. Biotechnology was one. I also mentioned an economics project on research evaluation of the North Central Station that served us very well and that was created with those kinds of funds. So the program could shift somewhat even in those days.

HKS: I suppose from the terms of just dealing with people it gave you a lever. Look, I'm dealing with Congress, we're going to have to make these shifts. That made the people in the field more receptive to things you wanted to do.

REB: Yes, but, taking people's jobs away from them is brutal.

More on Congressional Relations

REB: I want to comment a bit more about Congress. If I were to characterize effective relationships with Congress, it might go like this. They tend to be highly personal. Relationships develop that can be extremely effective, sometimes without regard to party. There are a couple of things that are absolutely essential, and one of those is trust. Congress tends to work with a handshake or a nod, and if you do anything to violate a trust you can be rendered ineffective.

HKS: How well do you get to know a member of Congress? Are you on a first name basis when you're out fishing and you're really friends in that sense, is that possible?

REB: Almost. Now that doesn't help with everyone, there are various degrees of formality. Bob Wolfe enjoyed unusually effective working relationships with many very senior members of Congress. Mansfield and many others.

HKS: He was an employee of Congress.

REB: He was with Congressional Research Service, Library of Congress.

HKS: Okay.

REB: Let me give an example. Bob Duncan was the Oregonian on the Appropriations Subcommittee in the mid-1970s when RPA came out. Duncan was the chief pusher that made budgets in the National Forest System go so far. I knew him and I think I called him Bob. One day Marie and I went to an Oregon state party in Washington, D.C., I knew the members of Congress from Oregon just have to show up. It's obvious that they divide those responsibilities, and on that day Duncan was covering for the Oregon delegation. He was bored to tears. He knew me from Appropriations, so he came over to visit. He asked how our new chief was doing? I said, "Fine." He said, "You know I'd like to visit with him sometime." This was just after Max had become chief. He said, "Why don't you give me a call?" So I told Max and Doug Leisz (then associate chief) that Bob Duncan would like to visit. I called Duncan, and he said come on down after work on a certain day. After we arrived, he turned to this staff assistant and said "Get a case of beer." We sat in Duncan's office for two or three hours drinking beer and talking about general forestry things. Those are the kinds of relationships that can be just extraordinarily helpful to the agency and pleasant besides.

There are some protocol questions that one just has be sensitive to. You have to be sensitive about the timing of announcements and opportunities to gain the member of Congress visibility in his district. If you don't do that; if you're not sensitive to that sort of thing it can arouse some...

HKS: If you develop a new lab in some district, they announce it, you don't announce it.

REB: Oh you seek opportunities for them to do it. You have to be very careful about feeding this to a Republican in a democratic administration and vice versa but there are often ways to get around this problem too. The system almost requires the interplay between the agency and Congress. Congress can't work in isolation and neither can the agency, it's essential that there be interaction.

HKS: Is there more interaction with the House or with the Senate?

REB: Probably the House because there are more members of Congress, and they have more time. But some of the Senate relationships can be extremely important too. Everybody in the Forest Service knows the heavyweights—Byrd of West Virginia, Leahy of Vermont, Hatfield of Oregon, and in earlier years many others. Talmadge and McGuire got along very well during those critical days of NFMA.

HKS: We all study civics and we learn how government works, but most of us don't really have any hands-on experience with how a bill really comes along. That's why I found Denny LeMaster's book so useful. It gave me a different slant on how Congress really functions. And you're saying the same kinds of things. I think it's very important to know what really happens, and the fact that so much is personal relationships.

REB: They're terribly important. Confidence in what you say and what you do.

BUDGET DISCIPLINE

REB: I would like to pick up one more point before we turn to EEO. I raised the question earlier about budget discipline, budget impoundments, and the Budget Reform Act of 1974. That changed in important ways the manner in which Forest Service research budgets were shepherded through OMB and the Congress. Up until the Budget Impoundment and Reform Act, it was possible for a station director with concurrence of the chief's office to lobby for some local activity such as a research lab or a program increase. With the Budget Impoundment and Reform Act, the authorizing committees of Congress are to be involved in the budget—heretofore the almost exclusive domain of the appropriations committees. The authorizing committees created the budget committee to set overall spending limits. The authorizing committees made recommendations to the budget committees which in turn sent their recommendations to the various appropriations subcommittees. The appropriations committees were supposed to work within those spending limits. That meant that the appropriations committees couldn't continually add to their budgets, which had been the pattern from the post-WWII period until 1974. This meant that we just couldn't tolerate anymore of those many requests for budget changes which I described to you in my run-in with Senator Bible's staffer. My office worked very closely with the station directors to say, sure we're going to take some risks; yes, we're going to reach out for budget support but if you go beyond our agreement and Congress provides additional funds, I'm going to ask you where in your station are you going to take the offsetting reductions. If you think about this kind of restraint on stations, it brought high degree of discipline to the budget process—unpleasant but necessary.

HKS: I think I missed the point in this amendment process. Amendments to what, to Congress, to OMB's, whose budget was amended? Who was Research amending? Which budget that you were using?

REB: The president's budget would be submitted to the Congress, and all of the amendments relate to the president's budget. The president's budget is the base document around which the whole debate occurs. When I say budget amendments, I mean particularly as modifying the president's budget.

HKS: But who's generating the amendments, you?

REB: In a formal sense the committees generate the amendments. If requested, the agency provides background information. What happened is that it led to a change in the long-term way we interacted with Congress. There were friends of Forest Service research who wanted to support our programs; we would encourage them to testify before the committee as outside witnesses. The appropriations committees, especially the House, have a day or two where people interested in that budget can testify. Tom McClintock was an especially good friend and supporter and Neil Sampson of AFA. There were two or three others who would usually with the RPA as a basic document. The committee staffs then come back and say this program has been suggested by such and such a witness; give us your capability to address the research programs that are suggested. These came to be called "capability statements." We would work out the locations and the nature of the project and return to the Hill. This was really significant during hearings because Yates would ask, what are your priorities, what do you want, where would you put the emphasis? Of course I couldn't testify outside of the president's budget but I could call attention to those capability statements. Now this is subtle, but it was terribly important in the way the budget was organized. That served as the basis then for amending the budget. Those capability statements really were ways of telling the committees our priorities. And, we weren't violating laws when we did it.

HKS: It's a common folklore, and McGuire said it's true. The Forest Service does very well at budget time compared to most agencies.

REB: I think that's right. What I'm trying to do here is address the most important single issue that a deputy chief has to be concerned about. Budgets are the one thing that no one else can deal with but the chief and, for separate program areas, the deputy chiefs.

HKS: Since 1915, research has been administratively separate from the rest of the Forest Service, joined only at the chief's office. Does that give the deputy chief of research a degree of autonomy that the other deputies don't have?

REB: Yes. It does give research a bit more autonomy for two or three reasons. One is that research isn't well understood by the other arms of the service and some of the issues are quite technical. Another reason is that the focus for action in the Forest Service are the national forests which make up about 90 percent of the budget. During the '70s and '80s, the chief, the associate chief, the deputy chief of the National Forest System, the deputy chief of programs and legislation, all essentially administering the national forests. Tom Nelson wasn't in charge of the national forests, he shared it with the others because that's where the problems and action were.

Essentially no one troubled me except when they had a few idle moments, and so I had a great deal of autonomy, and that reflected in the appropriations hearings. McGuire who came from research would defer to me because he didn't know some of the details. Max's main concerns were with the national forests. But, this issue goes back many years before my time as deputy chief. You may recall a memo that Earle Clapp wrote in 1937 in which he was protesting the way that research was being treated as he moved from the assistant chief of research up to associate chief. He was troubled by it. Harper was troubled by it. And Harper, as nearly as I can tell, was extremely aggressive at pursuing the interests of research. I've been told a number of times that McArdle was really quite uncomfortable with Harper's aggressiveness at times. On the other hand, if Harper hadn't been so vigorous at pursuing research budget interests, the research program in the Forest Service would has been much smaller. The relationship of research to the rest of the Forest Service has been an issue for as long as I can remember, or have read, going back to 1915. It's an uncomfortable relationship. Still, there are examples in my memory where the relationship has been outstanding and highly productive. Cases in point involved the Chippewa National Forest in R-9 and the H. J. Andrews Experimental Forest in the Willamette N. F.

HKS: How does Research interact with the National Forest Administration to know what they think their problems are. Technology transfer is a two-way street. You have RF&D meetings, that's an opportunity to talk about it.

REB: Yes, permit me to deal with the NFA relationship first.

HKS: Are the RF&D meetings an important vehicle for cooperation?

REB: Yes, they are. It's a more important vehicle for establishing a framework and environment for relationships than it is for identifying research priorities or joint tasks. But once that working relationship is established, it then sets the tone for people who actually do the work or set the priorities. You know, one of the most difficult questions that I've been asked through time is to describe how research sets priorities. Generalized answers are rarely satisfying. It's only when one deals with specific programs that answers become more concrete.

HKS: Sure.

REB: However, I would often turn that generalized question back to the national forests by asking how they decide priorities for the RPA or the forest plans under NFMA? In many respects, the processes are parallel. It's an imperfect integrative process where one samples the users and the public and then match funds, people, past costs, and historical patterns to determine what you're going to do.

BUDGET IMPOUNDMENTS

HKS: You wanted to talk about budget impoundments.

REB: The Budget Impoundment and Reform Act of 1974 changed substantially the way that we justified and mobilized support for budgets.

HKS: So that would prevent John Crowell, then, when you had your budget restored, of actually having it impounded. I mean he couldn't do that, could he?

REB: No, he could not.

HKS: But he could have under Nixon. I mean the rules changed.

REB: Nixon did it for a short time, but it quickly became a constitutional question involving division of presidential and congressional power.

HKS: I understand what the issue is.

REB: But after the Budget Impoundment Act, the administration could not withhold funds. That's probably one of the reasons why presidents today are now asking for line-item veto and balanced budget amendments to law and to the Constitution. So that they can reach into those budgets and eliminate specific items. Congress isn't about to give them that authority.

HKS: Carter to Reagan, Cutler to Crowell, you've talked quite a bit about that, but do you want to talk anymore about the shift between Carter to Reagan? Obviously it was abrupt and dramatic when it happened.

REB: Perhaps a sentence or two by way of summary. The Carter years were very favorable to research. The program was growing; it was accepted; but the Reagan/Crowell years had a trajectory that was just the opposite; it was a declining budget. The way that my office operated changed dramatically in the Crowell years. It was an almost 180 degree reversal in trends.

INDEPENDENCE OF RESEARCH

HKS: Independence of research, you wanted to talk about that. You sent me a letter that Max wrote to a state forester on the subject of reorganizing the Forest Service for research to dovetail at the regional level rather than the Washington level. I hadn't thought of that in terms of independence of research, but clearly it is.

REB: It will take a few minutes to develop this point, but independence is one of the most important issues to the research arm of the Forest Service. You may recall that Earle Clapp touched on that point in his 1937 letter, and every deputy chief that I am familiar with in the intervening years has been concerned about the question also, Harper, Dickerman, Arnold, all of them. There were few issues that were more important to me than for the ability of research to independently pursue questions, to analyze results and to bring them forth. In my ten years as deputy chief, no chief of the Forest Service, not John McGuire, not Max Peterson, ever told me what we could or could not do or ever challenged any of the findings that scientists brought forth. If they had said these findings are not appropriate for the time, you've got to change something, or withhold information, I would have been most troubled to the point of departing.

HKS: Do you know of any grumblings from the field that research has discovered something that challenges Forest Service policy and therefore research ought not to do that?

REB: I've heard that occasionally from lower level people in NFS, but never from the chief or his immediate staff. To carry this a step further, I had the feeling that John and particularly Max Peterson not only wanted an independent research group, but they were searching, they were eager for alternative ways of looking at the national forests. As I said, I did hear from subordinates but still senior people that the reserved areas for spotted owls are way too big, or there's no regeneration problem on Afognak Island or something like that. But I dismissed that sort of thing, and it was never a problem.

HKS: Could you elaborate on the example we used a few minutes ago about the letter from a state forester.

REB: Yes, and this occurred at a time when station and staff directors were concerned about how Max would view research. This wasn't a question with John McGuire who came from a research background but it was with Max. The concern was highlighted when Max consolidated Region 8 and the southern area of State and Private Forestry under the regional forester. There were lots of questions within the agency about where Max would seek the next consolidation. Then the letter came in from the state forester of one of the southern states who said essentially that the consolidation of Region 8 and the southern area seemed to be working, well how about consolidating research into that organization? Max passed the letter on to me, and we then visited for some time about how we would respond.

I prepared a draft response and gave it to Max for approval. Max's stand and mine too was yes we've got to streamline administrative activities and where possible share services and facilities, but research must remain independent of action programs. Max added one sentence to the letter saying in effect that if research is merged with an action or an administrative agency, inevitably the administrative agency captures research. A most significant policy statement concerning research. I shared that letter with the station directors and the staff directors. The issue of the independence of research within the Forest Service was put to at rest at least during Max Peterson's time, with that single letter.

HKS: The quest for independence on the part of research, isn't that a part of the tension within the Forest Service?

REB: I suspect that it is. Unfortunately, independence carries two meanings here. To the researcher it means independence to ask new questions, to frame inquiries differently, and to publish credible papers. That may not agree with current policy. To NFS and other users it often conveys the impression of being independent, often indifferent, to their problems and needs.

HKS: Can you reconcile these two points of view?

REB: Not as well as I would like. Personalities and different perspectives often exacerbate the problem. However, my observation was that researchers almost invariably are drawing from the same set of societal and natural resource concerns as were users; they simply needed the latitude to approach issues differently using the traditions and methods of science.

HKS: The '78 legislation on research, that doesn't address that issue, or does it? Independence of research.

REB: Not directly. But because it's a separate authorizing act addressed specifically to research, it reinforces the independence of research.

HKS: Will there ever be a concern that a member of Congress might be upset that the Forest Service gets all this money for research and turns out a paper that's critical of government action?

REB: I think that could be. One of the reasons our program in Alaska did not prosper in the late '70s and the early '80s was that Senator Ted Stevens felt that Forest Service research was in bed with the more active environmental groups. He felt that the researchers in Juneau were really handmaidens of the Sierra Club.

HKS: Do you think researchers who don't have administrative responsibility might tend to be more environmentalist?

REB: Perhaps so, they may tend to be more inclined toward the environmental side than the development side of natural resource issues but there are so many exceptions that I really wonder if one can generalize.

HKS: Seems to be an enormous advantage that Research has to fulfill its mission that the other parts of the Forest Service don't have. There's a lot of flexibility on hiring in terms of jiggering the job description around to go out and pick your person. You know how to take somebody off a roster, and you're much more likely to have successful programs if you can really pick the person that's fully qualified, rather than veterans benefits and all the other issues. But with affirmative action, research has that same load to carry that everyone else does. Is it true that Research has been more able to hire minorities and women and handicapped people than National Forest Administration because of that job description flexibility? You don't have to get somebody with a forestry background.

REB: I don't think it's so much because of job description flexibility as the fact that research depends on skills for which females and minorities are more likely to enter. Anthropology, sociology, geography, statistics, mathematics, and so forth. These are the kinds of skills that Research is recruiting and there tend to be more minorities and females in those skills than forestry.

HKS: Is it significant that research tends to be in urban areas, and more people would rather be in a university than out in the boondocks somewhere.

REB: Yes, rural America tends to be less friendly, especially to minorities. Maybe a little less so to women, and since research is located in more urbanized areas, some of the social aspects are easier to address.

HKS: Do you feel, did you ever feel, that research was being asked to take on more than its fair share of this to take the heat off of the agency?

REB: I didn't experience that but George Jemison certainly did. Every time I visit with George we talk about his experience with Assistant Secretary John Baker and hiring minorities.

DEALING WITH CONTROVERSIAL ISSUES

HKS: How do you deal with issues that involve science, but are strongly political in nature?

REB: It's never easy. You know one of the dilemmas that we've had for as long as I can remember is how does a person deal with a controversial issue, an environmental issue, a policy issue, without infecting his own value system to the outcome. My first encounter with this was in Grand Rapids, Minnesota, with Bud Heinselman.

I think an answer is that no one has been able to give an unequivocal response to that dilemma. Some scientists are better at it than others. Logan Norris, now department head in the college and used to be with the Forest Service, is one example. Logan Norris did research on environmental consequences of herbicides, such as 2-4D, 245-T, and other controversial chemicals. Logan was one of the most credible witnesses that the Forest Service ever had. I don't know his value judgments but he could deal with those controversial issues in such a way that he came across as a detached and thoroughly objective scientist. My admonition to scientists who are dealing with controversy is to put your evidence between you and your listener. That's one of those things the scientific method teaches—how to create a hypothesis and then to test it with detachment. But, still there's a very strong personal quality here. Some people simply are not able to separate their values and judgments from their science.

HKS: Presumably there is more detachment in research than in administration, but not necessarily. The kind of people that go into research, theoretically, are independent thinkers, but not always.

REB: But if one uses scientific principles, you can put your evidence between yourself and your listener.

HKS: I forget the name of the fellow who has been the head of the spotted owl study here, but he's certainly been under a lot of pressure.

REB: Jack Ward Thomas.

HKS: People want a certain answer out of that group.

REB: Jack Thomas came to the Pacific Northwest Station during my time here. He's a very forceful and able person, and I don't know that we could have chosen a better person to head the interagency scientific committee dealing with the spotted owl. I've been watching very closely to see how Jack handles these things. I have a suspicion that he feels, personally, a little toward the wildlife side. I think he's handling himself with great detachment. However, some of my industry friends here have raised some questions about Jack.

Jack did one thing recently that shows how treacherous this environment is. The *Oregonian* had an article about Lonsdale who was a challenger on the democratic ticket to Les AuCoin for the Senate seat occupied by Bob Packwood. There was a short sentence in Lonsdale's press release that said Jack Ward Thomas was a contributor to his campaign, the only contributor named in the release. Lonsdale is pro-environment and thinks that the national forests are being badly managed. My reaction to the *Oregonian* article was, Jack, how could you ever permit your name to be used in a political campaign? The issue was not that Jack couldn't contribute to a political campaign. The problem was that he permitted himself to be identified with one point of view. Fortunately, not much came of this issue, but it could have.

HKS: Yes.

REB: Thomas knew that that was a mistake. But so far as I know that's the only time he stumbled on this whole issue.

HKS: Well it's a tough one. We're all supposed to be objective.

REB: I have the feeling that scientists know that they have to find ways to separate their personal value system from their evidence. They too want to be independent and they know the only way they can be independent is to be objective.

HKS: Sometimes the questions go beyond the evidence, and since you're the expert you're asked to extrapolate from the data.

REB: Yes.

HKS: That's where your personal views can take over.

REB: Yes, and Jack Ward Thomas had imperfect evidence on the spotted owl and he and his committee was called upon to extrapolate. Still, he and his committee were, in my estimation, among the best available.

Person-in-Job

HKS: Let's talk about personnel. We talked about budget, and almost nothing gets done without budget, but nothing gets done without personnel either.

REB: Right. One of the things that I wanted to do when I came into Washington was to enhance the quality of research. That meant emphasis on recruitment and training and anticipation of movement into more responsible positions. I used to monitor statistically some of the demographic and educational characteristics of our work force—ages, level of degree training, and so forth. Of course I was encouraging station and staff directors to pay special attention to recruitment. I encouraged them on a number of occasions that if they saw a potential superstar, a person who clearly had research capacity, they should let me know and we'd find the money. The person was more important than the program in my estimation. I also made the same challenge for unusually attractive EEO recruitment opportunities, I suppose the most important thing, beyond recruitment, was the Person-in-Job career ladder.

The Person-in-Job concept is outlined in the Research Grade Evaluation Guide (RGEG) issued by the Civil Service Commission. Agencies had some discretion in developing operating procedures under the RGEG, and these were subject to much discussions and several revisions. The RGEG was especially resented by scientists whose research was at the application and extension end of science. Users, too, resented the guide because they thought it gave short shrift to technology transfer.

My own view was that the RGEG was intended to reward creativity and originality and that we tended to under-reward our most productive scientist and over-reward the less creative ones. Still, the Person-in-Job concept was better for science than any of the alternatives then available. I wish we could have found a similar system for technical specialists in NFS and S&PF.

Far and away the most crucial step in research productivity occurs on the day of recruitment. The next most important step was an attractive career ladder and reward system. Overlaying all of this were enhanced educational opportunities. So the Person-in-Job procedures were an important concern in all of my years as a research administrator. They remained important, they were important before I came on board.

COMPETITIVE GRANTS PROGRAM

REB: I saw some other ways, I thought, to enhance quality of work. The competitive grants program was one of those. The competitive grants authority in that 1978 legislation permitted research, singly or in combination with any agency, any public agency or private agency. The one thing that counted was the quality of the proposal. What I wanted was for Forest Service people, singly or in combination, to come forth with proposals that would be competitive in a peer-reviewed environment. That meant that our most productive and our most imaginative people would be successful.

HKS: I didn't know until this moment that those grants were available to Forest Service people. I thought they were for people at the Forest History Society or Oregon State University.

REB: You may be thinking of National Science Foundation grants which are not awarded directly to federal employees. USDA and Forest Service competitive grants are available to all qualified scientists.

HKS: Everybody.

REB: Forest Service people could singly or in a variety of combinations compete. We had a number of joint proposals—Forest Service scientist with a university person, Forest Service and an industry scientist.

HKS: Was it construed or could it have been construed, if you got a couple of those grants you were really upward bound, because the agency saw how smart you were?

REB: Not quite in those terms. But we recognized that these grants were highly competitive and reflected better science. That was also an important consideration in the Person-in-Job career evaluations.

HKS: So somebody who didn't want to be stagnant was under some pressure to apply for grants, to be active in this program. And to win occasionally.

REB: We encouraged Forest Service scientists to compete in order to build higher quality science, but also to foster cooperation with others through submission of joint proposals.

HKS: I'm trying to think of the management problems. Here is scientist X working on this long-term program and then he gets a grant, what happens to the program?

REB: In so far as we could during those years of declining budgets we tried to add these funds to those already available to the scientist. I viewed the competitive grants program as a powerful way to bring forth the best and the brightest.

HKS: We were told when I was at the station, and I don't know if it was official or not, that scientists were permitted to work up to 20 percent of their time on a project of their own choice. It's sort of like the farmer in the Soviet Union with his personal plot versus the commune farmer. A lot more productivity per acre on the personal plot.

REB: Yes, discretionary time is desirable and available, but it's not quite as simple as that. Planning is still an important part of the Forest Service research program—it's one of the strengths of the agency. I was restless under what I thought were excessively rigid planning requirements when my career started in the early '50s. But as time went on and I became deputy chief, I had the opportunity to discard formal planning procedures. I didn't do it because I thought it was such a hallmark of quality in the Service, and it is copied by many universities to be sure, some of the rigidities of planning. What the planning procedure requires is that you think through where you're going to go for the next five years. Give us your best estimate about where you see your program is going. Once we have agreement on that and all the signatures are on the project approval document, then the scientist proceeds. But almost as soon as they launch that program, they begin to modify their research, as they get feedback. So there's always some opportunity for the unanticipated or the exploratory research in that planning process, although it appears to be fairly rigorous at the outset.

HKS: Are some of these competitive grants large, like a million dollars?

REB: No, during my time they ranged from one hundred to two hundred fifty thousand dollars covering perhaps a three-year period.

HKS: That's the target. How did you administer the Competitive Grants Program?

REB: One of our dilemmas was that the first funding increment to the Competitive Grants Program came to the Forest Service budget. The Forest Service could not administer that program with objectivity. So we immediately transferred the funds to the Office of Science and Education to be administered by the Office of Competitive Grants. In other words, the Forest Service transferred, I think it was five million dollars, to that office so it would be administered independently of the agency—in which case Forest Service scientists could compete. Unfortunately the Competitive Grants Program in the Forest Service was suffering from serious budget erosion and the program eventually disappeared from the agency. We simply could not reduce our workforce and still move a big bundle of money into competitive grants. Anyway, competitive grants were one of the things that I wanted to do to enhance the quality of our research programs. I understand that competitive grants have been restored to USDA but with funds appropriated directly to the Office of Competitive Grants.

LOCATION OF RESEARCH

HKS: How do you choose the location of research laboratories? How does location affect the research environment?

REB: Location of the science workforce is another quality related issue although I'm less sure about unambiguous outcomes. About 60 percent of the seventy-five laboratories in the Forest Service are located on university campuses, and about 75 percent of the Forest Service scientists are at university locations. My predecessors and I wanted our workforce to be located where they could interact, not only with the forestry schools, but with other departments on campus. That's been a policy for a long time, and I tried to reinforce it. Of course, one of the best examples is here in Corvallis. The Forest Service today has thirty scientists and there are about fifty or sixty faculty members next door in the College of Forestry. EPA is located nearby with its science group and resources. And across campus are another forty to fifty scientists with interests in natural resources. And so, this is a rich environment for consultation and cooperation.

I had one hesitation about that generalization. I often asked myself, where are the most productive units in the Forest Service, and the paradox was that many of them were not on university campuses. For example, Peter Koch in Alexandria, Louisiana, Phil Larson in Rhinelander, Wisconsin, Jack Ward Thomas at Le Grande, Oregon, and Dave Marquis at Radnor, Pennsylvania. Those are not university locations. So it's a question that we still need to ask. My instincts tell me that we ought to be located in a scientific community, but somehow there's some stimulus that goes with other locations, perhaps proximity to forestry problems. In all probability the principal factor is people. Good people will be successful wherever they are located.

HKS: Do you know the chemistry that caused Oregon State to get this lab rather than University of Washington? There must have been some discussion at the time, both of them wanted it, or both deans would have been in favor of having federal money coming on campus plus the expertise, plus the prestige plus all the rest.

REB: I don't know first-hand the genesis of Corvallis. I suspect that it was part of creating work centers away from station headquarters, which came later in the West than in the East.

HKS: Corvallis is two hours from Portland, Seattle is four hours from Portland, I don't know if that made a difference.

REB: But you know the station has a location in Seattle, too.

HKS: That's right, but it's not anything like this.

REB: No, but it probably could have been. For years the PNW Station maintained a lease for a laboratory site at Sand Point Naval Air Station, a mile or two east of the University of Washington.

HKS: I just wondered what happened, if the congressional delegation got involved in that decision.

REB: Oh very much so, not only at Corvallis but at most laboratory locations. You know there were three phases in Forest Service presence here at OSU—not unlike the history of other university-based locations. Phase one was to be located within the forestry school starting in the mid-'50s. Phase two was constructing the first federal laboratory about 1960. The patron here probably was Senator Wayne Morse. Phase three was construction of the large complex, dedicated about 1976. The patron here clearly was Congressman Wendell Wyatt with whom I worked extensively.

HKS: There may not have been a champion in the Seattle district at that moment.

REB: That's right. Scoop Jackson and Warren Magnuson were never enthusiastic about the Forest Service and did relatively little to support it. Jackson and Magnuson built the defense industry in the state of Washington, but not natural resource programs. Phil Briegleb and I used to talk about that a lot. Jackson apparently was easy to talk to and Magnuson was too, but somewhere they were disenchanted with the Forest Service.

HKS: It strikes me as strange that when you see Portland, and all the western states where the Forest Service is, you'd think that all the members of Congress would be supportive.

REB: In general, western members of Congress supported the Forest Service, sometimes mixed with criticism. The support generally went to NFS because of the large number of national forests in the West. But, research derived support as well.

HKS: Why aren't more station headquarters on university campuses?

REB: There were often discussions about moving station headquarters to major university locations, including here in the Pacific Northwest. It never went very far here because it was so costly and so disruptive. Furthermore, Portland is an important forestry center in its own right and much closer to airports than Corvallis.

There are four station headquarters on major university campuses, the Forest Products Laboratory at Madison, the North Central Station in St. Paul, the Rocky Mountain Station at Fort Collins, Colorado, and the Pacific Southwest Station in Berkeley. The Northeastern Station, which moved several times over the years in the Philadelphia area, would have been the logical candidate for a university location.

HKS: Of course the Berkeley campus is four blocks away, but it interacts with the station.

REB: Yes. Ironically, the PSW Station will soon move to Albany, several miles away, because the present space is unsafe from earthquakes. Fortunately, the University of California connection will remain—and the station will be co-located with a major ARS laboratory.

FOREST PRODUCTS LABORATORY

HKS: Is this the time to talk about the Forest Products Lab?

REB: If you like.

HKS: To me, again as a civilian looking at the agency, the Forest Products Lab hardly seems to be a part of the Forest Service at all. It seems autonomous. I don't know why I feel that way.

REB: Your feelings are well grounded, but there have been many changes in recent years.

HKS: Does it have its own pipeline to Congress or what?

REB: It still does but with much more consultation with the WO. You know the origins of the Forest Products Laboratory, it was created in 1910. It was the first of the big highly visible research programs of the Forest Service. I think it grew up with a certain arrogance and an independence. FPL directors and their immediate staff worked hard not to come under the supervision of the Washington office. Al Hall was a director of both the Pacific Northwest Station and later the Forest Products Laboratory. He moved back to Portland after he retired. The first day I was on the job as PNW station director, Al visited. In his gruff way he said to me, "I hope they're not trying to tell you how to run the station from Washington, D.C." He was reflecting, as much as anything, his Forest Products Laboratory views.

That was of concern, particularly by Les Harper and his successors, that FPL was marching too much to its own drummer. I don't really think that problem was solved until the Dickerman/Arnold era, particularly when Dickerman insisted that the Forest Products Laboratory use the same reporting and planning procedures and organization as the eight regional experiment stations. We also have to recognize that Bob Youngs had a great deal to do with improved relations. Bob Youngs was an assistant director in Madison, came into Washington in the late '60s, went to New Orleans as Southern Station director, and came back to Washington in the early '70s as associate deputy chief. He was and is a very able and accommodating guy. Bob then went back to Madison as the Forest Products Laboratory director about 1975. Bob had a lot to do with removal of the hostile environment that existed for so many years at Madison. My own experiences with the Forest Products Laboratory were both pleasant and positive. The laboratory interacted well with my office and with the other stations. But Bob Youngs, and his successor John Erickson, deserve an enormous amount of credit. Another thing that helped was to have more personnel changes among FPL, WO, and the various stations. That issue that you described no longer exists.

HKS: It would seem to me that the Forest Products Lab would be very vulnerable when a Reagan comes in. Why is the government doing products development? That's what the private sector is so good at. That question must be asked from time to time.

REB: The possibility of that question was in my mind and notebook before every congressional appropriations hearing and every meeting with OMB. You recall, I gave you earlier several criteria by which we justified that Forest Products Laboratory research. All other things being equal, my top priority for strengthening research among the eight regional experiment stations and the FPL was the FPL. The directors were aware of my feelings and were generally supportive. In the real world of research budgeting, however, things were never equal.

HKS: What is the Lab's budget, roughly?

REB: When I retired, about fifteen million dollars.

HKS: Out of one hundred forty?

REB: Yes.

HKS: What's the budget at PNW, looking for a comparison.

REB: At the time I left, about twenty million.

HKS: I don't know if this station is bigger than most or not.

REB: The largest station was the Northeast, and the PNW was second but very close to the top. The smallest stations were PSW and North Central, and I was seeking ways to move funds into those stations.

HKS: The Lab doesn't get much publicity, and I don't know what more there is to ask about the Lab, other than it doesn't seem to fit the model very well. No where else in the agency is anyone concerned about what to make out of wood.

REB: The Lab is well known nationally and globally among specialists in the pulp and paper industry, wood preservation, and solid wood products industry. Its accomplishments over fifty years have been impressive indeed in almost every field of products research.

HKS: Was Crowell for or against the Lab?

REB: He was not excited about anything in research. I was with him at FPL one day. We were looking at research on creating thick panels of structural flake board using steam injection. He asked abruptly, "Why are you doing this research?"

HKS: I can understand the question.

REB: An hour or two earlier we had been looking at a display of conventional structural flake board, and he was marveling at how useful that had been to Louisiana Pacific, his previous employer. He said, "You know we built a major industry in Louisiana Pacific on the structural flake board made from aspen and other low-cost wood." Then two or three hours later he raised hell with me because we were doing this kind of research. I didn't have the temerity to say, "Where do you think structural flake board technology came from Mr. Crowell?" I certainly was tempted to do so. Crowell simply could not connect the generation of technology to public research.

SENIOR EXECUTIVE SERVICE

HKS: We got off personnel, but I had a personnel-related question, and that's the Senior Executive Service. I never really thought too much about the significance of that, it makes a certain amount of sense. I don't know what happens in other agencies when a new president comes in to change the administration. Do they just clean out the whole Senior Executive Service, or are those jobs fairly stable? They're fairly stable in the Forest Service.

REB: A proportion of the SES positions are reserved for career officers, and provided performance and conduct are satisfactory they're permanent. I don't remember what the proportions are, 60/40 percent or 80/20 percent with the larger proportion reserved for career people, by law and regulation the smaller proportion is available for political appointees. The Senior Executive Service was intended to provide more attractive career ladders for senior people, competitive with industry and the universities. It was supposed to provide a workplace of senior managers who could move from agency to agency and so forth. It hasn't worked out quite that way.

HKS: In the material you sent to help prepare the outline, you mentioned that you received a bonus or a merit, I can't remember the terminology. My interpretation was Max thought you were doing a damn good job. I mean it would have been his recommendation to get you that, right, no one else. He was your supervisor in the management sense.

REB: The Senior Executive treated me very well. Doug Leisz and I were the first recipients of Presidential Rank Awards. There were two, one was meritorious and the other was distinguished. Doug got the highest one, the distinguished award. He got it primarily because of his work in getting a hundred-thousand-acre tract donated to the Forest Service by Shell Oil Company, and Doug deserved it. I got the second one which was called meritorious. I was most pleased to receive it.

HKS: What's the nominating process? I'm trying to think of management. I'm assuming that Max was in the flow of paper in all of this.

REB: I don't remember the details, but it's a nomination that goes for review to the Department of Agriculture and then I think it goes to the Civil Service Commission for approval.

HKS: Each agency is putting up its candidates?

REB: Yes, and they are very limited in number. You may also have noticed that I was a recipient of a number of performance bonuses which were a part of SES.

HKS: There's a lot of evidence in your resume that Max thought you were doing a good job. How many other people...?

RELATIONSHIP WITH THE CHIEF

REB: Oh, quite a few got performance bonuses. I enjoyed working with Max. There were two or three brittle parts of that relationship. Max was anxious to have national forest people come into the research branch. There was resentment about researchers moving easily into the National Forest System, but not the reverse. Actually, a good number of NFS and S&PF people did come into research, but generally not at top levels. And he forced a position or two on me that I knew weren't going to work out, and they didn't. In many respects, those appointments were symbolic. Max would posture in public about technology transfer or some aspect of research, and the station directors and staff directors tended to resent that. But I would visit with Max in private conversations, and they were positive and meaningful. For example, I would give Max an example of new technology such as truss frame housing or press dry paper, and you know two or three months later he would use almost my same words when he described that process to Congress or somebody else. Max was one of those chiefs like McArdle and Cliff who had remarkable memory recall.

There was another aspect about my relationship with Max. During the days of retrenchment, the Crowell days, Max didn't put any premium on standardized organizations. That meant during those cutback deliberations I had a lot of discretion in taking advantage of retirements, transfers, and other vacancies. So the stations, which tended to have a standard organizational pattern, began to have some missing teeth. If one assistant director retired, we would mold the workload around the remaining staff. With this flexibility came opportunities to reduce the administrative work force without maintaining organizational symmetry. I was determined that the administrative side of research would reduce in proportion to the loss of scientists.

COMMUNITY OF SCIENTIFIC INTEREST

REB: I wanted to enlarge the community of scientific interest that was concerned with natural resources, and again this was a matter not of originality but of emphasis on my part. One of the first things I did as deputy chief was where we had senior vacancies and no obvious candidates from within, we reached outside to recruit senior people. We brought in the deans of several forestry schools directly into senior positions in the Forest Service—Bob Dils, dean of Colorado; John Gray, dean in Florida; George Marra, who went to the Forest Products Laboratory, came from Washington State University; and George Brown, now the dean here. Others were Dave Thorud, now dean at Washington; Ross Whaley, now president of SUNY; and let's see, Hank Montrey from Weyerhaeuser. These were people that I came to know for a variety of reasons including work in science and education in USDA. I wanted to bring people into the agency, some permanently if they would stay, some from sabbaticals or in temporary positions. And it worked out well. We were lucky—we got some of the best. All were senior people who moved easily through the agency and were very effective.

HKS: How about those people who might ordinarily have been next in line?

REB: In almost every case there was no obvious successor. You know if you've an in-house candidate who is clearly able to do the job, you don't go outside. But that option isn't always available. The two-track career system also complicated successional possibilities.

HKS: Of course it's happening a lot throughout the agency now. We're talking to the Forest Service about preparing training documents, because so many people are coming in at senior levels who don't know anything about the Forest Service.

REB: Some of that's brought about by EEO and diversity requirements. Others in the Forest Service brought in senior people as well. Einer Roget, for example, who was the deputy chief for State and Private Forestry, came out of the Soil Conservation Service, and Jerry Miles, deputy chief for administration, was a special case. Jerry went from the Department of Agriculture to the newly formed Department of Energy as chief administrative officer. He was fired by Schlesinger and came back to the Forest Service. Jerry was outstanding and contributed much to the Forest Service. The point I would make here is that these external senior recruits gave a marvelous account of themselves and added a great deal to the perspective of the agency. Unfortunately, during the Reagan years, this was viewed as non-competitive, probably with some political overtones. It became much more difficult to bring people in from the outside under the Intergovernmental Personnel Act, or other procedures. Much of the movement I described occurred during the Carter and preceding years.

HKS: I was going to ask, and maybe you just answered that, was it difficult at all dealing with the civil service requirements to come up with job descriptions and to demonstrate that they qualify?

REB: Not particularly difficult during the Carter years. But very difficult during the Reagan years. But anyway, that recruitment from the outside, I think with only one or two exceptions, was most helpful to the agency. If I were still with the Forest Service, I would encourage more of it. We talked earlier about the pioneering research units, and I voiced some reservation. The progenitors of that idea, the Agricultural Research Service, also did. But we wanted to find ways to recognize people who were unusually strong performers, and we used less rigorous, rigid ways of doing it. For example, Don Marx, an outstanding mycorrhiza researcher, was considering leaving the Forest Service. I asked what it would require to keep him, and his supervisor replied that he probably would appreciate a little bit of discretionary money and it might be appropriate to give his work a little more recognition. So I went back to Washington and dug up some money, twenty-five thousand dollars, and informed the staff that we were going to retitle Marx's work, Institute for Mycorrhizal Research. Don spent the remainder of his career with the Forest Service and in 1991 was recipient of the prestigious Marcus Wallenberg Prize.

Kent Kirk at Madison, Wisconsin, was doing marvelously creative work on the biodegradation of lignin, understanding the enzymatic relationships that cause lignin to decompose. He too was a winner of the Wallenberg Prize, and is now a member of the National Academy of Sciences. We wanted Kirk to have more visibility, so we created an Institute for Microbial and Biochemical Technology. That was a way of according visibility to our most creative people.

A major quality that I was seeking, and others before me have too, was recognition of Forest Service scientists by the superscience agencies like the National Science Foundation and the National Academy of Sciences. Kirk has been elected to the National Academy of Sciences, but I was always disappointed that more Forest Service scientists were not. George Hepting of the Southeastern Station and Kirk, I think, are the only two, although I think others are now in the pipeline. The paradox is that Forest Service, particularly the Hubbard Brook research group in New Hampshire, have led collaborators into the National Academy of Sciences—Gene Likens and Herb Borman. They were elected to the academy based on some of their cooperative work with Forest Service research.

HKS: I always assumed it was the same problem you have on the university campus, getting tenure for a forest economist because the econ department chewed them up because they're not pure enough. Forestry is a conglomeration of stuff, and I would have thought the National Academy of Sciences has the same hangup.

REB: Election to the National Academy is probably the premier science recognition, short of a Nobel or Wallenberg prize, in the country. The academy has come under a lot of criticism, incest among others. Harvard tends to elect his own and so forth. The fact that some of the smaller institutions just don't have anybody to sponsor them and push, so it's an imperfect process.

RESEARCH COLLABORATION

HKS: How does Forest Service research fit in the larger field of science in general?

REB: I think increasingly better. A major goal of mine, upon returning to Washington, was to enlarge the scientific community that dealt with forestry research. I described to you the planning process where the sixty-one forestry schools under the McIntire-Stennis Act and the Forest Service joined together in the late 1970s to prepare a coordinated research program. That was a joint activity all the way, and I think did a great deal toward building the confidence and cooperation among the forestry schools and the Forest Service. We tried to do other things. I made a special effort to collaborate with the Natural Resources Group in the Cooperative State Research Service and with the Department of the Interior agencies—the Park Service, the Bureau of Land Management, and the Fish and Wildlife Service. We used to have frequent meetings concerning their research programs.

I wanted to lay the foundation for more collaborative and cooperative work. One of the products of our efforts was to prepare under one cover a directory of research in all of the agencies so that it would encourage contacts among our field people. I'm holding here just such a directory. It was fairly difficult to establish collaborative research with the Interior agencies, partly because they were smaller, party because their programs tend to be more volatile in terms of people and budgets. I also wanted to link with the industry much better. I borrowed heavily from a mechanism used by the Forest Products Laboratory.

Each year the Forest Products Laboratory would invite industry representatives in to comment on their programs, and I thought that procedure was working well. I used to join that meeting in Madison whenever I could. I thought, why don't we try this at the regional experiment stations. So I encouraged industry/experiment station committee meetings, and it worked reasonably well. The forestry schools and the Forest Service would join me in inviting industry groups in. If it didn't foster a lot of collaborative research, it certainly was a mechanism by which people got to know each other and could visit and exchange views; maybe that was the most important output. I don't know how the efforts fared after my retirement. It takes a great deal of time and energy to keep them going.

HKS: Isn't RPA a rationale for this very thing?

REB: Yes, it is.

HKS: The Assessment.

REB: Yes, the Assessment, but also the Program. The RPA program was the vehicle around which we built a lot of these efforts. I also used to on occasion join the state foresters. Each year the state foresters would meet under the auspices of State and Private Forestry. I wanted them to become much more aware of our research programs and be involved in them. The state foresters set up a research committee, but I retired shortly after that came into being.

Let me try something else out. I felt there were a number of policy issues, of research issues, that needed an independent voice. The organization that I had my eye on was Resources for the Future. This started during John McGuire's time. He was sympathetic to the notion and he also knew the then RFF president, Frye, from his California days.

HKS: Right.

REB: Frye was succeeded by Emory Castle, who came from Oregon State University. I wanted to commission some research for the newly forming forestry program within RFF, and over the years sent something like a million dollars to RFF for a variety of things, including the scientific foundation for multiple use management. John Krutilla, an economist with RFF, took the lead in that work and published a book as a result of it. This was another effort to enlarge that scientific community and to give the Forest Service an independent source of information.

HKS: That was before the competitive grants.

REB: Yes, it was. I didn't want the Forest Service to be inward looking. I wanted it to be outward looking, and I wanted to be aggressive in pursuing that goal. If I were still there I would still be tracking that goal.

HKS: It's my stance that the National Forest Administration could use, at least historically, a little more of that broad-minded view.

REB: Yes. Probably right, but we also need to keep in mind that the national forests are the focus of most controversial issues in the Forest Service—and NFS is the lightning rod. John McGuire was especially active in trying to maintain outside contacts.

HKS: Its butt's in a sling very often because it's not paying attention.

REB: I still need to reinforce that point.

University Research

HKS: When I came to the station in '62, I didn't realize that McIntire-Stennis had just been passed and that's really what was happening. But I wasn't aware of why it was happening. There were some complaints among people I worked with about the quality of research that university faculty were doing, because they used the Forest Service money and then they'd support grad students. There were different agendas. That was the reason they put Forest Service scientists on the campus so they could watch the Forest Service money a little better. This is what I was told. Was there truth to that, was that a significant concern? Every professor has to support graduate students, that's what the university's all about, but Congress didn't say to support graduate students. They want to get research done. Has that been an issue or a difficult thing to unravel on a campus?

REB: I don't know the details of this circumstance that you're talking about, but many criticisms probably were unfair or uninformed. Today those concerns are largely a nonproblem although mutually, generally uninformed criticism still exists. This location, Corvallis, is a case in point. Several of the Forest Service scientists here sit on OSU graduate committees, and occasionally serve as chairperson of that graduate committee. Some of the Forestry Science staff members teach, not so much repetitive undergraduate courses, but special lectures at both undergraduate and graduate levels. The collaboration here is so close that sometimes you can't tell the parent organization of individual scientists. There are always exceptions of course.

HKS: The criticism was that the quality of research that the average professor does wasn't up to Forest Service standards.

REB: That issue cuts both ways.

HKS: I suppose. I mean it could have been professional jealousy, I don't know the motivation behind the comments.

REB: Jealousy and competition sometimes exist. And there is often a tendency to deal in cliches and stereotypes—which usually are uninformed and unfair.

HKS: Yes, I know.

REB: But today the working relationships are excellent. You know, much depends on the personalities of the titular heads of the two organizations. Carl Stoltenberg, dean here for over twenty years, set the stage for collaboration; George Brown, his successor, has followed up. The Pacific Northwest Station also has encouraged these relationships. But there were other locations where personalities got in the way; cooperation became strained. Forest Service research would have liked to have been at Penn State, or at Purdue, but the deans of those two schools and the Forest Service just didn't see eye to eye.

HKS: So the deans weren't against it in principle, it's just the chemistry didn't work.

REB: Interpersonal relationships had something to do with it, but I think maybe they were against it in principle also.

HKS: Why would that be? They're always looking for new and better things. What made it a problem to a dean of having an institution like this on the campus?

REB: I can only hypothesize. Maybe competition for attention. Maybe a hostility toward federal involvement.

CONTRIBUTION OF SCIENCE

HKS: Bob, tell me about a paper in *Science* that influenced you to the extent that you're still citing it.

REB: You know that the agricultural and the forestry research system in the United States comes under periodic assault and question. Outmoded, pedestrian, unimaginative, and a whole series of other pejorative adjectives. I used to look at those reports, particularly the Glen Pound report of the early '70s, a National Academy sponsored study.

Pound was criticizing agricultural research, and indirectly forestry research, because it's organized in similar ways. I kept asking myself, why is it that American agriculture is setting the standards for agricultural research everywhere in the world. How can productivity increases be in the order of 6 percent per year, if it's so outmoded, pedestrian, and so forth. I didn't find a satisfactory answer to that question until I saw a paper in *Science* (vol. 205: 1101-1107) in 1979, a paper by Evenson, Waggoner, and Ruttan, which described the economic benefits from agricultural research. That paper went something like this:

The authors reviewed thirty or thirty-five case studies about the cost-effectiveness of agricultural research. Sure some methodology problems, but the overwhelming evidence was the returns on invested dollars from agricultural research were in the neighborhood of 30 to 35 percent per year, some as high as 100 percent rate of return per year. Then, the paper went on to characterize agricultural research, and let me add parenthetically, indirectly forestry research as well, to characterize the reason the system was so successful. The authors cited two reasons: one is that the agricultural system is highly decentralized and second it's strongly interconnected. What they meant by decentralized is we have fundamental research, we have applied research, we have field experiment stations, we have experimental forests, and so forth. All of which carry the problems closer to the user groups and provides a delivery system for the information. So it's a strongly decentralized system. The second ingredient is that it's a strongly interconnected system. It's articulated, in the words of the authors. It means that scientists know what their peers are doing.

We see this in forestry with such interconnecting mechanisms as the Western Forestry and Conservation Association here in the Northwest, the Society of American Foresters, IUFRO, and others—all of which provide those linking mechanisms among technical foresters and researchers. As a result of that *Science* paper, I became more and more certain that the decentralized, interconnected system of forestry research of the U.S. was a source of strength, not a weakness. It wasn't pedestrian. Sure, it could be improved at the edges, but it wasn't pedestrian; it wasn't outmoded; it was the proper way to go. Consequently, after that time I became very much more reluctant to close field locations or to close stations. Sure, we made some corrections and some changes in a few locations, but I felt that a regional system of experiment and satellite locations was terribly important to the well-being of forestry research and provided a highly effective way to interact with user groups.

HKS: This sort of analysis of success of agriculture research must go well with Congress at budget time.

REB: Oh, I think so, but understand that Forest Service appears before Interior committees and Agriculture appears before the Agriculture committees. Still there were enough parallels and enough forestry examples that I used them before our committee hearings.

HKS: That's true, but the research is cost effective. You give us one million dollars and we'll give you two million dollars worth of benefits.

REB: Yes, and those cost-benefit studies, many of which came out of that research project at St. Paul, Minnesota, in the North Central Station, were used in budget hearings. But you know, explanations of the kind that I've just given really served fairly well in Congress. If you know where you're going and why and can give reasons, you really can quiet a lot of criticisms—and, in fact, gain understanding and support.

HKS: Congress must like the Forest Products Lab.

REB: Mixed bag. Yes, they do, people who know what goes on there like it. People who say we've got to make cuts ask, why are you doing this research. Relating to what I said a moment ago, the Lab suffers somewhat from centralization. It's not well-known among potential users and supporters but tries hard to maintain regional contacts. This was one of the benefits we tried to achieve in creating better relationships among the Lab, the WO, and the eight regional stations.

HKS: But the kinds of things the Lab does where it's providing engineering and all that to the private sector to...

REB: We always encouraged key people to visit the Forest Products Laboratory. That was almost the first on the agenda for policy makers. Assistant secretaries Crowell, Cutler, and Peter Mayer were there, members of Congress have also visited. It provides a fascinating way to demonstrate what technology can do.

The technologies that have come out of the Forest Products Laboratory, and there are many of them and they are major, were sometimes easier to describe than were the biological and social sciences. I remember one occasion when we were presenting the first stages of the budget, this time before the secretary of agriculture. The chief and his deputies were sitting across the table from the secretary, Bob Bergland, and some of his budget staff. It was obvious that Bergland was bored to tears. He was yawning, and he was looking at his watch, and casting his eyes around. I had some wood samples with me that had come out of research at Forest Products Laboratory. I said, "Mr. Secretary, these are some examples of recent technologies coming out of our research program"—yellow poplar studs from a new sawing process, other samples of reconstituted wood and reconstituted 2 x 4s. The secretary's eyes brightened and conversations must have gone on for another half hour or forty-five minutes. Finally, he had some kind of a social engagement, it was after six o'clock and somebody was summoning him to leave. But he went from inattention to full-fledged attention, and was so fascinated that the discussions kept going. The story traveled around the Department of Agriculture overnight and the next morning, because all the agencies were undergoing these kinds of budget hearings, and the wags had it that the only way to get the secretary's attention now was with a 2 x 4. [laughter]

HKS: I guess secretaries are human too.

REB: Anyway, I did use a lot of artifacts and samples, and that made for a somewhat easier entree in the appropriations process. It lent an element of concreteness to something that otherwise appears to be fairly abstract.

HKS: Does the Public Affairs Office come to research for stories to tell?

REB: Yes, and a fair number of research stories did go through Public Affairs. We could have done a lot better job of that. The journalists used to tell me that science is one of the easiest stories for them to tell. But it was always a chore to supply the stories, scientists were so occupied with their own day-to-day work.

HKS: Taxol. Journalism is running way ahead of science, as it does so often.

REB: There are substantial funds here, right now, some of them from the drug companies, and some I think through appropriations, to understand more about the Pacific yew and its ecology and management. It's big business here at Corvallis today.

HKS: I see a yew outside the entrance here. I thought it's lucky it hasn't been cut down.

REB: I'm not sure that it's a Pacific yew, because there are many species, some introduced.

International Forestry

HKS: Talk about International Forestry.

REB: Another of the goals that I brought with me to Washington was to strengthen International Forestry. International Forestry within the Forest Service has a highly volatile history; on occasion it could be very important, at another time it could subside to virtually nothing, depending on what posture the U.S. had toward other countries at that time. I came into Washington when International Forestry was at a very low ebb. That staff group was down to about six people; there was just no support anywhere. We were looking for ways to strengthen that program.

About that time, tropical forestry was beginning to move up in the world agenda, including within USAID. But, it was also a time when budgets were being cut back and ceilings were being lowered. A related issue here is that a long time friend and colleague, Dr. John D. "Jack" Sullivan, moved from head of natural resources in the Cooperative State Research Service to the science and technology group in USAID. Jack was an old friend and a shaker and a mover and knew how to get things done. When Jack went to USAID there was already some ferment about the Forest Service being the technical repository for forestry skills useful to USAID. I noticed in your book, *Changing Tropical Forests*, credit is given to Dan Deeley for initiating some interest in science and technology in USAID. I want to add Deeley was important in stimulating interest within USAID, but so were Jack Sullivan, Jack Vanderyn, and Nyle C. Brady. Nyle Brady headed science and technology in USAID and was very supportive. He had just come back from the Philippines where he had been director general of the International Rice Research Institute, one of the premier world agricultural research centers.

What this was leading to was that USAID asked the Forest Service to be their source of information on forestry schools. That led to the Forestry Support Program, FSP. And it's still there and it's growing. USAID provided funds for staffing and the Forest Service developed skill rosters and background information that can be called on for forestry programs anywhere in the world. The upshot was that International Forestry started to grow, and then attracted still other funds and more activities. My office directed some additional money into it as well. When I left the agency in 1986, International Forestry had gone from six to twenty-five people. Since then it's gone up another fifteen or twenty and by 1991 achieved separate deputy chief status.

HKS: Let me step back a bit. Why do you think it was in Research as opposed to State and Private Forestry in terms of the kind of assignment that International Forestry had? State and Private Forestry is the outreach part.

REB: Somewhere I read why that happened. It had been one of those opportunistic things.

HKS: I suspect State and Private Forestry didn't care for it and someone in Research did, and that's why it's there.

REB: All of those things could have been true. But in fact, the volume of business in International Forestry was greater in Research than it was in any of the other deputy areas because most of the exchanges were technical and scientific in nature, and that brought the research group into play. So I had two portfolios, Research and International Forestry. As an extension of your question, there was some discomfort, especially in the National Forest System, about why Research had this activity. They felt that they weren't getting a fair shake. I tried hard to dispel that concern, but I don't think I succeeded very well.

HKS: Some of the experts you would use would come out of the National Forest System.

REB: Yes, and if it were appropriate we went out of our way to involve them, disaster assistance was a case in point. The forest management seminars for foreign forest administrators, started in the early '80s, also was assigned to NFS.

HKS: Bob Spivey, do you know Bob?

REB: Yes.

HKS: Well, he was working in Honduras, that's National Forest Administration providing...

REB: That happened after I retired but as I understand the job, it concerned forest administration, which certainly should have been supplied by NSF or S&PF.

The movement toward an independent International Forestry program began, partly through congressional action, partly by some interested non-governmental people in Washington, to create a separate deputy area for International Forestry. That happened with the 1989 Farm Bill. International Forestry was separated in 1991 with Jeff Sirmons the first deputy chief.

HKS: I realize that foreign aid in general takes a beating whenever the economy is down, because why export money when we've got jobs here that need it. International Forestry faces that at budget time. How did Congress view International Forestry? Was it challenged?

REB: Foreign assistance is not a favorite with Congress or their constituents. However, the House Appropriations Committee I think was favorably disposed toward International Forestry because tropical deforestation was and is a current issue. Yates used to question me about it. It seems to me that we did get some budget increases for Hawaii and Puerto Rico.

HKS: The Third World. My specific knowledge is pretty primitive, but it seems to me that expertise from the Soil Conservation Service and all throughout Agriculture would be in great demand. Was there ever a conflict within the department about the Forest Service assignments?

REB: Yes, there was. It happened late in the Carter administration. OICD, Office of International Cooperation and Development, had the principal responsibilities for international programs. Several agencies within USDA had international activities, Agricultural Research Service, and other agencies. During that period international programs were taken from those agencies and placed in OICD headed by a Joan Wallace. The Forest Service program was so small and so diffuse that nobody within that agency reached out to pick it off. The role that OICD has served in intervening years is to mobilize resources, that is people and skills within the department, to address problems around the world. Despite the fact that International Forestry remained with the Forest Service, OICD has been a pretty good supporter. ARS was deeply resentful of having lost International Agriculture. One of the things I attempted to do was to keep our international program fairly diffused and with a low silhouette so it wouldn't be consolidated with OICD. One of the agency's concerns today should be that the higher profile of International Forestry risks this consolidation.

HKS: Agroforestry can be called a kind of forestry or a kind of agriculture. It depends on where you're standing when you describe it.

REB: Yes. The Forest Service has an agroforestry project in Lincoln, Nebraska, today. The questions that you are raising are difficult to answer concisely, because agroforestry represents many different combinations of trees and woody shrubs on one hand and various agronomic, horticultural, and pastoral systems on the other.

HKS: They may not be important.

REB: But the question about the support for International Forestry is an extremely volatile one. Right now, overall, international programs are one of the least popular things in Congress. The exception is global forestry including deforestation, global warming, and biological diversity. The 1992 UNCED Conference in Rio de Janeiro reinforced this need.

HKS: The president is going to give away one hundred fifty million dollars for it.

REB: For global forestry. As I read this message, one part of the international programs, forestry and natural resources, are currently in favor. What the implications are for the Forest Service, I don't know.

About the volatility in international programs, let me go back to the Renewable Resources Research Act of 1978. We wanted unambiguous authority to do international forestry research, because the previous authorizations were vague. I was chairing the meeting where we were drafting legislation for the new bill. There were other people also providing background. The committee decided it should seek clear authorization but with a low-silhouette for International Forestry in that legislation. If you read the paragraph about International Forestry you'll see some vague references to the fact that the bill authorizes cooperation with industries, universities, foreign governments, and so forth. What we wanted to do was have the authority but with extremely low silhouette so it didn't provide a hang-up point in congressional deliberations. If I were to write new legislation today I would give it high visibility, but it was a risky approach in 1977 and 1978.

HKS: What's the timing that put last year as the year to create the deputy chief's job in International Forestry?

REB: It was a consequence of the Farm Bill of '89, the most recent authorizing legislation for both agriculture and forestry. One title in the farm bill had to do with forestry. If memory serves me correctly, that's what created an independent international forestry program in the Forest Service.

HKS: The Forest Service has been involved in international forestry for a long time, but there was some threshold that was crossed.

REB: Yes, this was an important threshold. As I understand it, the chief congressional support came from Congressman Vento of Minnesota. His staff assistant, Jim Bradley, who's a forester, came from the Forest Service. Vento went to Puerto Rico, met with Frank Wadsworth, and was taken up with Frank's charm and ability. Vento became very much interested in international forestry. So that was the congressional side. On the constituency group side, Warren Doolittle, who's now heading the International Society for Tropical Forestry, and other groups lobbied for more visibility for International Forestry. That's how it came about as I understand. Others can comment in more detail about how that came into being.

HKS: I'm going to be interviewing Frank later in the year. He sent me this huge resume, thirty-five or forty pages, very detailed. He visited, officially, about thirty different countries. How was Frank a part of International Forestry? You said you were down to six people in International Forestry, but you had a bunch...

REB: We had six people in Washington, D.C. We also had small programs in Puerto Rico, Hawaii, and the Forest Products Laboratory, plus cooperation with Mexico and Canada at several stations.

HKS: Frank would have been an important part of that.

REB: Yes, Frank and the Institute of Tropical Forestry under the Southern Forest Experiment Station. The appropriations for Puerto Rico, Hawaii, and FPL came out of the regular budget.

HKS: Puerto Rico is tropical, and because of the language situation, it's easier to disseminate something in Spanish in the Third World.

REB: Exactly. Puerto Rico was always important in the Forest Service scheme of things because it was a window to Latin America. The institute had Spanish speaking skills, and I'm told the second largest tropical forestry library in the world.

HKS: Where is the largest?

REB: Oxford University. For me, separating Research from International Forestry as a separate deputy area would have been difficult because I was fond of both jobs. But on balance, I think that breaking it out as a separate deputy area is a good idea at this time. I'm not sure how durable it will be because the total Forest Service budget for International Forestry today is in the twelve-fourteen million dollar range. That compares to one hundred fifty million dollars in Research, one hundred or more million dollars in State and Private Forestry, and nearly two billion dollars in NFS.

HKS: A station director has a larger budget than International Forestry.

REB: Yes, and when one of those inevitable efficiency issues comes around, they're going to ask why do you have a fourteen million dollar program with a separate deputy chief. So the Forest Service will undoubtedly be called upon to defend that program.

HKS: I assume that Sirmon's hunting license is to make it grow and make it permanent.

REB: He probably will do that, but the main influence of that program is going to come from its leverage on other programs. A better way to say that is that the technical skills needed for international work resources resides not with the Forest

Service, but with its university and state and private cooperators. But USAID and the State Department are going to have one hundred fifty million dollars from Bush's initiatives. How are they going to spend the money? They are already calling on the Forest Service for ideas. So it's the leverage the new deputy chief can assert that probably represents his greatest potential impact. (This exchange occurred before President Bush's defeat in 1992.)

HKS: Five years from now that money is going to be gone. I'm assuming that one of Sirmon's responsibilities is to make a permanent program out of this and not soft money from the president.

REB: I have the impression that Jeff sees it in that way more as a catalytic role than as a full-fledged operating program with a large budget. I know that's the way I would view it. But just think of the influence that the program could have. The World Bank is a major contributor to the international forestry programs. UNDP located in New York has another major forestry program. USAID, I think they had a hundred or so million dollars in forestry, even before Bush's pronouncement. So that's where the money is.

HKS: That's right.

REB: But the skills are in the Forest Service and its partners.

HKS: And there is a precedent for the Forest Service carrying out these other assignments, like for AID.

REB: So international forestry programs, with the support of AID and some people in both the Forest Service and AID, began to grow. One of the things that I did early on was to invite Nyle Brady, senior administrator of science and technology, Jack Vanderyn, Jack Sullivan, all of USAID, to join in some round table discussions with Max Peterson. Max was very willing to join in those discussions. I think these informal meetings had a very positive influence on the working relationships with USAID.

The second aspect of International Forestry began to develop about the time when I became a member of IUFRO's executive board, and especially in 1981 when I became vice president. IUFRO was searching for funds to do a series of research problem analyses around the world, and USAID was very supportive in underwriting some of those. In fact, so much so that we invited Nyle Brady to be a keynote speaker at the Ljubljana IUFRO Congress in 1986, and that was another reinforcement in working with USAID. I do want to talk about the IUFRO connection in a little more detail. But it was closely linked to some of our developments domestically.

HKS: You talked earlier about Research's outreach. What did the forestry deans think about International Forestry? Some schools are strong in international forestry. N.C. State is.

REB: The short answer is generally positive—Yale and the Universities of Washington, Minnesota, N.C. State, and many others. Essentially all of them have international forestry programs of one sort or another. Here at the College of Forestry in Corvallis, about 40 percent of our graduate enrollment, and this is one hundred sixty students, are from other countries. Oregon State is not nearly as active as some of the forestry schools. But international forestry is a part of virtually every curriculum. The Council of Forestry School Deans and Leaders has a committee on international forestry.

THE PEACE CORPS

HKS: There's a lot of interest by the American students at the Duke School of Environment on international issues. These are the popular courses; save the tropical rainforest and all of that. Are there jobs for First World people? If you graduate today in America with a specialty in international forestry, what can you do? I understand you can train a Third World person to go back and do a job in their own nation.

REB: The employment prospects for forestry are really not very good. I'm confronted with that question almost weekly here at OSU. Both American and foreign students ask, "Where can I go to work?" The problem is not that the need isn't there, the problem is the budget deficit. The most meaningful way for young people to enter into that arena is through the Peace Corps, which is eager to have foresters. I encourage young people to consider that route. It's a tough life, but it gives a person two or three advantages. One of them is a language skill, sometimes more than one language. And the second one is that it gives them cultural and technical exposure to a country or region of the world. A third advantage is that they have some employment rights that go with the Peace Corps assignment. The Peace Corps right now is far and away the most attractive employment possibility. It's a tragedy, so much interest, so much need, and so little opportunity.

HKS: I was wondering about the students who want to work in the Third World, but at First World salaries.

REB: It's difficult. But you know the Peace Corps has been around for long enough that earlier participants are beginning to occupy middle and upper-level management positions both in public and private pursuits. Dave Harcharik, who's assistant deputy chief in International Forestry, is a Peace Corps graduate.

The interesting thing to me about the Peace Corps volunteers is that life is never the same after they return. Have you noticed that among Peace Corps graduates? They view the world differently. The Peace Corps is providing background and international skills for the U.S. much the same as the colonial services did for Britain and France.

IUFRO

Executive Board, 1976-81

HKS: You've been talking about IUFRO and how it's related to international forestry, so let's continue on that. Tell me more about IUFRO involvement and its significance to forestry research.

REB: As I've said before, I came into Washington in 1975 and was appointed deputy chief in 1976. I was elected to IUFRO's executive board at the Oslo Congress in 1976. That began what turned out to be nearly a twenty year involvement with IUFRO. It strongly reinforced and fortified my interests in international forestry. My first executive board meeting was in Nigeria, hosted by Domonic Iyamabo, one of the IUFRO executive board members from that country. I was to visit with Domonic many times over the next fifteen years. It was in 1976, and I must say that perhaps was the most significant international involvement I had up to that time. It's the first time that I'd really seen tropical forests with all their forestry problems, including economic and social consequences. The president of IUFRO was Walter Liese of Germany, another person I came to know well.

Liese appointed me to the Finance and Planning Committee of IUFRO, which really was the inner committee of the executive board. I had a couple of jobs at that time. One of them was that I was chairman of the Honors and Awards Committee, which is a fairly significant one. I also was involved in monitoring the administrative activities of the Union, including the secretariat in Vienna.

Administrative Problems

REB: The story of the secretariat is a complicated one, and I won't go into it at this point except to say that IUFRO moved its secretariat from Rome with FAO to Vienna, Austria, in 1971. Liese was having serious problems with the secretariat including questions of performance and confidence. I couldn't tell at that time whether this was a conflict resulting from Liese's Prussian and Teutonic tendencies, or whether it was the Austrian tendency toward laissez-faire. But Liese talked with me about it repeatedly. One also has to understand that one of Liese's sons was dying with a long term illness, a brain tumor. And so Liese was under great stress, and he involved me in some of the administrative chores of the Union. The question of the secretariat for IUFRO remained until I became president in 1986.

HKS: Right.

REB: A second issue came up during that time, for which I was more an observer than a player. It was the China problem. The People's Republic would never accept Taiwan as a separate country. For that matter, Taiwan wouldn't accept mainland China. I remember that Liese and Tai Satu, the vice president from Japan, developed an accommodation that gave representation on IUFRO principal governing body, the international council, to the People's Republic of China. They also worked out an arrangement where Taiwan, the Republic of China, would have observer status. It was a tender, fragile relationship, but at least it avoided confrontation at the Kyoto Congress, and it carried in a very uneasy way up until 1990. In many respects this first term on IUFRO executive board was a precursor for problems that I would address later on. Liese was serving as a treasurer for the Union. His office was sending bills to many countries of the world, and with currency exchange problems, delinquent dues, and all of that, an incredible workload developed.

HKS: Because the secretariat wouldn't do it?

REB: Because the secretariat couldn't or wouldn't do it. At that time levels of trust also were low. The secretariat was handling IUFRO news and some routine membership questions, but the treasury responsibilities stayed with the president. Liese and I had some discussions about Union finances. Amy King of the Forest Service, who served as an assistant to George Jemison when he was president of IUFRO, said, "Why don't you appoint a treasurer?" And I asked myself, why didn't I think of that? So Liese and I talked about creating the office of treasurer, a shared perception, I believe. So late in my first term, we proposed the creation of the office of treasurer.

We had preliminary discussions with a very able research administrator from Switzerland, Walter Bosshard, who was director of the Forestry Research Institute of Switzerland. Walter was willing to take on that job. Furthermore, as director of the Swiss Institute, he volunteered to underwrite the costs of the office. The incoming president was Dusan Mlinsek from Yugoslavia. Yugoslavia, was a soft currency country and didn't have a reliable banking system. Mlinsek just didn't want to have anything to do with the office of treasurer. So office of treasurer with Walter Bosshard as treasurer was timely and has worked out well. This brings us then to the Kyoto conference of 1981, which was hosted by the Japanese.

SPECIAL PROGRAM FOR DEVELOPING COUNTRIES

HKS: I went to that one.

REB: Then you know what an outstanding event that was. At the Kyoto Congress, John Spears of the World Bank and Marco Flores Rhodas, assistant director general of FAO for forestry, presented a paper to IUFRO which said essentially—why doesn't IUFRO bring its resources and skills to bear on forestry problems in the developing world. There were some discussions following the paper—in the end, IUFRO accepted the challenge.

In the meantime, I was elected vice president of the Union which involves chairing the program committee. At our first executive board meeting following Kyoto, I was asked to take the lead on the challenge from the World Bank and FAO. I must say there are very few assignments that I relished more than that one. I prepared a position paper suggesting how IUFRO might respond. We called a special meeting of the Finance and Planning Committee (a small policy subcommittee of the Union) in Vienna in March 1982.

At that time, the executive board agreed with my assessment that we couldn't handle this assignment on a volunteer basis, that we needed to have a person to give our leadership to the developing countries program. We agreed that we would advertise globally for that position which had financial support from the World Bank. There were about fifteen or twenty applicants. The Finance and Planning Committee met again in Zurich in November, as I recall, to assess where we were. We went through the list of candidates and the work load, and we agreed to recommend to the full board recruitment of Oscar Fugalli. Oscar Fugalli had recently retired from FAO in Rome where he occupied a senior position in their forest resources department as a silviculturist.

We agreed that to recommend Oscar Fugalli as coordinator for the Special Program for Developing Countries, SPDC, to the full executive board. I had some ideas about what the SPDC would do, and so did Oscar. I wanted to put the early emphasis on problem identification—what are the problems in the developing world for which science could make a substantial contribution? I didn't want to start with organizational questions to assess researchable problems. That began a series of workshops involving Asia, South America, and Africa underwritten by, among others, USAID, World Bank, and UNDP. In fact I think there was something like ten or twelve donors that ultimately helped underwrite the ten workshops.

The purpose of these workshops was to ask what are the most important problems that we ought to be addressing, with as much input from the scientists from the region as possible. The first workshop, held at Kandy, Sri Lanka, was concerned with multipurpose tree species suitable for the Asian/Pacific region. We came out with a planning document with Keith Shea and Les Carlson of Canada doing the secretarial tasks. Oscar Fugalli was at this workshop and assumed increasing responsibility for others to follow. We held another similar workshop in South America and two more in Africa all aimed at multipurpose tree species. For Africa, one was concerned with the sub-Saharan region; and the other for the southeast Africa Division Five, under the leadership of Bob Youngs, took on the question of problem identification for utilization research in each of the three continents. An eighth planning document, this one in cooperation with the International Food Policy Research Institute, addressed natural resource policy questions.

HKS: What did you do with these problem analyses?

INCOFORE

REB: As we moved into these regional problem analyses, we began to ask: what kind of an organization do we need to strengthen global forestry research in the developing world? All of us were very much aware of what agriculture had done with the big international research centers, and we wanted something like that for forestry. It was time to deal with the organizational questions. I prepared a paper for the 1986 Ljubljana Congress where I outlined a concept called INCOFORE, International Council for Forestry Research and Extension. I wanted to use these problem analyses as a basis for the organization.

HKS: Are those analyses well distributed? Do most forest school libraries have them?

REB: No, I don't think most forestry schools do.

HKS: It's a fascinating document. I wonder who has access to it.

REB: They've been sent to donor agencies, and they're available in limited numbers from the IUFRO secretariat.

HKS: Small press runs.

REB: Let me give you a later paper which describes in detail what we did and the origins of this program. I think it's an important reference document. (IUFRO. 1989. *INCOFORE: A Research and Extension System for Tropical Forestry.* 35 pp. IUFRO Secretariat, Seckendorf-Gudent-weg 8 A-1131 Vienna, Austria).

HKS: Thank you.

REB: INCOFORE was an organizational concept to serve as a starting point for forestry research on high priority topics in Asia, in South America, and in Africa. These contributed to discussions that were also underway among other major players.

Bellagio Conferences

REB: About this time, FAO, the World Bank, the World Resources Institute, and UNDP were trying to stimulate interest in global forestry, including not only research but also action programs. There were two parallel efforts underway. One was called Tropical Forests: A Call for Action sponsored by the World Resources Institute, UNDP, and the World Bank. Another was the Tropical Forestry Action Plan (TFAP) under development by FAO in Rome. Both were to look at overall global forestry, of which research was a part. The TFAP was issued in 1985, and it was soon merged with the World Resources Institute effort into a single program called TFAP.

The Tropical Forestry Action Plan was a high visibility program. It was obvious that forestry research had piggybacked on that process. The TFAP served as a basis for a conference on global forestry at the Rockefeller Conference Center in Bellagio, Italy, in 1987. Bellagio was very significant because it was here that the system of International Agricultural Research centers came into being about 1970. In any event, the conference in 1987 dealt with the issues raised in the Tropical Forestry Action Plan, including research. The Bellagio I conference came out with about ten recommendations, two of which concerned research.

A second conference was convened in the U.K., but it was called Bellagio II. This one dealt only with research. Oscar Fugalli and I attended because of our IUFRO connections. There were about thirty or forty donors at Bellagio II, chaired by David Hopper of the World Bank. A task force proposed several alternatives to the participants and recommended paralleling the INCOFORE concept where forestry would be a stand-alone international forestry research institute similar to the then status of the International Council for Research in Agroforestry (ICRAF), located in Nairobi.

HKS: What happened next?

REB: The Bellagio II conferees, very senior people, rejected the stand-alone institute and said instead we must consider forestry research for incorporation into the CGIAR system. CGIAR, Consultative Group for International Agricultural Research, consists of about forty donors that underwrite the International Agriculture Research Centers, a two hundred fifty million dollar a year undertaking. They didn't reject many details presented by the task force, but they said you need to be in the larger agricultural research system. From my point of view, this was a much more favorable outcome than we had dared to expect.

HKS: What happened to IUFRO's role after the CGIAR group accepted forestry research in its mandate?

REB: In many respects the action was taken away although IUFRO was never the sole player in the game. And as deliberations moved forward, the action shifted more to the CGIAR group. The donors indicated about 1990 their intention was to incorporate forestry research into the CGIAR system. Actually deliberations went on for another two or three years before forestry research was formally accepted into CGIAR. One of them was ICRAF, with global rather than African mandate. The second, now named CIFOR, Center for International Forestry Research, to be located in Asia. ICRAF would handle agroforestry research and CIFOR most other forestry research. Small parts of forestry research, such as policy and plant genetic resources, would be dealt with in other of the existing CG centers. Australia was appointed as the executing agency to bring this new entity into being, and we're now in the final stages of those deliberations.

HKS: Rockefeller's supporting it and other foundations?

REB: There are forty governmental and foundation donors. The story of the CG system is a great one leading to the Green Revolution in wheat and rice. It's the place where forestry research always wanted to be. If I were ten years younger I'd have my hat in the ring for the director general's job of that new forestry research institute in Asia. IUFRO was a significant player in this achievement, but like all victories, you know there are dozens of fathers.

HKS: Let's follow up on what you just said that forestry research now is positioned where it always wanted to be in this structure. I don't completely understand the significance of what you said.

REB: It's significant for several reasons. The International Agricultural Research Centers (IARCs) are the premier research institutions to approach agricultural problems in the developing world. Forestry joins that group. Because the IARCs are so visible and they've done so well, they're the recipients of major funds, two hundred fifty million dollars a year for the thirteen centers then in existence. There is no funding source anywhere in the world to equal those. What that means is that some of the individual institutes have annual budgets of more than twenty million dollars a year, rice, wheat, and so forth. You know that Norman Borlaug won the Nobel Prize for his work on wheat, and that's the family of institutions that forestry joins. The system has been enlarged to eighteen institutions now, two of which are forestry, ICRAF and the one yet to be located in Asia called CIFOR, Center for International Forestry Research.

The role of the international centers is to do upstream or more basic research. They address the more fundamental obstacles to greatly increased production of X, you name the commodity. They do the pioneering work, generally beyond the capacity of developing countries. Forestry obviously has an equivalent role in that kind of research. Some of the early problems tentatively identified for the forestry institutes include genetic improvement, seed source studies, tree breeding, husbandry techniques including fertilization and insect and disease protection, irrigation or drainage, and whatever else is required to impose the performance of tree crops. It's the ecological equivalent, really, of what has happened in wheat and rice. You can imagine that tree species involved are going to include some of the pines of North America, eucalyptus of Australia, and acacias from several places in the world. Other things that have been identified as obstacles in forestry include soil relationships, including mycorrhizal studies, other soil symbionts like bacteria and things of that sort, and nutritional requirements. Still another area tentatively identified for accelerated research is utilization. One of the quickest payoffs is to use wood more effectively to make it last longer and burn it for cooking and heating more effectively.

The points that I want to make are that IUFRO was given the charge for special programs for developing countries, the job was given to me as vice president in 1981; it was an assignment that I relished. When I became president of IUFRO in

1986, I didn't surrender that job, I kept it. Now I think maybe that brought about some resentment in the executive board, but I was so intrigued by that program that I wanted to keep it.

HKS: When you say utilization, will this generate work for the Forest Products Lab?

REB: It could. Even without the new international forestry research center, FPL was and is playing a substantial role in forestry utilization around the world.

HKS: This is a change of direction for IUFRO, which historically has been a gathering place for people.

REB: That is correct. And this was a major concern to some members of IUFRO's executive board during the early years of SPDC, who wanted to keep the SPDC entirely separate from IUFRO regular programs. Ironically one of the later criticisms was that the SPDC was not integrated sufficiently with the regular program.

HKS: But no action plan. Now IUFRO is taking advantage of the talent it has to solve a world problem, directly.

REB: Yes, to all of the above. We knew that the traditional way of doing IUFRO business, very much voluntary—you came, you participated, you brought your own travel funds and so forth—that simply wouldn't work in the developing world. IUFRO, and especially some of the European members of the executive board, were very hesitant about this departure from the traditional IUFRO role, as you can imagine.

HKS: I'm sure.

REB: But there were enough supporters at those two meetings, one in Vienna and one in Zurich, that the board hesitantly said go ahead with your proposals for a special program for developing countries. I must say that today the IUFRO executive board and others in IUFRO who are familiar with it are proud of what the Union has done, and they don't want to relinquish this program under any circumstances.

HKS: You're still involved in it?

REB: More now as an observer than as an active participant. So this was a major development during my time, and this concept has been endorsed by the IUFRO Congress in Ljubljana and strongly endorsed and reinforced again by the IUFRO Congress in Montreal.

IUFRO PRESIDENCY

REB: Permit me to backtrack for a moment. My vice-presidency ended in 1985, and then came the question of who would be the next president of IUFRO. Early on, I had no special interest in being the next president of IUFRO until I became so involved in the SPDC. I began to ask myself, can I have more influence on this program as a president than otherwise, and of course the answer was obvious. So I announced that I wanted to be a candidate for the next president of IUFRO. I know that there were folks that were unhappy about that. Walter Liese was one. There was the issue that I was American. After all Jemison had been the president just a short time before. A nominating committee was appointed. The other candidate was Walter Bosshard who was director for the Forestry Research Institute in Switzerland. Walter and I agreed that whoever was elected, the other would support fully.

I am told that the Nominating Committee noted the presidency of IUFRO had been in a German speaking country for several terms and that the Union had only one president outside Europe, and I received the nomination. The International Council approved. There were a lot of things that I could do as president that I couldn't do otherwise. Although Fugalli and I tried to keep board members informed, I think some may have been unhappy because they thought we kept that work pretty close. I was in weekly phone contact with Fugalli in Vienna about fast-moving events, but the board met only once a year. In any event Fugalli and I left our jobs late in 1990, he to be succeeded by Loren Riley of Canada and me by Salleh Mohammed Nor of Malaysia as president, with overall supervising of the SPDC to Jeff Burley of the U.K. who was incoming vice president for programs.

THE SECRETARIAT

HKS: Earlier you mentioned problems with the secretariat in Vienna. Could you elaborate?

REB: It became obvious to me in the 1980s that the secretariat in Vienna was simply not functioning well, and in fact was in decline and that the concerns that Liese had a few years earlier were genuine. Our secretary at the time, Otman Bein, knew that there was a lot of concern about his performance. I suspect that he felt that he had a mentor in me who would encourage his continuation. We had known each other socially and on two occasions we hunted together in Austria. In the meantime Bein had a serious accident. Both of his arms were in casts. He knew then that I was a candidate for the presidency, he wrote me a letter and said that because of poor health he was contemplating retirement or resignation from IUFRO. And I am almost certain that what he anticipated is that I would come back with a letter and say we can't spare you. I wrote to him saying that his first consideration was to his health and to his family and that we would accept his resignation. I knew I had to do something about the secretariat, so he made it easy for me. And we, with the assistance of the Austrian government, replaced him with Heinrich Schmutzenhofer. Heinrich has provided all the qualities of leadership and vision needed for IUFRO's continued growth. It was never a question about the previous secretary's dedication to the Union, it's just that events and times passed him by.

HKS: Now Jemison had been president, Harper had been vice president, so there wouldn't have been any eyebrows going up in the Washington office of all this extra duty you'd taken on.

REB: No problem at all. In fact I think encouragement. I announced in mid-1985 my intention to retire early in 1986. But before I was elected president of IUFRO in 1986, I visited with Max and Dale Robertson about post-retirement support for my IUFRO activities. There never was any question. I drafted a letter for Max's signature and got Dale's approval on it saying yes, we will support your office and travel expenses during your time as president, and we will also support your past president's costs. The Forest Service has been unstinting in supporting those out-of-pocket costs. Otherwise all of IUFRO's activities have been done without compensation. I was a volunteer.

HKS: What authority does Max have to make that agreement? Are you part of a volunteer program?

REB: Yes, I am. While I am unsure of any other specific authorities, the Forest Service frequently uses its funds to support work that furthers its aims at strengthening domestic and international forestry. They provided me office services (pointing to this building); they provided me secretarial help and office space and things like that. But I think over that four year period I probably gave half my time to IUFRO activities, and I still give a quarter...

HKS: So your office in this building is because of IUFRO, not because of deputy chief emeritus or something.

REB: Primarily. I also have an office in Peavy Hall where most of my teaching responsibilities are met. To finish IUFRO administrative issues during my presidency. The treasurer and the secretariat were by now fairly well taken care of. Understand that Liese's presidency to mine, membership doubled, going from seventy-five hundred individual scientists to fifteen thousand. By 1990, one hundred six countries and seven hundred research institutions were members, so the workload was growing enormously. When I became president, I wanted to gather up all the administrative activities in the Union, which were scattered among various members and committees of the executive board, so I asked Jim Cayford of Canada to serve as the chairman of the administrative activities of the Union. Keep in mind that at that time the vice president was in charge of programs. The six divisions and the sixty subject and project groups were under the overall guidance of the then vice president, but the administrative side—treasurer, secretary, publications, and dues—really didn't have any counterpart. I asked Jim Cayford to take over the leadership of that cluster of activities so that it would give me as president more time to be involved with both committees. Jim took on that job. But what I was really aiming for was a second vice president of IUFRO. We proposed this to the International Council at the Montreal Congress and they approved. And so a second vice president was created. One dealing with the administrative affairs of the Union and the other dealing with the scientific affairs. That's the organization today. As I pointed out, the president today is Salleh Mohammed Nor of Malaysia, assisted by Jeff Burley, vice president for programs of the U.K., and Jim Cayford of Canada, vice president for administration.

The China Problem

REB: By way of summary, what I'm recounting here are the administrative and organizational changes of the Union that were addressed during my time as president. Let me touch on a couple of other issues in IUFRO. One was the China problem. Keep in mind the non-political, non-governmental nature of IUFRO, and it all came to a head in conflicts between Taiwan (ROC) and mainland China (PRC). I mentioned earlier that Liese and Satu had found a temporary accommodation to the China problem, but in the meantime Taiwan was getting restive as hell. Taiwan actually has more IUFRO participants with its twenty million people than China does with a billion people.

I recalled reading several years ago, in a foreign affairs journal, a discussion of the conflicts between People's Republic of China and Taiwan. You know this has been a bitter debate over the years. That foreign affairs paper mentioned only one exception where PRC China recognizes Taiwan, and that's in their joint affiliation with the International Council for Scientific Unions, ICSU, located in Paris. ICSU had developed an accommodation by which both Taiwan and China could be represented in the same international bodies. Somehow that stuck in my mind. So as we were headed for a confrontation, China wanting to expel Taiwan from any administrative role in IUFRO, I recalled the ICSU precedent. I said, the solution here is for us to adopt the principles of ICSU, which provides a home for scientific unions all over the world, and IUFRO was one. So we recommended to the international council in Montreal that IUFRO follow the rules of ICSU, and they adopted it. I think it would have been interesting for you to sit in on the debate.

HKS: I'm sure.

REB: The secretary and I met with representatives of the two Chinas in Montreal, and I thought we had an agreement about how to resolve the matter. PRC China would hold the seat in the current International Council meeting in Montreal but that both Taiwan and mainland China would sit at the table in the next International Council meeting, in keeping with ICSU's procedure. PRC China obviously came to Montreal with instructions from their government that under no circumstances would Taiwan sit at any table or be a co-equal with PRC.

HKS: At the Fifth World Forestry Congress in Seattle in 1960, I was McArdle's chauf-
feur. The Forest Service sent me to Seattle because I knew my way around. He was
having all these problems. They had a Friendship Grove for sixty-four nations.
And some country, let's say Czechoslovakia, had a brand new flag, but no one
knew it. The United States government presented them their old flag to carry in
the parade, and they were upset. I imagine in IUFRO you have the same problem.

REB: A parallel situation, in both cases the implications are far more significant than
whose flag flies or who has a seat at the table. I remember a big ceremony where
one flag was marched out of the arena and out of the display area. I can't remember but it could have been the Taiwan flag, so these international protocols
are extremely sensitive. In any event, the PRC delegate challenged these arrangements in the International Council meeting, contrary to what I thought was a pre-
arranged agreement. I didn't want to have a public debate, but finally we did. As
chairman of the council, I took a fairly active role and explained the circumstances and what the alternatives were, and I was terribly concerned that the ICSU
precedent was going to fail. But when it came to vote, it was fifty-five for the position that I was proposing and one opposed.

HKS: You mentioned the larger significance of the debate. Could you explain?

REB: The China question in its own right is a major one in terms of international
relationships, but the larger significance of this is that it was a powerful and dramatic reaffirmation of the non-governmental, non-political role of IUFRO. One
has to think about it for a moment, but that's an important point in international
relationships, and it served IUFRO extremely well. IUFRO enjoyed, for example,
East/West and North/South contacts, because it was non-aligned and non-
governmental in nature. IUFRO could move easily between western Europe and
eastern Europe and the Soviet Union. There are many other examples about how
critical it is for the Union to be non-political and non-governmental.

HKS: The history group put on a conference in Zwolen, Czechoslovakia, about three
or four years ago. Someone you may know, Dariosh Voshmgir, was there to
represent the secretariat.

REB: Dariosh? He is an Iranian working in Vienna.

HKS: I guess you know him pretty well. We stayed at his hotel when my wife and I
went back to Vienna with him. When we were driving in Czechoslovakia he
showed us his passport. Because of IUFRO he had some sort of generic passport,
it was almost like being a member of the State Department or something. He
didn't have to worry about visas and border crossings the way all of us did. I
thought, isn't that interesting that IUFRO has enough stature that he could enter
and leave the East Block, which in those days was petty damn hard to do without a
lot of rigmarole when you're driving. My esteem for the organization went up a
notch. Gee, this really is recognized as a significant body.

REB: Yes, it is. It gives a legitimacy to inter-country interactions that often could not be achieved governmentally. The irony is that many governments recognize that they need this kind of body to carry out their business. For example, the U.S. government tends to be very supportive of IUFRO activities because it fulfills a need that governments can't meet. But it came to a head on that China question and there was, I think, a substantial reaffirmation.

An additional significance of the two China debate was that it gave us a generic mechanism to deal with shifting alliances among and within countries under a wide range of political circumstances.

LATIN AMERICA

REB: Now a couple of other things that came out of my presidency. I very much wanted to draw Latin America more into IUFRO. Latin America was the most reluctant continent of any, and IUFRO has more trouble in Latin America. We, with the help of our Spanish representatives, were now publishing IUFRO news in Spanish, the Spanish are doing it, and we adopted Spanish as a fourth official language of IUFRO.

HKS: I want to ask you one question about Spanish. Since Spanish is not a traditional scientific language, was there any reservation to do that?

REB: Spanish imposes some additional costs on IUFRO. IUFRO still conducts most of its business in English. And so, informally, English still serves as the principal working language, even in Spanish countries. Still, if we wanted to reach Spanish-speaking groups, both in the new and old worlds, official recognition of Spanish was important both for its substance and its symbolism.

HKS: Right, when we put on our conference in Costa Rica, we had money from Rockefeller. They said, you have to have someone to do translation. We were dealing with the humanities and not with science, and most people there really couldn't speak English very well. It was an interesting experience for me.

REB: A couple of IUFRO research planning workshops were held in Latin America, where the papers and the discussions were in Spanish.

HKS: It was more a courtesy. You go into someone's country, you've got to speak their language.

REB: Absolutely, but informally, English still is the principal working language of the Union.

EXTENDING IUFRO COOPERATION

REB: One of the things I wanted to do with IUFRO had a parallel with what I was trying to do domestically, and it was to build bridges and to enlarge the groups that deal with forest sciences. In this regard I asked that an executive board meeting be held at the headquarters of FAO in Rome, I think in 1987. The relationships between FAO and IUFRO had been uneasy for a long time.

HKS: I know.

REB: I wanted to say IUFRO cares. I think that meeting had a fairly positive outcome. I also wanted to use IUFRO to symbolize East/West relationships, and early in my presidency we arranged for an executive board meeting that would be split between Vienna, which was then western Europe, and Prague, which was eastern Europe. It was to show these joint relationships. Wouldn't you know it, about six months before the meeting came off, the Iron Curtain collapsed. [laughter] But that didn't detract at all from the joint meeting venue. As symbolic bridge building, it worked out as well that the Iron Curtain was down. It just changed the nature of the discussions.

Several leaders of forestry research in eastern Europe, the USSR came, including Alexander Isaev who was then chairman of the State Committee of Forestry for the USSR. He spent the whole week with us. Now Isaev is a scientist in his own right, he's an entomologist and has written several books on entomology and was active in IUFRO before Gorbachev appointed him as chairman of the State Committee of Forestry, a ministerial level post. So that bridge building meeting worked out well, and IUFRO has now adopted a practice where we try to have a split venue in meetings. We'll meet in, say Chile, but we'll visit Brazil before we go to Chile. Or, we will meet in the Philippines, but we'll spend a couple of days in Taiwan. What I was trying to do was extend the reach and the concerns of IUFRO.

In terms of bridge building, in 1988 or 1989, the executive board was scheduled to meet in Beijing. We were going to stop over in Taiwan before we went to Beijing and had to go through Hong Kong in order to get our visas cleared. But Tianneman Square conflicts intervened, and most of the board would not have come to Beijing because it was such an anathema to what people believe. So we still went to Taiwan unofficially and then on to the Philippines for that formal meeting, but it was another example here of trying to cast a larger shadow for IUFRO.

IUFRO AND GLOBAL SCIENCE

REB: I have tended to deal with administrative and organizational questions in this discussion, but we need to keep in mind that for IUFRO these are only means to an end. The real reason for IUFRO's being in place is the extension of science, to share results, to cooperate, to anticipate the next generation's problems. IUFRO is a robust organization. As I said earlier, there are more than sixty subject and project groups many of which have working parties beneath. More than fifteen thousand scientists are affiliated with the Union. As of a few months ago about one hundred six countries, but there are going to be more now that the Soviet Union is breaking up into separate countries. The reason for IUFRO's existence is not administration, it's to provide a forum for the exchange of science. It's a networking organization, and I'd guess I'd like to think it's been very successful.

HKS: It's certainly broadened my horizons, I've gone to more than a half dozen IUFRO conferences.

REB: During my four-year presidency, there was something like two hundred symposia, workshops, and colloquia held in various parts of the world, dealing with the subject content of IUFRO, about fifty a year. Fifty percent of those workshops and meetings were held in Europe, 25 percent in other industrialized countries, and about 25 percent in the developing world. I think this is a partial manifestation of IUFRO's growing interest in developing countries. I suspect that in the next term of Salleh's presidency, a still larger proportion of these will be held in the developing world.

HKS: Bob, do you want to add some additional comments on IUFRO?

REB: I'd like to conclude our discussions on IUFRO with a few impressions that occurred over those eighteen years with which I've been associated with the group. The first one concerns the globalization of IUFRO's efforts. Prior to World War II, IUFRO was very much a European organization. The U.S. was there in observer status, but not terribly active. After World War II, IUFRO was reconstituted, and the U.S. and other industrialized nations began to participate. Les Harper was the vice president, and later George Jemison was the president, Tai Satu of Japan was a vice president leading up to the Kyoto Congress. But still, in those early days, Europe was the dominant guiding influence on the Union.

I became president in 1981 and in 1986 the second president from other than Europe. But by that time other regions of the world, North America, Canada and the U.S. in particular, Japan, Taiwan, were becoming much more active. At the time my presidency ended in 1990, 28 percent of IUFRO membership was from Europe, 28 percent from North America, about 14 percent from Asia and the balance from the other continents of the world. I was followed by a non-European president also, by Salleh Mohammed Nor, who's the first from a developing country. The point that I'm making here is that Europe was the major influence in IUFRO until sometime after World War II, but they've created a robust group of children, and IUFRO has really become global in outlook. I think Europeans view this with some ambivalence. They hate to see their children leave home, but they're also proud as can be of what they've created, and I think properly so.

Another observation concerns the subject matter content of IUFRO. Very much development oriented—silviculture, watershed, genetics, and things of that sort—until recent years, but with a growing environmental content in the program. Concern about air pollution, biodiversity, ecosystems studies, and topics that wouldn't have been very germane fifteen or twenty years ago. The Union has responded in many ways. I should also point out that the things that are not directly research-oriented but otherwise support forestry have been enlarged and they include your own activities, forest history, library services, computer sciences, and a whole set of facilitating and supporting activities for science have become a major part. IUFRO has also extended its scope substantially. There's now a very active boreal forestry group, and the tropical groups. In fact there are many tropical groups, and this would have been unheard of only two or three decades ago, so IUFRO has really expanded in many directions.

OTHER FOREST SERVICE ACTIVITIES

RPA AND PLANNING

HKS: Do you want to go back and look at RPA in terms of research planning?

REB: RPA at its inception was largely administered by the chief's office and by Max Peterson, the deputy chief for programs and legislation. The roles of research really were two. One was to provide technical backstopping, that is the use of computers and computer programs and algorithms and other technical skills. These were drawn upon heavily in managing the database and gathering information for RPA. Research also was the principal provider of information for the Assessment. So there was a lot of technical support and technical backstopping for RPA.

The second aspect was that research had an important part in RPA because it was one of the programs that was under evaluation and review. As I've said already, we gave a great deal of attention to justification for research, especially early in my time as deputy chief. Now I want to introduce a thought here that may be more important by way of hindsight than foresight. I was terribly uncomfortable in the formative years with RPA and with the National Forest Management Act—data acquisition and data handling were getting out-of-hand. There was far more information available than one could assimilate. My concerns gradually formed into this notion: that we were attempting technical solutions for what were essentially political problems.

I've reflected over that many times. That the people recruited to help both the RPA and land management planning within NFMA were some of the best and most able of the Forest Service. They did incredibly well in developing and in fact changing state-of-the-art on the various data gathering and analytical tools. But the issues here, I realize more clearly every year, were not technical, they were political. I keep asking myself, if we were to start over again, if we had had a better comprehension of that, might we have organized the display of the information differently? Might we have acquired data in different ways? Probably in larger chunks so that we could display them more concisely. Did I mention this, that we in research were presented one day with something over fifty thousand options?

HKS: No, I hadn't heard that before.

REB: We were. By the time you take eight or nine alternatives, and eight or ten program alternatives within those, and then build in still other ways, you can generate just enormous numbers of alternatives, beyond comprehension of anyone. Anyway, the principal point I want to make is that had we viewed this as a political process we might have organized and displayed information differently. To its credit, RPA has been slimming down, at least the public displays have been slimming down in more recent years and I find them much more attractive.

Research Planning in USDA

REB: Now let me back up to 1982. In 1982 the second farm bill was passed, and it broke the science and education groups in the Department of Agriculture into a separate agency called the Science and Education Administration (SEA), and a new assistant secretary was created. This took Science and Education away from the assistant secretary who was also looking after the Forest Service and the Soil Conservation Service. Orville Bentley was appointed the first assistant secretary for Science and Education. The 1982 Farm Bill required the science and education agencies to do for science essentially what the RPA had required for the Forest Service. The 1982 Farm Bill required four things: an assessment of the problems confronting the agricultural and forestry research sector; a five-year plan; an annual list of priorities; and an annual report of accomplishments.

Orville Bentley, presiding over the Joint Council for Food and Agricultural Research, asked me to take the lead in preparing the documents to satisfy those legislative requirements. Of course I had help from some very able colleagues and guidelines from the Joint Council. I began with the same perception that I've just given you for RPA and NFMA, it's essentially a political process, not a technical one. I started out by saying that none of these documents were going to be more than fifty pages long, shorter if possible, and the audience is going to be informed laymen and members of Congress. The documents that came out of that process are here (displayed on the table).

HKS: While we're talking about these four documents, briefly summarize how they're significant, in terms of research planning and USDA.

REB: The four documents are an equivalent to the RPA process, that is an assessment, a five-year plan, and an annual listing of priorities for science in the Department of Agriculture. The priorities would be the equivalent of the policy statement that the president is required to send to the Congress under RPA. The fourth document is the annual accomplishments for USDA research extension and education including forestry research. In many respects, this was my way of saying, maybe there's another way to approach the RPA.

Let me make another observation in that regard. If you look back on the history of not only Forest Service research but the Forest Service itself, you'll see a whole series of planning documents, the Capper Report, the Copeland Report, the Timber Resources Review, a National Program for the National Forests, and so forth. RPA in most respects is a later generation equivalent of those earlier planning efforts. My view today is that the next chief of the Forest Service would be well served to put a whole new face on the RPA process, the half-life of the current approach is already past. Now the RPA is required by law, so we can't lose all of the identity to it, but it desperately needs a new approach.

HKS: We hear about the amount of number crunching; each national forest had to come up with a plan. It was a terribly demanding requirement that Congress put on the agency.

REB: I think a lot of these were self-inflicted wounds.

HKS: Self-inflicted.

REB: I remember hearing about running the FORPLAN plan for the Deschutes National Forest. I think the computers ran for thirty hours [laughter] and then aborted.

HKS: Sure.

REB: But you know science has some responsibilities here too. We need not only to develop some of the technologies but we need to offer advice and counsel about the limitations and management of numbers, databases, and so forth, so it's a shared responsibility. Research didn't do as well as it should.

HKS: For RPA, you have to go outside of the agency for the Assessment. You're dealing with a lot of different kinds of data, different parameters. You've got to put it all in a big box and shake it up and come up with an answer. It's enough to do it for the national forests, but to do it for all lands.

REB: Actually the Assessment had lots of antecedents, and the Assessment was not the difficult part of RPA.

HKS: Is that right?

REB: That's right. When I say we have lots of experience with Assessments, the timber surveys were going on even before the McSweeny-McNary Act of 1928 and the

Copeland Report in the early 1930s. And there have been a whole series of timber assessments, each more refined than its predecessor, and each raising more questions for the next one.

HKS: Right.

REB: Now the difficulty with the Assessment in the RPA days is that we were required to get information on other resources—wildlife and water and things like that. Those were more difficult, and the information not nearly as mature as it was for timber, so that gave an apparent imbalance to the RPA. Those issues will be corrected as time goes on.

    The most difficult problem in my estimation came with the program side, and it was generating alternatives among dozens of options. The combinations and permutations became huge. There was a companion problem with the forest plans required under NFMA. That's where new tools like FORPLAN and others came into play. Here too we were generating all of the data that people could gather and computers would handle. Excessively so in my estimation. The upshot was that the information generated under NFMA for the individual forests could never be very well integrated with the national programs. Consequently, Congress would deal with Forest Service budgets in terms of the traditional line items and not in terms of the plans themselves. Now I know Max wrestled with that question at great length and never came to a useful conclusion, and I didn't help him very much either. But imagine the detail that was amassed for each of the 153 national forests, and this was supposed to integrate upward into that program. You couldn't amalgamate them, so there was a discontinuity between the forest planning and the national program.

HKS: There's been some litigation that has refined our understanding of what the law really said, right?

REB: Oh, is there? I wasn't aware of that.

HKS: Lawsuits over forest plans, and the judge says this plan is not adequate.

REB: I wonder if the inadequacy is based on other legislation than the internal requirements of RPA, for example the Endangered and Threatened Species Act, or the procedural requirements under NEPA.

HKS: I'm not sure. The forest supervisor for the forests of North Carolina is quoted in the press as saying, "We did the best we could but the courts say we've got to go back and do it again." That's the sum of my knowledge.

REB: Those lawsuits are generally based on procedural shortcomings of such processes as the National Environmental Policy Act, NEPA, and the Endangered and Threatened Species Act, and other legislation. Not the internal implementation of RPA.

HKS: Complicated times we live in.

REB: Yes. My concluding observation about the planning in the Science and Education Administration is that Orville Bentley and the deans of the agricultural schools and all of us that participate in the science and education planning would take these documents to members of Congress and to other interested groups. In many respects it quieted a lot of the external concerns. We knew who we were, what our goals were, and how we were going to accomplish them, and we displayed them in a relatively concise way.

HKS: How does this translate to the guy sitting in a field research station in Bend, or Grand Rapids, Minnesota? How is his life different because of this planning process?

REB: Excellent question. In terms of the planning process, not much. Because he already knows what his colleagues and his contemporaries are doing. They generally know where to go with their research program. So this process isn't for them. Detailed planning is not for internal use, it is to satisfy constituencies and the gatekeepers outside the agency.

HKS: Like in financial matters, you leave a paper trail so that the auditor can follow closely what you did. It's the external review process.

REB: This overall process has been repeated many times in the Forest Service to give coherence, visibility, and timely justification to a program for those who can influence its well-being. In one form or another this process is and will be repeated many times over.

HKS: If you don't do that, you don't get the budget so people can do their research.

REB: That's right, that's the way the ability to do the hands-on science is affected. You asked me about the world as a deputy chief sees it. What I just said is a statement about the program part of those three activities, that a deputy chief deals with. You may recall that the other two were, first, budgets and second, personnel.

HKS: Let's turn to some conceptual questions. What were the top three scientific developments by Forest Service researchers when you were deputy chief? Things that you're really proud of; here's a turning point in scientific knowledge, or watershed studies, or however you might want to characterize that.

REB: There were a lot more than three to five important scientific developments, and most had their roots before I became deputy chief. However I'd like to think that scientists during my time set in motion research that is providing those same kinds of outputs today, ten and fifteen years later.

HKS: Sure.

TRUSS FRAME HOUSING

REB: Let me give you a half a dozen examples. Best Opening Face, that is the use of computers in sawmills, work by Hiram Hallack at the Forest Products Laboratory. Truss-frame housing, the use of engineered structural members around which you frame and sheath a house, work of Roger Toumey of the Forest Products Laboratory. That was another step in a long series of things that came under the Lab having to do with more efficient engineering of wood. Did you know that the amount of wood used in a house today is about half of what was used in a house at the time of World War II?

HKS: No, I didn't know that.

REB: Yes, just enormous economies in wood efficiency and it comes from a variety of sources. But one of the most important is improved engineering of wood, treating wood as an engineering material. Floor systems are cast differently than a pre-World War II house for example, and sheathing is entirely different.

HKS: I remember in forestry school we were taught that houses are way overbuilt structurally. That it's simpler to put the extra stud in than for the carpenter to figure it out. So what you're saying is that they have dealt with that issue?

REB: Yes, and it's a series of developments, one of which was truss-frame housing. Building a house in its entire cross section, instead of just the roof truss. Serendipitous things come out of findings like that. For example, in truss-frame housing, the work crew could go from a foundation to an enclosed ranch-style house in one work day. You would space these trusses on the foundation just like spacing slices of bread. Once the trusses were in place workers could take reconstituted wood or plywood panels and sheath the house before the workday ended. It's always interesting to see the serendipitous relationships that come out of things.

DECAY ORGANISMS

REB: If we turn to the biological side, Kent Kirk came through with a much better understanding of the mechanism by which natural organisms cause wood to decay. He was working on the lignin part and a colleague in Sweden was working on the cellulose component in wood. They were the winners of the Wallenberg Prize for that achievement. You can see the importance of understanding how wood breaks down. The obvious possible application is biological pulping, or the breaking down of lignins into other useful organic compounds. You know it's still short of application, but it was a marvelous fundamental discovery. By the way, Kent Kirk was elected to the National Academy of Sciences on the basis of that research.

HKS: It has implication for landfill problems, biomass disposition, accelerating the rate of decomposition.

REB: Or the conversion of that material into other useful products. It's a fascinating area, and there's a tremendous effort going on around the world in that today. There were other major accomplishments in insect and disease research. One that I mentioned to you earlier occurred right here in Corvallis. Roger Ryan's finding the parasite/predator complex that would control the introduced larch case bearer, solved that problem with very little fanfare. And there have been many others.

ACCUMULATION OF KNOWLEDGE

REB: Permit me to focus on what I think has been the major contribution of Forest Service research. It's not the spectacular and the highly visible technologies, as important as they are and as attractive as they are to the press and the media. It's the accumulation of knowledge that backstops major fields. If you look at the literature cited in a major monograph, on any of the important tree species in the U.S., on questions of seed and nursery practices, on insect complexes, on utilization practices, you'll see hundreds and hundreds of citations that came out of Forest Service research. Small bits and pieces and increments of information. Some of the classical handbooks, reference sources, that you'll see on the libraries of forest practitioners anywhere in the U.S. or in the world came out of Forest Service research. The *Woody Plant Seed Manual*, the two volume *Silvics Manual*, a manual on silviculture of the important forest types of the United States, insect and disease handbooks, wood engineering references, and on and on. This, in my estimation, is the major contribution, not spectacular, not flashy, but it's the accumulation of knowledge that backstops a whole discipline.

HKS: Sure, because it gives people something to build on in any direction.

REB: In my days of justifying research budgets, especially during times of cutback, that was the most difficult part of the budget, to defend and to justify, because it doesn't have those catchy handholds and immediate intuitive appeal. We searched for more attractive terms like continuing or foundation research which accounts for like 75 or 80 percent of the research of the Forest Service. This is the body of knowledge that permits a profession to deal with new or unanticipated problems. But it's difficult to maintain.

HKS: You used a term earlier that was new to me, upstream research. Is that basic research?

REB: It can be basic or it can be applied. It's the anticipation of the problems that are likely to come along. It is used frequently in international agricultural research to describe the basic or complex applied research needed to solve a problem— research that often is beyond the capacity of indigenous science.

So when a Douglas-fir tussock moth outbreak comes along you reach into the literature and call on the experts, and within a short time you have skills and people address that program. You can apply that test to almost any field: fire, insects, disease, silviculture, ecology, hydrology, wood science, and so on. An extremely difficult program to justify, not only in the Forest Service but in the Department of Agriculture as well. It's also the kind of research that invites pejorative adjectives like pedestrian, outmoded, outdated, and unresponsive.

HKS: I was surprised when I worked for the station here how much of the Forest Service manual comes out of research. People in the field don't know that and they probably would be upset, but the researcher has the technical information at hand to write procedures for fire danger rating and that sort of thing. So research really is pervasive in its influence.

Non-Forest Service Forestry Research

REB: One of the goals during my time as deputy chief was to straighten the non-Forest Service component of forestry research in the U.S. The McIntire-Stennis Act authorizes those expenditures to be up to 50 percent of those of the U.S. Forest Service research, with the implied understanding that the matching funds required in McIntire-Stennis would make the forestry schools equal in research effort to those of the Forest Service. At the time I went to Washington as deputy chief, I suppose the forestry schools research expenditures from all sources combined were approximately half those of the Forest Service. In the following years, the Forest Service research budgets did not increase, but funding sources increased elsewhere. The Forest Service was an important contributor to university research, but the National Science Foundation, NASA, Department of Energy, and Department of the Interior agencies all began to put money primarily into the forestry schools. I think today that earlier goal of approximate parity in research programs has been achieved, although not directly through McIntire-Stennis contributions as originally envisioned.

HKS: How about compatibility, cooperation with industry research, like Weyerhaeuser's research program and Westvaco's? Is there any link there other than a collegial link?

REB: Yes. It tends to be a volatile kind of cooperation in the sense that it ebbs and flows depending on the priorities of the two groups. The cooperation is far more effective on the land management side than it is on the products and the marketing side.

HKS: Because of proprietary interest.

REB: That's correct, and it's the research inside the plant gate where the industry can capture the benefits. I think it's obvious that the collaboration on the resource side is not nearly as easy to convert into a patentable or proprietary interest, and so there's a lot more collaboration. By the way the industry's support of land management research is really relatively modest. Weyerhaeuser and Westvaco are the biggest performers. That's the area where the public agencies, including the Forest Service, do the bulk of their work. The research inside the plant gate is primarily the industry side. When you count research and development including process improvement, new products, market development and things like that, the industry probably spends several times as much as the Forest Service and the forestry schools combined. Even on products-related research, I found a surprising sharing of information among industry scientists and counterpart researchers at the Forest Products Laboratory.

ADMINISTRATIVE STUDIES

HKS: When I was at the station I heard the term several times, in reference to National Forest Administration, administrative studies. That the regions actually carried out certain kinds of research of a very broad nature. I'm assuming that still goes on.

REB: Yes.

HKS: Why is that? Is it they want a faster turnaround time than the stations would...?

REB: Administrative studies are simultaneously a source of strength and a source of considerable tension. There's an intellectual curiosity on the parts of people who work in the National Forest System and in State and Private Forestry that should be encouraged. Administrative studies were intended to deal with site specific issues with questions of scale and operational procedures. It was the A part of an RD&A sequence. But it hadn't quite worked out that way. In most cases administrative studies have a terrible track record. Many of them are poorly planned, poorly executed, and almost never recorded for later reference. Keith Arnold and Ed Schulz, deputy chief for NFS, felt very strongly about that situation and worked out an agreement where administrative studies required approval of Research. As deputy chief, I didn't follow up aggressively on that point, because it was such a point of contention and friction.

There are many examples, however, where Research and NFS have worked together to the advantage of both. One of the best examples in the country in the H. A. Andrews Experimental Forest here, where awards have been given by Research to the district people because of their work with the application side. The Chippewa National Forest was another innovating one.

RESEARCH BY STATES

HKS: How about the states, other than the state universities? The city and county of Los Angeles have done a lot of work, administrative studies, on fire, of I assume rather high quality. State nurseries would be an example too.

REB: Let me comment on research in the states. I tried to draw state foresters into research activities and participated in their annual meetings on a couple of occasions. I don't know what the follow-up has been since then. The story on the states in research is extremely uneven. It ranges from zero to fairly substantial. The states with research programs that come to mind are Georgia and Texas. When you add up all of the state forestry research activities, they probably amount to less than 5 percent of the total research expenditures in the country. The big performers in mainstream forestry are the Forest Service and the forestry schools. Today about three hundred million dollars total expenditures for the two groups combined, maybe a little bit more. Industry's research is estimated to be about three hundred million dollars also, but that's mainly product development, process improvement, and so on, so the total enterprise is about six hundred million dollars. States are very uneven in what they do, but they are very willing cooperators. For example, we have experimental forests on state lands.

OTHER FEDERAL AGENCIES

HKS: How about other federal agencies? The National Park Service must do recreation research. BLM must do some range studies. Is there coordination?

REB: Yes. Yes, there is It can be improved, but it's better than meets the eye. For example, here at Oregon State University the Fish and Wildlife Service and the National Park Service have cooperative research units in place. The Bureau of Land Management has just enlarged its program; they will have about five or six scientists here at the forestry school, and of course the Forest Service has thirty or so scientists here. So the collocation of several federal research groups leads to information sharing and collaboration. As George Jemison noted, Corvallis is a center of excellence for natural resource programs; I very much agree with him.

HKS: In general terms, is the quality of the scientists in the other agencies of the same caliber as the Forest Service? The Forest Service has been in the business a longer time.

REB: Today, essentially yes. There may be some minor differences in background and training, but it's not significant. The biggest difference among several of the agencies is the way they organize to do research. Forest Service research is largely independent of the action programs, going back to 1915 when Earle Clapp became the first assistant chief for research. Much of the research of the three interior agencies comes under the action programs. In other words the national park superintendent has overall control of the research of the park. The Bureau of Land Management administrator has control over research and so does the Fish and Wildlife Service. The exception to these generalizations are the co-op units, such as those at OSU.

HKS: So continuity is always in question in the other agencies.

REB: Not so much a question of continuity, but that's important. It's the ability to ask the visionary and independent questions. Because when research programs come under close supervision by the action program, they tend to do day-to-day problem solving for the administrator and understandably so.

HKS: When you're testifying in Congress on the budget, do they ever say well we just gave some money to BLM, why don't you use their research?

REB: One always has to be prepared for those questions. In fact, I would have welcomed them. I was trying to build bridges to the Interior agencies, and I knew fairly well what their programs were. Now I need to make one point in fairness to the other agencies. I said that the bulk of their research comes under the close supervision of the line officer, such as the area director of BLM, the park superintendent, or the refuge supervisor. A part of each of those agencies' research now is done through cooperative units. That's a Department of the Interior exercise where small teams of scientists are located generally on university campuses to deal with long-range research questions. That's the mission of the co-op units here at OSU. Once again, a part of those research programs do have some independence, but the bulk of them are under very close surveillance of administrators of action programs.

HKS: BLM has had problems of managing or not wild horse herds and the damage or not they cause in competition with the wildlife and all this. I'm assuming some of their research is very specific, how to transplant wild horses or...

REB: I'm not sure that that program is backstopped by very much research.

HKS: It's one that gets a lot of media display. You can buy a feral horse or whatever they're called. The BLM has auctions and all of that.

REB: I don't know how much research is being done, not very much. The Intermountain Station had a small piece of that in the '80s. I don't know whether they continued it.

Upstream Research

HKS: The next question I guess, to use your terminology, deals with upstream research. Overall, what's the track record of Forest Service research anticipating the problem, so when the problem does appear, when it becomes publicly known, the Forest Service is ready with some kind of an answer?

REB: The track record, of course, is mixed. People who need solutions to problems, advice on problems right now, tend to be fairly uncharitable about our ability to anticipate those questions. I think we do better on it than meets the eye. I've already given you an example of the Douglas-fir tussock moth outbreak of the early '70s, and we had emerging research on a pheromone and a virus plus other entomological and ecological research that had accumulated since the previous outbreak. So we could mobilize that information fairly rapidly. Even with that vast accumulation of information on silviculture, insects and disease, fire and so forth, we still get caught off balance.

HKS: Let me ask you a specific question that maybe is not a good example but it's the one I was thinking about when I asked that question about upstream research. The red-cockaded woodpecker was listed as an endangered species in 1968 or 1969, and it's still a problem. Is this a problem biologically, or is it a problem in terms of management decisions? The problem hasn't been solved, it's still with us.

REB: We knew something about the red-cockaded woodpecker before its listing; I suspect more from ornithologists than foresters. Mike Lennartz, who was located at the Southeastern Forest Experiment Station at that time, took on the problem. Mike and his associates learned enough about the breeding and the feeding requirements of the red-cockaded woodpecker that they thought we could have timber and birds too. The breeding requirements required a small cluster of old growth pines that had decay in the center so that the birds could nest and breed. The foraging needs required far more area but were less restrictive; the birds could forage in managed timber. When I left Washington, I felt we had some workable solutions. But I suspect that what is involved here are value judgments and political issues where science really is of only marginal consequence.

HKS: About two months ago I got a call from a friend in Louisiana, suggesting that we do a study of the history of the woodpecker controversy, said it's really a hot issue. So I did a little bit of research on the history of the literature, and I was amazed how much was known, and how long it's been known.

REB: I think we need to ask ourselves the nature of the problem. Is this scientific and technical or is philosophical, depending on the value systems of individuals. The spotted owl out here is some of both. Oftentimes people call for research on essentially a political and value-driven problem as a way of prolonging the debate or avoiding a difficult political decision. We try to guard against that. I think some of that is involved in the spotted owl controversy.

HKS: It's a long story, but I'll just cut through to the punch line. I talked by phone to a guy, a USDA scientist at Beltsville, who was getting close to retirement. He worked on taxol twenty years ago, and he, I think, could best be characterized as disgusted at the hoopla about taxol now, reinventing the wheel. All I can go by is what he said, the understanding of the chemistry of taxol was very sophisticated twenty years ago and was put back on the shelf. But there's upstream research.

WHOLE-TREE UTILIZATION

REB: But that happens to a lot of science, and I think we ought to view that as a legitimate part of the process. Let me give you an example from the Forest Service. In the early 1970s the pulp and paper industry said that wood supplies are getting short and we want to look into whole-tree utilization. They said it would be far more efficient for us to chip the whole tree in the forest, put it into a van, take it into a mill, and then separate the bark from the chip. This process became known as the BCSS, Bark Chip Separation and Segregation. And John Erickson, now the director of the Forest Products Laboratory, was the project leader at Houghton, Michigan. The Forest Service had much encouragement from industry including support for budget increases from Congress.

It was a problem intuitively appealing to the Forest Service as well. John Erickson and his team were successful in removing enough bark that the chips met the standards required by the pulp digesters. But in the meantime, the industry found that there were some unanticipated disadvantages to bark chip separation. For example there was enough grit and sand in the chips that caused excessive wear on the tubes and the pipes in the paper mill. There were a couple of very large trials, but so far as I know no mill today is using it exactly as developed by Erickson and his team, although portions are in use. It's a technology that for the most part is sitting on a shelf, and I'll wager that we'll be back to it. That's what happens with much technology or piece of technology. It sits there until it fits into something else. I'm not surprised about the taxol story. I recall reading that there was interest in taxol for other reasons a good long while ago, and that its advocacy for cancer was a later development which renewed the interest in taxol.

HKS: I guess it's a very complex molecule.

REB: That's right.

HKS: That's one of the difficulties, the side effects from the ingredient that dissolves it so you can inject it is part of the problem, isn't it? Taxol is not toxic, but the carrier of taxol...

REB: Could well be. I read in *Science* a short time ago that there's something like fifty or sixty or seventy laboratories trying to synthesize the organic molecule called taxol. This laboratory has a substantial budget for the ecology and the management of Pacific yew, in fact there are survey crews now inventorying Pacific yew in the Cascades.

HKS: We always thought that the only practical use of the Pacific yew was to make bows.

REB: Do I make myself clear? That one tries to target research so that the payoffs can be predicted but it's at best an imperfect...

HKS: RPA sort of forces an upstream thinking.

REB: Yes, it was intended to do that. But our vision is terribly imperfect. Then serendipity comes along. I mentioned the truss frame house. The reason that Roger Toumey was looking into new construction techniques for wood frame houses was to avoid the failures that were reported around the country from wind and earthquakes. He began to look at the failure points in a wood frame house. The failures occurred where the studs meet the floor and the roof. So he said to himself, if we make the union between the wall studs and the ceiling and the floor truss differently and stronger, we'll avoid those problems. He didn't anticipate a different construction technique for houses, but when we put the house together, imagine the roof truss like this [gestures] and the studs that anchored in there with nail plates, and yes you can take the wall studs and anchor them to a floor truss as well. Imagine that you have a cross section of a house in a single large truss. This solved the first problem but it also altered the construction techniques. You fabricate these trusses in a nearby fabricating plant, put them on a flatbed truck, move them out, and set them up one right after another like an erector set, then sheath the structure, and there's the house.

HKS: It has an impact on building codes.

REB: Actually the adoption of truss frame has been fairly slow. But we can be sure that it or variations of it will evolve into construction practices. Let me give you another example of serendipity. The pulp and paper industry, which is far more advanced in forestry research than the solid wood products or the wood preservation group, suggested to the Forest Products Laboratory that they do research on the bonding mechanisms of wood fibers. Very basic upstream research. Vance Setterholm of FPL took on that job, and he began to form sheets under heat and pressure and under constraint to keep the fibers from squirming. The sample papers were much stronger. Then he began to do that with pulp furnishes from short-fibered red oak and other hardwoods. He was finding paper strengths that exceeded those of long-fibered softwoods. That process came to be called pressdry. Vance didn't set out to do that; he wanted to understand the bonding mechanism, but one of the incidental byproducts was that if we form a paper mat differently than we had, it will permit use of tree species not utilized before, and in some respects, achieve superior products.

Now the industry has not adopted in any significant way pressdrying, but they've moved toward it. Pulp and paper mills all over the world, I am told, are using a process called extended nip. You know paper is squeezed rolls. But if one can extend the contact that occurs at that time, more densification of paper results and higher strength properties occur. This is called extended nip. In many respects, that's an example of how research manifests itself. Our ability to predict where it will take us is at best imperfect, but we went from basic research to fiber bonding, to a new way of forming paper, to a modified system in pulp and paper mills that takes advantage of some of those processes.

ORIENTED STRAND BOARD

HKS: Last fall in south Georgia I toured for the first time an OSB plant. I understand there's a German patent, all the equipment comes from Germany...

REB: Could well be.

HKS: The impact on land management has to be extraordinary.

REB: Enormous.

HKS: There's a market for hardwood, hardwood not of furniture grade. Hardwoods have been a problem, and now suddenly they're out looking for hardwood.

REB: Technology has to be a major strategy in the management of forests. Let me make another comment on reconstituted wood, structural flakeboard. As I understand it, the genesis of that work came out of Canada, and I think it may have been an industrial researcher who worked on this concept. It was picked up, particularly at Washington State University by George Marra who later became deputy director of FPL. George did a lot of work with reconstituted wood and the Forest Products Laboratory also did some of the modifying and the adaptive work on structural flakeboard. So here was a technology that had its roots and its reinforcement in several different places. Structural flakeboard today is a major growth component of several forest products firms.

Even here in the plywood region of the West, we're seeing structural flakeboard. Now the next generation of research had to do with orienting the strands, this is called oriented strand board. Actually I don't know where the seminal technology occurred. I suspect that some of it came from British Columbia because the Wallenberg Prize was given there for developing the concept of oriented strand structural board. Oriented strand board is in contrast to the flake board, where the flakes are not oriented. The orientation gives greater strength properties. But that was a predictable next generation in the way that we go. What have the consequences been? The plywood industry is in serious decline because an alternative was found. On the other hand, the big trees that formerly went into plywood now can go into lumber.

HKS: Earlier we talked about the legitimacy of a Forest Products Lab, with the federal government being engaged in products development. A related question in my mind is, in Canada there's FERIC, the Forest Engineering Research Institute of Canada.

REB: Right.

HKS: Which is a federal/private enterprise. Why isn't there something analogous in the U.S.? It seems such a logical thing when you watch what it's doing in Canada. We've already accepted the philosophy, having a Forest Products Lab. Were there ever suggestions that we get into equipment analysis and testing and development?

REB: There really are two questions here. Let me take the second one first. The FPL does do product analysis testing and development, especially where a strong public interest can be demonstrated. For example, FPL had major leads in developing new lumber standards, in evaluating wood preservatives for efficacy and safety, and in developing fire safety standards for wood houses. The Lab often is uncomfortable with these tasks, feeling that industry or other agencies should do the repetitive work.

   The second question has to do with prioritizations of part of the research. I was never comfortable with the Canadian approach to FERIC (equipment engineering) or FORINTEK (forest products research), or for that matter similar efforts in New Zealand or Australia. For example, in Canada the public was expected to provide part of the funding and the private side the other. I suspect that the public funding was fairly stable and the industry contributions volatile. Furthermore, the industry contributions have a way of strongly levering the public funds to the advantage of the industry.

   There have been new developments in the U.S. since my retirement to accomplish some of the same ends—new legislation to deal with cooperation and patent procedures that observe the proprietary nature of some research. Questions about these new approaches should best be directed to FPL. I have the impression that there are better alternatives for the U.S. than privatizing of public research laboratories.

HKS: I was thinking of development costs. Feller-buncher technology is so important to the South. They're expensive gadgets, and the small landowner or the modest size landowner can't buy one to test it out. It's the practical part, and FERIC seemed to be such a logical solution. I would suppose some American enterprises, John Deere and Caterpillar, didn't want the government ranking them, sort of a *Consumer Reports*.

REB: The Forest Service is sensitive to that applications side for their own internal use. They have two equipment development centers, one in San Dimas and one in Missoula. The equipment development centers were intended to test and modify equipment that would be useful to the Forest Service, mainly fire fighting and transport equipment, nursery and tree planting machines, safety equipment, and so forth. My impression, however, is that much equipment development has been fairly well addressed by the private sector, especially in North America, in central Europe, and in the Nordic countries.

INTERNATIONAL NETWORKS

REB: I want to comment about international networks. One of the great advantages, as I see it, in IUFRO and other international networking associations is the sharing of technologies. For example, Europe is ahead of the U.S. in matters related to worker health and safety. Much of the technology developed there is now used by U.S. workers—safer chain saws, protective clothing to prevent injury, and reduction of noise and equipment vibrations.

I think we owe credit to Europe, and especially the Scandinavian countries, for some of the developments that led to things like the feller-buncher and the various kinds of forwarders. That occurred after World War II. The reason for this was because of their labor shortage and rapidly rising labor costs. This led to research and development programs in Scandinavia to substitute machines for labor. The U.S. and Canada also took part, but major contributions came from Europe. IUFRO has a very active work group on forest engineering that collaborates, especially with Europe but with other countries as well on various harvesting and transport systems. So some of that technology that you were seeing in the South, I'll bet had its origins in parts of Europe.

HKS: A guy in industry was explaining to me that workman's compensation insurance has just about eliminated chain saw work. The cost per hour per unit production for operating a chain saw is greater than the cost per hour of operating a feller-buncher because of workman's compensation liability. OSHA regulations, everyone has the ear plugs on now if they're around noisy equipment and all the rest of that.

REB: Yes, but much of that technology came out of Europe. Chaps to prevent chain saw injuries to legs, I think that came out of Europe. Some of these hard hats with the ear muffs.

HKS: Right.

REB: In the case of Scandinavia, and the U.S., we're oftentimes substituting capital for what at one time was labor, you know hand felling and so forth. This ability to anticipate, to substitute among the various factors of production, is very important in the private sector, and should be in the public sector as well. In fact, one of the things that should guide a research program is how to combine these production factors—land, various forms of capital and labor—more effectively.

INDUSTRIAL RESEARCH

REB: I'm going to make another observation. The Forest Products Laboratory has had an annual meeting with industry groups for more years than I was in Washington. As I mentioned earlier I used to join those meetings. It was interesting to observe how industry and the Lab interacted. There are all kinds of subtle and subliminal signals that flow back and forth in those meetings. Three groups participated; one from wood preservation; the second from the solid wood industries, that's plywood and sawmilling; and the third was from pulp and paper industries.

You could see in those discussions various degrees of sophistication and awareness of science. Pulp and paper was far more advanced scientifically than either of the other two. And the pulp and paper industry would ask the Forest Service to put emphasis on the basic or the upstream research, and they would do the adaptive work. The solid wood products groups were mixed. I remember a Weyerhaeuser representative saying essentially the same thing as pulp and paper, you do the basic studies and we'll do the adaptive research. But the solid wood products companies that were less well backstopped by research would ask the Forest Service to do the more adaptive and the more applied research. The wood preservation groups had essentially no research program of their own. They kept encouraging the Forest Service not only to work with various preservative questions including new chemicals, toxicity, and things like that, but actually to evaluate and field test the treated wood—from more basic research all the way to application.

The question often arose in my mind about the proprietary research the industry was doing, what that meant for a publicly supported research program. Lab scientists often could tell what the industry was working on by what they told them directly and what they discouraged them from working on. Still another source of information came from the keen interpersonal relationships among industry and FPL researchers. There would be private discussions with pledges of confidence which were invariably observed. But the industry had a way of sending those subtle signals about what they thought was important and what they thought you shouldn't be working on.

HKS: I was talking by phone to people at TAPPI in Atlanta, they're sort of the bibliographer for the industry. They don't know what the industry is doing. They said by the time we find out and include the research in our database, it's already been patented and sometimes even obsolete.

REB: The temptation is always to suggest an inventory of what we're doing and that will identify the voids and the gaps. However, this simply is an unworkable way to go with industry. They simply won't respond, and it's understandable why they won't. I always discouraged that approach when we worked with industry because it lacked realism. The thing that does work are these informal exchanges where information flows back and forth in a variety of ways.

HKS: A dramatic advance in technology, maybe dramatic is not the right word because it took place over a twenty-year period, this paper mill is right on the highway out here, just north of here on the outskirts of Albany.

REB: Willamette Industries.

HKS: Twenty years ago you'd be on I-5, you'd have this tremendous volume of crud coming out of the top. I drove by the other day and I couldn't see anything.

REB: Oh, I think the air quality standards have really been improved. It was a fascinating mill. I came down I-5 a number of year ago with Dave Mason. Dave was on the station advisory committee, meeting in Corvallis. He was in his eighties if not early nineties. He told me a story about that Willamette mill. It was built in the 1950s under the guidance of Ira Keller. Keller said he could build a mill for eight million dollars and it would use wood residues not round wood material. You may remember twenty years ago that building didn't have exterior walls; it was just a steel skeleton with a big digester. Dave said that that mill came in on budget and it set in motion the growth of the pulp and paper industry here for the next thirty or forty years. The pulp and paper industry here has grown on wood residues, not round wood—wood that formerly went into wigwam burners and hog fuel.

MISSIONARY WORK IN RESEARCH

HKS: One of the questions that your colleagues wanted me to ask you was a little bit surprising. Is there need for missionary work to advance research within the general field of forestry? I could see that fifty years ago and thirty years ago; is it still an issue?

REB: Yes.

HKS: Is it because the general practitioner just doesn't think research is necessary or...?

REB: I don't think it's different whether it's research or anything else, you simply have to champion your product. You have to give it visibility, demonstrate its worth and its quality. That's what I was trying to do when I waved a 2 x 4 in front of Bob Bergland. Some people are far more effective at it than others. Some of the big names in science, Carl Sagan, Paul Erhlich, others, have the ability to articulate a science program. It surely has to be one of the important considerations in recruiting a leader of a research program.

HKS: The *Journal of Forestry*, over the years, has received a lot of criticism from the run-of-the-mill forester in the field because there's nothing useful in it. I have no idea what percentage of foresters feel that way, but you've heard the criticisms.

REB: Yes.

HKS: SAF is always fiddling with the journal to make it more appealing to the guy on the ground.

REB: And the audience is so diverse that they can't satisfy very many. For science and technology they've come up with *Forest Science*, which was a Steve Spurr contribution of forty years ago, and now three volumes of applied forestry.

HKS: Regional.

REB: Regional journals, and so the technical content of the *Journal of Forestry* has gone way down.

ECONOMICS RESEARCH

REB: There was a cyclical quality to that issue. I was looking at the annual report of the Pacific Northwest Station during my first year as director. Timber supply was the big issue in the 1970s, and we featured it in the annual report. The incremental changes in our research program were tilting in that direction with Project FALCON, another program involving close timber utilization and so forth, which was then in response to an economic issue tempered by emerging environmental questions. Later in my time here the emphasis shifted very strongly to environmental concerns, and the research moved in that direction. I used to think that if you watched a research program for ten years, you'd come right back to where you'd been earlier. I think that that's normal and reasonable. You never drop everything you're doing and go on to the next fad, it's always an incremental change with a strong cyclical quality.

WHAT WOULD I DO DIFFERENTLY

MORE RESEARCH DECENTRALIZATION

HKS: What would you have done differently? You know now the things that worked and the things that didn't work.

REB: You know I've reflected on that question in the seven years that I've been away from Washington, D.C. One of the things that I would consider is further decentralization and slimming down of the administrative structure of the research branch. Some of that slimming down was occurring during the Crowell days as budget declined. I was determined that we were going to make as many reductions on the administrative side as we were required to make on the science side. The work was still there, but the remaining people would be forced to triage their time. Often I was surprised at how little loss occurs with neglect or abandonment of lower priority jobs. Also I would be looking to push to lower levels decisions affecting science. Fewer assistant directors at the station level; the place that I would place more accountability and responsibility is at the project leader level. And I would say to the project leader, you have now more authority to make decisions, to do things right and to do them wrong. Those are risks we're prepared to take, but you do it. It takes a mature organization to do things like that, which the Forest Service was with its limited recruiting and declining budgets of the 1980s.

I would also increase accountability. I would say to the scientists, you've more authority to make decisions, to plan your programs, and to allocate budgets, but you also have to be accountable for what you're doing. That means I would tighten up the Person-in-Job procedures. Decentralize, but enhance accountability.

HKS: Where we're sitting right now as a case study, is there more administrative structure in the forestry school than there is in the research branch? You've got a dean, you've got one or two assistant deans of some kind, and a director of graduate studies and several department heads.

REB: The administrative structure of a forestry school is different because of the nature of the workload. Student advising and teaching, both at the undergraduate and graduate levels, drives that work. I don't think it's fruitful to compare the administrative overhead in the two organizations. I have the impression that both organizations work equally hard with about equal efficiency, but the administrative workload is different. Both organizations need periodic review and belt-tightening.

HKS: It seems to me that in university research, professors don't need the dean's permission in the review process, they are more autonomous. There's less concern about continuity, about dissemination of the research except via "publish or perish." This is the perception that I have and it's not based upon anything other than very casual observation.

REB: Oregon State is not typical of forestry schools, but it's a highly regarded one. Carl Stoltenberg was and George Brown is providing strong leadership to the departments. But university professors are not nearly as autonomous as you might think. Part of the rigor and the discipline that goes into a university appointment are the demands of the classroom and students. A university tends to be organized around teaching programs and that in turn strongly influences research programs. Universities, and Oregon State is one of the few exceptions, tend not to do much team research. If they do, they often create quasi-independent institutes to do it.

One more observation. People both in the Forest Service and the forestry schools, especially those who do not have close personal and professional working relationships with the others, tend to characterize the other by stereotypes and cliches—generally unfavorable and unfair ones at that. A pity!

HKS: Right.

REB: One gets the impression that university professors come and go as they see fit, and in some respects they do, except when they have to meet requirements such as teaching, committee assignments, and student advising. Their schedules often are more rigorous and demanding than those of a Forest Service scientists, especially when a quarter or semester approaches an end.

HKS: Observing at Duke, five years ago they had a professor in what's now the School of Environment who was interested in Canada. He left, Canada dropped off the radar screen. That's how you were worried about the pioneer research, your concern that the continuity ends when that superstar retires and there's nothing to carry that on.

REB: I'm not sure I fully understand the question.

HKS: I'm trying to compare university research to Forest Service research in terms of continuity, that the university professor has more autonomy to pick and choose the subject. They have to do good jobs on what they choose, but there's no research plan for the school, no agenda, no RPA-driven concepts.

REB: The university research program is more sensitive to a broadly-based teaching program and to external forces than is the Forest Service. I don't know whether that's evident or not.

HKS: Grantsmanship.

REB: And that's part of the reason. Most university research funding comes from external sources. Oregon State University has one of the largest state supported research budgets in the country. I think it's about $2.5 million per year, and the other six or seven million dollars in their annual budget comes from external sources. External sources can strongly influence the direction that research takes. It's amazing how responsive university professors will be to external sources. But we need to keep in mind that recruitment to a faculty is more often driven by teaching needs.

HKS: The University of Washington when I was a grad student, a lot of Atomic Energy Commission money was available for soil study because of the bombing in the atolls in the Pacific. I think half of the forestry graduate students there were working on Atomic Energy Commission fellowships.

REB: Sure. It was enormous opportunism. The university professors require freedom, but they can be tugged around by money sources, actually more readily than the Forest Service. A point that you're making though is one that I would build into what I said previously—decentralize and slim down. Where greater discretion and greater freedom can be given, people will react with responsibility. There will always be exceptions and this is one of the problems with public employees, especially federal employees. We tend to gear our administrative procedures to our losers, not our winners.

HKS: This leads into what is my final question—your new career. Your vision of academic research must have shifted a bit since you've been in residence in this fine university. That's sort of what I was leading to with the previous question.

REB: Would you please permit me to stay with the previous question for a minute. You asked me what I would do differently, and the first thing I said was that I would decentralize, I would put more responsibility farther down in the organization, greater freedom to make mistakes but also greater freedom to be innovative. The second thing that I would do, and I would argue this with the chief and his deputies, that the Forest Service would be better served if there were greater autonomy among the various program areas. It's almost like the profit centers in an industrial group.

I would argue that the programs of the Forest Service are major sources of strength in their own right, and I would seek greater visibility and some independence of action for the three, now the four program areas, including International Forestry. But I would also insist to the point of indignation that there be greater mutual support among the four programs of the agency. Greater respect for each other's mission and greater support for the work of the others.

HKS: I experienced that when I was doing research on the history of the Forest Service. Bitterness might be too harsh a term, but it's close to correct for some of the retirees, who felt that timber management and fire were the glory guys all during their long, forty-two year careers. They always were dealt the nickels and dimes, timber management got the millions, and I guess that's human nature.

REB: You probably can't dismiss the problem by fiat. But do you see the philosophical point that I would make? For example, Max Peterson and his predecessors insisted that they were the line officers directly responsible for nine station directors, including Forest Products Laboratory, eight regions, two areas of State and Private Forestry. When you add WO staff reporting directly to the chief, he is responsible for the performance of nearly thirty people. And he can't do it.

HKS: That's too many.

REB: The chief goes through the motion of trying to monitor the performance of nearly thirty people. It makes him a mechanic. It takes time away from the longer range visionary things plus external contacts that only he can do. I would argue that the deputy chiefs ought to deal with the performance not only of their Washington office staff but also for their line officers in the field, subject to audit and review by the chief and the associate chief. The deputies would be held accountable for the performance of their line officers in the field; they in turn would be responsible directly to the chief. Of course, field line officers would continue to have direct access to the chief and associate chief. But there is more to the change than who reports to whom. I think this would allow for more innovative thinking not only in the chief's office but also in each of the major program fields.

NEW PERSPECTIVES—NEW RESEARCH?

HKS: Somebody told me recently that the Washington office is an anachronism, and since New Perspectives, it's the last outpost in the Forest Service of functional thinking. The question I'm asking, it must have an impact eventually on research. If we now truly have interdisciplinary management, the kinds of questions that are coming out of the field are going to be different.

REB: I can approach that question either as a cynic or an advocate, depending on the time of day. In a positive vein, I made the point several times in this interview that a new leader has to reassess the environment and, when appropriate, put a new face on things. Sometimes the new face really is a reaffirmation of existing practices or only slight modifications. I see some of that in New Perspectives, New Forestry, and now Ecosystem Management. It's putting a new face on some things that have been around for a long time. To be sure today, especially on federal lands, we need better integration of resources. That's the positive response. The negative response is that it's a gimmick. It's a way of trying to quiet criticism, and not very much is likely to come out of it. The next time we have a change in leadership we'll have a new gimmick.

HKS: I was thinking back to your statement that one of the most significant contributions of research is a solid body of basic grunt work literature. Definitive, authoritative, evaluative, on very arcane slices of knowledge. Bibliography of Frasier fir silviculture or some such thing, the building blocks. New Perspectives says well that's not the kind of building blocks we want any more. Or does it?

REB: The basic building blocks would be required no matter what direction management practices go. The place where that body of knowledge is lacking is in some of the newer fields of science: recreation research, social forestry, wildlife and fish habitat, hydrology, and ecology. These are newer entries into the field and a body of knowledge is far less developed for them than it is for the older ones. Another point. Research aimed at integration is far more costly than unidisciplinary research.

HKS: And in a sense it might be more ephemeral.

REB: It will certainly be ephemeral if it doesn't observe some fairly basic principles of experimentation and experimental design. Too much of the ecological research in the West is observational and deductive, not experimental and inductive.

HKS: It's harder to generalize in integrated research.

REB: By the way, that's one of the major shortcomings of this thing called administrative studies. They just don't have a memory. One of Arnold's and Dickerman's major contributions were to give a lot of support and emphasis to integrated study, like the 3-Bug program, SEAM, CANUSA, and FALCON. That was a significant contribution.

I really wanted to make those two points. Decentralize and a respect for more autonomy among the various program areas, and I think that would serve the Forest Service very well.

HKS: Who would have been the resistor to this proposal? If you had come to chief and staff meetings and said, "I want to decentralize research even more than it is," who would have said, "Wait a minute"?

REB: I don't know, I think that's a question you ought to ask others. I have the impression that this is not an agency requirement; it is more the proclivities of individuals. It's true that there's some agency chauvinism built into that issue. We're the Forest Service, the premier federal conservation agency. I think we can still be a premier federal agency with an enviable mission. Do you know of any other agencies in government that have a more comprehensive mission than the Forest Service?

HKS: No, in sense of identity or of esprit de corps. I suppose the FBI at times since past has been, the agents have been very proud employees.

REB: Agency elan was there for the FBI, but law enforcement is shared by several federal agencies.

HKS: One of the criticisms I've heard of research, hell those guys don't even have the uniform. You've got to belong.

REB: To many within and outside the agency, the national forests are the Forest Service—all other activities are in support of this program. The Forest Service has a strong legislative mandate to manage one hundred ninety million acres of land, to cooperate with state and private forestry, to conduct a research program, and to participate in international forestry. What other agency of government has as comprehensive a mission as that? Unfortunately, in the minds of many, that mission focuses largely on the national forests. If one thinks about it, the contributions to American forestry from Research and from a strong S&PF program could exceed those of the national forests.

HKS: Most agencies don't have land to administer.

REB: I don't know of any forestry agency in the world that has a more comprehensive mission than the Forest Service, and a legislative foundation to support it.

HKS: When you look at the Forest Service philosophically, the existence of the agency or not, it is, as McGuire once called it, a grand experiment. The Forest Service is inconsistent with the general sweep of American history, with having the federal government having a hands-on role in day-to-day operations out in the field. Owning the land and managing it. We don't have federal farms.

REB: No.

HKS: But we have federal forests.

REB: But that large and comprehensive role, not only for national forests but also for other programs, is a source of vulnerability. Perhaps with the benefit of hindsight, my desire to enlarge the participation in science programs was a way to recognize a monolithic organization with a dominance in the field, as a source of vulnerability, but also a loss of access to other sources of innovation.

Along these lines, as I've said before, I would encourage more senior exchanges between the Forest Service and its external cooperators—but with one important caution. It is important that the best people possible be chosen for these assignments, not second choices. And when the Forest Service sends its people to other agencies and departments for short term assignments, they too must be first choices.

HKS: Is part of it the complexity of comparable health plans and retirement benefits?

REB: Yes. We underestimate the importance of the fringe benefits. Forest Service research was simply not competitive in the early post-World War II period with universities. Consequently, some of the best and the brightest of the Forest Service people left. They became the major professors and the deans of the forestry schools. The most important reason was economic. Universities were paying more. You remember those days.

HKS: Sure.

REB: Then Eisenhower came in, a Republican who said that there ought to be equal pay for equal work. That brought the comparability issue into the federal government, and in a short time, the Forest Service became fully competitive, and in fact as university salaries increased more slowly in the '70s and the '80s, the Forest Service became more than competitive. Better salaries plus Person-in-Job meant we could pull people away from universities, industry, and other organizations. So we often underestimate the economic and the fringe benefits, including retirement benefits. Until recent years, Forest Service people were captives of the retirement system; they simply couldn't afford to leave.

HKS: Yes, I understand that.

REB: Now that's been liberalized and made more flexible under the Carter administration with the Civil Service Reform Act which liberalized retirement options so that one could leave and come back.

HKS: If it fits I'm going to ask Keith Arnold about that, because there's someone who has popped in and out from universities to Forest Service and back a couple of times.

REB: Keith did all of that when the retirement system was not so generous, so he did it at some personal cost. Maybe that's why he's still working. [laughter] Does that answer your question? Economic and fringe benefits are rarely talked about, but they're important.

HKS: Our expectations as individuals have changed a great deal in the past generation in terms of retirement benefits. Health care is something else, you can't get two people together for an evening without health care costs coming up like it did for us last night. You have to address the issue.

POST-RETIREMENT THOUGHTS ON PERSONNEL

REB: This relates, I think, to the next set of questions. Let me just comment about my retirement. I moved into the deputy chief's job at a fairly young age, I was forty-eight. My tenure could have been as long or longer than Harper's and still retired at that informal retirement age of sixty-two that the Forest Service observes for its senior people. It wouldn't have been as long as Clapp's, which went about twenty years. But I didn't want to stay that long. The signals about when you ought to retire are really terribly imperfect. I was unsure about how much more I could contribute. It was time for me to retire and to give someone else the opportunity to do the job. And so in January 1986, I retired after ten years in the job.

The exact timing of my retirement was also strongly conditioned by relaxation three years earlier of an onerous and punitive pay cap on senior federal employees. Federal retirement is based on one's highest three-year salary, and I couldn't afford to forego the additional retirement benefits.

HKS: Let's be sure to talk a little bit about the selection of your successor, to the extent that you understand the mechanism and are willing to discuss it. You're not quite sure how you got to be deputy chief, the mechanism, because no one talked to you about it, you weren't interviewed for the job. Is this normal?

REB: I didn't have a keen appreciation of the process at that time; I do now—although I'm not sure we keep prospective candidates any better informed. During my years as deputy chief, I tried to maintain informal rosters of potential candidates for all senior positions in Forest Service research. It was a useful device for me to discuss successional questions with station and staff directors as well. While the position of deputy and associate deputy were not on the rosters, they were very much on my mind and came up frequently in executive discussions of chief and staff. I don't remember all possibilities but there were four names I would discuss with Max and other deputies at our executive sessions.

One of them was Ross Whaley who had come back to the Forest Service in the late '70s and left five years later to become president of the School of Forestry and Environmental Sciences at Syracuse. Another one was Dave Thorud who had been in the Forest Service for six or seven years and is now dean of the College of Forestry at the University of Washington. Internally there were two candidates. One of them was John Ohman who was deputy chief for State and Private Forestry, who had been director of the North Carolina station, and wanted to come back to research. And the fourth was J. B. Hillman. J. B. had been the director of the Southeastern Station, came in as associate deputy chief with me about 1976, lasted in that job about two weeks when, I think it was McGuire said we needed him more in the National Forest System. So J. B. became associate deputy chief for NFS and stayed in that job for several years. He was highly regarded by the National Forest System. So those were the most plausible candidates, two from the outside and two from the inside, that I was talking with Max about. There were others who could have handled the job but because of family reasons, newness in their current job, or anticipated retirement did not figure strongly in the selection.

John Ohman had been deputy chief for State and Private Forestry for several years, and I had the feeling that he was getting weary. That S&PF job is one of the most trying of any of the deputy chief activities. John really wanted to come back to research as my successor. However I thought that J. B. Hillman, of the internal candidates, was a strong contender. He was an outstanding director of the Southeastern Station, and as associate deputy chief in NFS worked very well. But he was eligible for retirement in a few months of the time that I was eligible. He adopted a different agenda. He wanted to go back to his rural roots in western Virginia, and he just didn't want to be a candidate. John Ohman became the next deputy chief for research.

HKS: You have this list in front of you of people of significant accomplishment by any measure, yet four or five of them stood out. Can you articulate what caught your eye? What did you look for?

REB: It is a complex calculus that came from consultations among people who were supervisors of potential leaders. It came from my own personal impressions and observations including how well does this person handle himself in public; how well does he articulate his ideas; what is his vision of the future; and how quick are his thought processes, among others.

HKS: Physical stamina strikes me as significant. You guys travel an incomprehensible amount compared to us civilians, and that's hard work.

REB: I said that this was a calculus, it was a complex integration of a whole set of factors, and you also have to add family considerations. Are there teenagers in high school, how well will they take a move, can we postpone a move or should we advance it? What is the next move beyond this one? If we bring a person into a staff position, what is their potential beyond that? And I've mentioned several times the complication that the two-track, Person-in-Job career ladder proved for research. But, like all such forecasting, there's an error term. I used to seek all the information I could, I'd layer my own judgment but occasionally you make mistakes. Occasionally means mistakes no more than half the time. [laughter]

HKS: McGuire said one of the most important jobs the chief has is to provide for succession. You've got to keep the people coming at all levels of the agency.

REB: That's right.

HKS: Who's going to be the next regional forester, who's going to be the next forest supervisor, you've got to make sure this happens, otherwise everyone retires and there's nobody waiting that's ready to go.

REB: That's right. You remember the three big jobs of the deputy—budgets, programs, and personnel. Now I've given you some details about the succession to my job. John Ohman was deputy chief for research for three years and retired for a number of reasons.

Jerry Sesco became the present deputy chief for research. Jerry worked under me as budget coordinator for several years and did very well. He went to the Southeastern Station as assistant director, later becoming director of the station. He stayed in that job a relatively short time and moved to the WO as associate deputy chief for research, one of the two associates under the deputy chief for research, then John Ohman. When Ohman retired, Jerry became deputy chief.

HKS: Sure.

REB: That's how I dealt with questions of succession. I've omitted many of the details of internal discussions, but you can probably guess the nature of them.

HKS: Does the chief make a recommendation to the secretary?

REB: Yes.

HKS: The secretaries pretty much rubber stamp these recommendations? I mean, can you count on it?

REB: More so at senior levels below the chief/associate chief. That's why every chief that I know, especially Max and John McGuire, spend a great deal of time thinking about their succession and the timing of succession. I'll wager that was a major part of your discussion, a significant part with both Max and John.

HKS: In your case, where you postpone your retirement, postpone is not quite the right word, but you had an incentive to stay on a few more years. This succession calculus is changing daily.

HKS: Yes, that's correct.

HKS: Because if you'd retired this year rather than that year, there'd be somebody else in the wings.

REB: Right! Politically the succession for the chief's job is an order of magnitude more sensitive and more complex than it is for any of the subordinate positions. Sometimes the associate deputy chief can be sensitive, and sometimes one or more of the internal positions can be sensitive.

HKS: I suspect that not all personnel matters you dealt with were rewarding.

REB: It's fun to deal with positive personnel things, but all of us in those jobs have to deal with adverse ones as well.

HKS: Absolutely.

REB: The adverse personnel activities never get the publicity, nor should they, but they were among the most painful and time consuming jobs that I had to deal with.

HKS: Sure.

REB: Downgrading a senior scientist for nonperformance, personal indiscretions, fiscal malfeasance, and dealing with difficult personalities. All of us I think have etched in our minds vividly those many encounters.

HKS: It's tough, and the list of grievances gets longer with all the affirmative action...

REB: The legal foundation for adverse action really has tilted the benefit of the doubt to the employee. That's probably the way it ought to be. The thing that always used to trouble me were those who finessed those things, to move a substandard performer into a different job, generally horizontally, sometimes vertically, in order to get a nonperformer into a different position. I did so on two or three occasions, but never felt good about it. I would do those things differently today. For those three staff positions in the deputy chief's office, I always tried to choose people who had a clear capacity to move up, but occasionally things were so unsatisfactory elsewhere in the Washington office that I would occupy one of those positions with a marginal performer intending to move them out at an early opportunity. It was a mistake. Marginal performers don't become better performers with a changing job. But sometimes the situation in the antecedent position was so unsatisfactory that you just had to do something.

HKS: It might have been just that chemistry, get them with a different mix of personalities elsewhere.

REB: The courageous thing to do is just simply to confront the issue directly. It's a performance question, what are we going to do about it and how. One of the things that a senior administrator cannot do is deal with all of your personnel problems at one time. You tend to address them incrementally and sequentially. Where are they; what do I need to address first? What's second? What's third? As soon as you resolve the first three problems, you expose still others.

HKS: Sure.

REB: It requires enormous energy to deal with the adverse side of things, and you just don't have enough of it to do all of the things at one time.

HKS: It's a matter of priorities, you deal with the worse first. Are we at a good time to move into your new career?

New Career

REB: Yes. This is going to be short.

HKS: All throughout your time in the Forest Service you were working cooperatively with the universities. Now that you've had quite a few years of hands-on experience as a university professor, have your perceptions changed? Would you have done things differently in your relationships with the universities if you had known just a little bit more what you understand now, having been here on campus for so long?

REB: I really don't think so. Maybe minor and marginal changes, but I really don't think so. I personally view my working relationships with the university system positively, and one of the most satisfying. I knew Oregon State University from my days as director of Pacific Northwest Station. I knew it through the eyes of George Jemison, who joined this faculty when he retired. I knew Carl Stoltenberg for a whole lifetime. I knew George Brown, now the dean, who earlier joined us for a year in Washington. So I knew a lot about OSU. But I knew a fair amount about the other sixty or so forestry schools. So, I don't think I would have done things very much differently, except to reinforce what we were already doing. More collaboration, more cooperation, more confidence in what each other was doing.

The thing that I have come to appreciate more here though were the day-to-day operations of a university, not the longer term strategic view but the daily operation. It's different. In earlier days I might have wondered why professors are sometimes so relaxed and casual; today the follow-up question would be, how far along is the quarter or the semester? How many undergraduate and graduate students are you advising? How many committees do you serve on?

HKS: Yes.

REB: University professors can work day and night as a quarter or a semester approaches an end, mainly because of the teaching burden. They have to relax between semesters, and I am far more charitable about that.

I think the most satisfactory aspect of university life, and other professors tell me the same thing, is working with students. I've had several graduate students, all of them with foreign experience, but I also serve on a number of graduate committees. Most of my students are upperclassmen or graduate students. I may see a slightly more mature group. It's fascinating to learn about where they come from, their hopes, their aspirations, to assess their potential, to encourage their development, that's been the most rewarding part of the job.

HKS: Do you have formal classroom responsibilities?

REB: I teach, yes, but it's diminishing. I was responsible for the Starker lectures for the first four years here, and I still teach two seminars on international forestry, and I'm a guest lecturer in dozens of classes.

HKS: What textbook do you use on international forestry?

REB: I don't use textbooks. I use my own material.

HKS: Jan Laarman and Roger Sedjo have a book out.

REB: I want to read the book. Their textbook would be more for undergraduates. I use a lot of my own visual material in teaching and a series of references, contemporary references about issues. There are some personal satisfactions that come with being here, a good forestry school and reasonably well supported. But equally important is that this is where our family is. Our two sons and two daughters are graduated from Oregon State University. Three of them have two degrees from this institution. There are grandchildren here, very close at hand, they're all in Oregon. So there were some very personal reasons for coming.

HKS: I'm impressed by the vision or the finesse of Carl Stoltenberg. I'm impressed that Jemison and you are here. It strikes me that having you two guys at bat is a real coup. Both of you guys added a cosmopolitan element that's missing in a lot of forestry schools. I would wonder if other deans are a little bit jealous of Oregon State for having you guys around, if there'd be a perception that a lot of federal money must come this way because you're here.

REB: That part of it didn't turn out to be true. I suspect that your assessment is correct. Carl has indicated that he always wanted to have some kind of a senior or old-hand presence here to interact both with faculty and students. I'm never sure how well that works out, but I spend lots of hours visiting with faculty members, and especially with graduate students. I think that's what Carl wanted. Carl was and is a very astute observer of people, his ability to size up capacity and potential is amazing. I don't say that because of George Jemison or me, I say it because of the quality of some of his other faculty recruitments here.

HKS: In 1969 I wrote to Carl. I was finishing my Ph.D., and I asked him for a job. I wanted to teach forest policy. He wrote back to me, he said, you know the Forest

History Society is moving to the west coast from Yale, and one of its potential sites is here in Corvallis. He said if the Forest History Society will pick up your salary half-time, I'll pick up the other half. So I wrote to the Forest History Society and introduced myself, that's how I got into the Forest History Society. We went to Santa Cruz instead.

REB: That's the way Carl operated. He oftentimes changed the rules of the game after you got here. I can imagine that later on he would have searched for ways for your salary to be paid by the Forest History Society. He was very imaginative about those relationships, and keenly appreciative about how important that affiliation would be to this institution. He was also a tough administrator. He could make some very hard decisions. But Carl also was sufficiently astute that he was generally right. He more than anyone else made Oregon State one of the leading forestry schools in the country.

HKS: So you're winding down on IUFRO.

REB: I will serve on the executive board as past-president for an additional two years.

HKS: Until the '95 meeting in Finland. Are there roles that you play in IUFRO after that?

REB: Minor or none at all. Are you asking what am I going to do next? I'm going to be a truck driver.

HKS: That's right, you mentioned that last November when I was here. You want to get one of those big belt buckles. And on that bit of whimsy, let's stop. Thank you for an informative and thoughtful interview.